LIGHT WITHOUT HEAT

LIGHT WITHOUT HEAT

THE OBSERVATIONAL MOOD FROM BACON TO MILTON

DAVID CARROLL SIMON

CORNELL UNIVERSITY PRESS

Ithaca and London

First published 2018 by Cornell University Press

Printed in the United States of America

Library of Congress Cataloging-in-Publication Data
Names: Simon, David Carroll, author.
Title: Light without heat : the observational mood from
    Bacon to Milton / David Carroll Simon.
Description: Ithaca : Cornell University Press, 2018. |
    Includes bibliographical references and index.
Identifiers: LCCN 2017048069 (print) | LCCN 2017052392
    (ebook) | ISBN 9781501723414 (epub/mobi) |
    ISBN 9781501723421 (pdf) | ISBN 9781501723407 |
    ISBN 9781501723407 (cloth : alk. paper)
Subjects: LCSH: Literature and science—England—
    History—17th century. | English literature—Early
    modern, 1500-1700—History and criticism. |
    Observation (Scientific method)—England—History—
    17th century. | Philosophy of nature in literature. |
    Empiricism in literature. | Bacon, Francis, 1561–1626—
    Influence. | England—Intellectual life—17th century.
Classification: LCC PR438.S37 (ebook) | LCC PR438.S37
    S56 2018 (print) | DDC 820.9/004—dc23
LC record available at https://lccn.loc.gov/2017048069

For Jerry

Across the parlor *you* provide examples,
Wide open, sunny, of everything I am Not.
You embrace a whole world without once caring
To set it in order.

—James Merrill

CONTENTS

 **ACKNOWLEDGMENTS**

I am grateful to my mentors at the University of California, Berkeley, who exemplify the open-ended, generous, patient expectancy this book calls "the observational mood." Victoria Kahn has been an important source of counsel and encouragement since the very beginning of the research that led to the writing of this book. Timothy Hampton made the literature of the French Renaissance come alive for me, and I am thankful for his ongoing engagement. Judith Butler helped me think through some of this project's conceptual knots; her searching practice of reading is an inspiration. I owe particular thanks to Joanna Picciotto, with whom this book is often in conversation: a necessary friend and treasured interlocutor. My thinking in these pages also bears the imprint of illuminating conversations with the late Janet Adelman, Anne-Lise François, Kevis Goodman, and Barbara Spackman, as well as formative experiences from my undergraduate years at Brown University—in the classrooms of Susan Bernstein, Elliott Colla, Nicolás Wey-Gómez, Meredith Steinbach, Arnold Weinstein, and Esther Whitfield.

During my time in Berkeley and in San Francisco, I also incurred great debts to Andrea Gadberry, Amanda Jo Goldstein, Lily Gurton-Wachter, and (a friend from previous lives as well) Tristram Wolff, not only for talking my ideas through with me but also, more recently, for reading and responding to portions of this book. For creating a happy and lively world of intellectual exchange (and for much else) in the Bay Area, thanks are likewise due to Corey Byrnes, Colin Dingler, Katrina Dodson, Tom McEnaney, Julia Otis, Lealah Pollock, and Toby Warner.

My colleagues in the English Department at the University of Chicago have shaped this book in many ways, and I am grateful for the seriousness of their engagement. I owe special thanks to Joshua Scodel, who has read every chapter with care. The book is much better for it. For their generous responses to pieces of the manuscript, I am grateful to Bill Brown, Maud Ellman, Frances Ferguson, Tim Harrison, Mark Miller, Michael Murrin, Larry Rothfield, Lisa Ruddick, and Richard Strier. For our ongoing interchange of ideas about affective flatness (among many other topics of shared concern) and for her friendship, I am grateful to Lauren Berlant. Thank you to the first friend I made in Chicago, whose ongoing solidarity has been crucial to the completion of this project, Adrienne Brown. For thought-provoking conversations about this book's historical stakes, I thank Bradin Cormack and Jim Chandler. Three department chairs helped sustain the steady, collegial, and supportive environment in which I wrote the manuscript: Elaine Hadley, Frances Ferguson, and Debbie Nelson. In the context of a different research project altogether, Sianne Ngai made a decisive impact on my thinking about Montaigne—one that has also left traces on this book. I am thankful for the friendship and conversation of Tim Campbell, Rachel Galvin, Heather Keenleyside, Julie Orlemanski, Zach Samalin, Kristen Schilt, Richard So, Justin Steinberg, Forrest Stuart, Chris Taylor, Sarah Pierce Taylor, and Sonali Thakkar. Thank you as well to friends with whom I've shared time in the weird borderland between literature and science, Patrick Jagoda and Benjamin Morgan. I have also had the pleasure of sharpening my thinking about the works I discuss in this book in seminar conversation with graduate students; particular thanks are due to Beatrice Bradley, Sarah Kunjummen, Jo Nixon, Allison Turner, and Michal Zechariah.

I was lucky enough to share pieces of this project as works-in-progress with audiences at the University of British Columbia; the University of Missouri; Princeton University; the Renaissance Seminar, Chicago; Columbia University; the University of Maryland, College Park; and Northwestern University; as well as meetings of the Renaissance Society of America and the Shakespeare Association of America. I am grateful for their queries and observations, which were of great use as I composed and revised this book. I am especially grateful to Adam Frank at UBC for his reflections on this book's (still) unarticulated psychoanalytic stakes.

I am thankful for permission to reprint portions of this book that appeared first as journal articles. A part of the third chapter appeared as "Andrew Marvell and the Epistemology of Carelessness," *English Literary History* 82, no. 2 (2015): 553–88. A small portion of the first chapter appeared, much altered and repurposed, as "The Anatomy of Schadenfreude; or, Montaigne's Laughter," *Critical Inquiry* 43, no. 2 (2017): 250–80.

Many other Chicagoans deserve thanks: Thom Cantey, Joshua Chambers-Letson, Pete Coviello, Harris Feinsod, Andy Ferguson, Julia Fish, Andrew Leong, Emily Licht, Richard Rezac, and Kate Schechter.

I thank the American Council of Learned Societies for a year of fellowship funding that gave me the time to finish the manuscript.

At Cornell University Press, I owe special thanks to Mahinder Kingra for his interest in this project and his thoughtful feedback on the manuscript. Thank you as well to Martyn Beeny, Karen Laun, and Bethany Wasik. Many thanks to my copyeditor, Irina Burns, whose sharp eye made a big difference. I am also grateful for the extraordinary encouragement and helpful suggestions for revision I received from the press's two anonymous readers.

I thank Leah Beeferman for meeting me at the intersection of "nonchalance" and "cold color," and for her cover art. Thank you to Amberle Sherman for expert proofreading.

Thank you to my family: my (identical!) twin brother and best friend, Matt Simon, whose conversation makes me smarter; Danielle Williams; Dan Simon; Maria and Lou Passannante; and Lyndsay, Patrick, Gabby, and Ellie Adesso. I want especially to thank my mother, Candace Carroll, and my father, Len Simon—lovers of ideas and of literature. Finally, I thank Jerry Passannante, my best reader and companion in all things. This book is dedicated to him.

 LIGHT WITHOUT HEAT

Atmospheres of Understanding

Scientific Emotion and Literary Criticism

This book proposes a new interpretation of the scientific imagination at the threshold of modernity. It describes a vision of inquiry that belongs to philosophers, moralists, essayists, and poets across the seventeenth century. In the works of Francis Bacon and his first successors, the observer achieves freedom from obsolete premises and distortive passions by embracing an experience of carelessness. Indeed, mental laxity is the condition under which discoveries are made. In advancing this view, I present an alternative to a consensus that encompasses both intellectual history and literary studies. In its Baconian moment, the story goes, scientific investigation owes its credibility to the stringency of its procedures; the New Science is said to earn its modernizing descriptor by at last imposing the strictures of method on the errant speculations of medieval natural philosophy. This version of events, however, tells only half the story. Methodical rigor is chief among the values late modern scientific culture inherits from early modernity, but cognitive waywardness reveals distinctive qualities of the seventeenth century's scientific sensorium. By adopting the approach of the literary critic—by attending to the formal dimensions of a variety of written works, including contributions to natural philosophy—I recapture a forgotten perspective on what it means to know the natural world.

In the spirit of the apologists, evangelists, and practitioners of Bacon's reformed philosophy of nature, I define "carelessness"—which also goes by names like "disinvoltura," "nonchalance," and "indifferency"—as *dispassion without labor, including the labor of self-discipline*.[1] To distinguish a scientific inflection of the term, which correlates affective weakness with perceptual amplitude, from its many other (often flatly disagreeable) senses, I speak throughout this book of an "observational mood": an attitude of calm, outward-facing awareness. Though the etymology of "observation" risks confusion, suggesting as it does "the action of following a rule or practice," ordinary usage rightly conveys the experience of impressionability I have in mind.[2] Another potential liability of the term is the emphasis it lends to vision. This book explores all manner of sensory—and even strictly mental— perceptions enabled or intensified by casual indifference. In the end, what tilts the balance in favor of "observation" is the idea of passive witnessing.[3] My terminology signals an absence of eager expectancy: the slightness of desire when all it wants is to see what happens.

When indifference is careless rather than staunchly impassive, it can only be relative: the sensation of feeling less rather than nothing. Indeed, it's impure by necessity; only the most assiduous effort to protect indiffer- ence from the passions could even pretend to the status of perfection (to say nothing of actually achieving it). That's what distinguishes the observa- tional mood from the not-yet-ascendant paradigm of "objectivity," which insulates sober judgment from emotional interference (or at least aims to do so).[4] My argument runs in the other direction: the happenstance of diffuse emotion is a good occasion for understanding.[5] To borrow one of Bacon's metaphors, a useful point of comparison for the observational mood is the heightened sensitivity a person achieves with a simple willingness to listen. When human ambitions do not control the scene of investigation, Bacon explains, Nature takes the opportunity to speak unprompted.[6]

The Observational Mood

In showing the importance of an affective vocabulary to Baconian philoso- phy, no one succeeds as well as Lorraine Daston and Katherine Park. Their searching account of the "passions of inquiry" brings the psychology of sci- ence back down to earth (from its stubborn reputation for near-supernatural detachment), unfolding narratives of desire and gratification within an expe- rience of laboratory trial too often misconstrued as coldly technical.[7] My own wish is not to exchange the philosopher's dispassion for hotter affec- tive climes but rather to reject the misconstrual of indifference as the cessa- tion of feeling—and attend instead to the experience Michel de Montaigne calls "nonchalance" and locates on the spectrum of ordinary moods. The morphology of Montaigne's term, which suggests the absence of *chaleur,* or emotional "heat," is responsible for this book's title. *Light without Heat* refers to the power of illumination Montaigne, Bacon, and their intellectual kin ascribe to effortless "cool." In these pages, I grant pride of place to affective blues and greens, arguing for their importance to the conduct of scientific inquiry.[8] Though a compelling precedent, Daston and Park's study cannot be the right point of departure for my investigation. This is not only because "curiosity," which they locate at the heart of experimental science, radiates the heat of fervid desire; they also cast the seventeenth century's positive reappraisal of that concept, which now loses its suggestion of sinful errancy and unseats the medieval paradigm of "wonder," as a manifestation of the ethos of relentlessness from which I take my distance.[9] For Daston and Park (as, indeed, for the Aristotelian tradition), "wonder" is the affective name for a question that needs answering, and understanding extinguishes it.[10] With a similar singularity of purpose but a redoubled intensity, "curiosity" suggests an analogy between the pursuit of knowledge and the acquisitive thirst of a social class with money to burn, resembling the propulsive force of "greed" as it drives the scientist onward in the quest for understanding.[11] In this book, I turn down the emotional temperature on the history of Baconian philosophy and its cultural analogues. Granting priority to cool sensuality, I both explain early modern reservations about impassioned momentum as an engine for "advancement" and demonstrate the importance of exactly the opposite claim to the scientific imagination. Without the directionality of desire, the mind loses its way—but cognitive disorientation reanimates the sensorium. Thinking without feeling *without having to think about unfeeling* is a crack in the dome of received opinion: it lets in light.[12]

Adapting Daston and Park's terminology to my theme, I exchange majestic Dispassion for less exalted "dispassions of inquiry": gentle moods of seeming inconsequence that pervade the practice of science and scientifically minded literary experiments without having to earn their legitimacy as guarantors of credibility. This book assembles a diverse collection of literary and philosophical scenes through which readers are invited to inhabit the observational mood: a stroll through a vineyard, a fishing journey, a sleepy bedside experiment, and an exploratory conversation about the cosmos—to name only a few examples. The dispassions that animate these episodes do not goad readers onward with curiosity's promise of imminent understanding; nor do they pose wonder's question about the causes to which a given phenomenon should be attributed. This is certainly not to insist that these *words,* "curiosity" and "wonder," play no significant role in Baconian writing; it's only to suggest that their established meanings obscure the atmospheric

pleasure at the center of early modern scientific inquiry. Inherently anticlimactic, the dispassions know little of triumph—beyond the slim margin of
enjoyment inherent to grateful awareness. Instead, they facilitate immersion
in a practice of investigation in which different features of the perceptible
world come gradually but steadily to light. By placing diverse cases under a
single heading, I call attention to a family resemblance: in addition to their
soothing ease, what such varied instances of the observational mood have in
common is an incidental power to intensify receptivity.[13]

By embracing an attitude of multidirectional interest, the inquirer savors the
cognitive benefits of an experience for which no one claims responsibility—
which explains my preference for the irresponsible inflection of the term
"careless" (as opposed to "carefree").[14] In early modernity, it's exactly this
unlabored quality that most distinguishes the observational mood from
extant models of indifference, moderation, and tranquility.[15] Consider, for
instance, the case of Epicurean *ataraxia,* or "imperturbability": the transcen-
dence of worldly cares.[16] Of the several relevant philosophical contexts for
the observational mood, this one proves especially revealing. Throughout
this book, I return with some frequency to the arguments, images, and topoi
of the Epicurean tradition. These often prove irresistible to expositors of
the observational mood: not only the triumphant escape from anxiety into
peace but also the rhythm of steadily manageable gratification, the image
of the contemplative garden, the view of the turbulent sea (and the victim
of shipwreck) from the safety of the shore, and the thesis (which gains new
traction in the seventeenth century) that the world is composed of atoms.
Yet the reason I have not produced anything like a study of Epicureanism is
that the observational mood is an exact reversal of one of its basic tenets. For
Epicurus and his great evangelist, the poet Lucretius, coming to understand
the world's true nature brings about a desirable state of tranquility. Like
Stoicism, Epicureanism is a therapeutic program. The Epicurean journey
from knowledge acquisition to inner peace is crystallized in the following
promise: If you retain no superstitious beliefs about the afterlife, you have no
reason to live in fear of death. As Virgil in his *Georgics* (29 BC) puts it, chan-
neling Lucretius: "Happy the man who has been able to discover the causes
of things, to trample under foot every fear."[17] With respect to this model, the
protagonists of this book have exchanged cause for effect: an experience of
tranquility enables a newfound understanding of the world. The world is not
the way to contentment; contentment is the way to the world. One corollary
of this point is among my central themes: the easygoing pleasure inherent
to Baconian inquiry is ordinary, arising from quotidian experiences of mental drift. The difference from the Epicurean program is therefore decisive.

As James I. Porter puts it, "[Epicurean] ataraxy for humans requires a kind of effort that forever keeps them at a remove from the divine."[18] Such an ethos of exertion is contrary to the observational mood, and it implies what the protagonists of this book do not accept: that true peace of mind is the special privilege of the gods.

Another revealing point of comparison in the history of philosophy is Pyrrhonian skepticism. It too frames ataraxia as the well-earned reward of wise self-cultivation. In *Outlines of Skepticism* (second century AD), Sextus Empiricus portrays skeptics as those who "are still investigating," as opposed to dogmatists who leap hastily to conclusions.[19] Bacon and his followers show this same appetite for open-endedness—but Pyrrhonian inquiry, unlike the observational mood, is a program of self-management.[20] Interestingly, Sextus deviates from his view that tranquility is the deliberate "aim" of skepticism (and so more precisely anticipates the observational mood) when he describes emotional calm as an accidental byproduct of suspended judgment.[21] In recounting the story of Apelles, he makes ataraxia the unforeseen outcome of an abandoned project. The painter struggles unavailingly to represent the "lather" on the mouth of a horse, and, at last, in his frustration, throws a sponge at his work-in-progress and thus produces the desired effect; in just this way, Sextus explains, "when [the skeptics] suspended judgment, tranquility followed as it were fortuitously, as a shadow follows a body."[22] Though the story might be taken as a perfect parable of the incidental, Sextus uses it to recommend the suspension of judgment. He transmutes an accident into a deliberate course of (mental) action. For the inheritors of Bacon's vision, however, the observational mood really is a fleeting "shadow." Transmitting techniques for sponge throwing misses the point.

Over the course of this book, I return several times to this distinction between eager aspiration and contented description. There is danger of confusion, however, in both over- and understating effortlessness. I do not insist that the inquirer has nothing at all to do with the achievement of the observational mood. Indeed, I affirm the nonparadoxical possibility of a successful effort to attain it, as long as the pursuit is casual, implicit, subordinated to some other end—such as knowledge of the natural world. The feeling of easygoingness would then follow from comportment as likelihood rather than effect—as a possible corollary of action rather than its definite purpose. The advantage of taking this subtle difference seriously is that it allows us to appreciate the sense of luxurious abandon that pervades scenes of observation and trial in the sphere of Baconian science. Those thinkers who most value the observational mood amplify exactly this dimension of it: they narrate experiences in which leisurely indifference is an extravagant gift of circumstance. Yet there is still another wrinkle. Some of my cases describe apparently strenuous steps effectively taken in order to achieve the observational mood, but they nonetheless downplay exertion by casting the very experience of taking those steps as recreational pleasure. This latter perspective would be easy to brush off as mystification, a willful insistence on the counterfactual, but there are good reasons not to do so. Some courses of action really do take effort without feeling very much like they do: playing a game, for instance, or learning to play an instrument. I make no attempt at a metaphysics of action; I do not render a verdict on what should count as evidence of agency. In this respect, my approach is consistently phenomenological and psychological rather than ontological. What matters is how seventeenth-century authors describe the scene of inquiry: what they say about how it feels and how they take their emotions to matter.

My emphasis on both fortuitousness and slightness explains my fondness for the language of "mood." Thinking now of our late modern habits of speech, what I most value about this term is its suggestion of an enveloping "atmosphere" that comes and goes.[23] The metaphor emphasizes the transient relationship between the self and what it feels, and it renders affective experience situational rather than expressive of inner life. This sense of the indirectness or contingency of emotion is also conveyed by idioms such as "I'm in a good mood" and "I'm in a bad mood," which refer to affective backgrounds rather than foregrounds—unless we want to say that mood enters the foreground in such phrases but remains underexamined, vague, out of focus. Notwithstanding these good reasons to favor the vocabulary of "mood," my preference for the term—like my use of emotion-words throughout this book—is a matter of rhetorical efficacy; I appreciate the aptness to my theme of some of "mood's" ordinary implications, and yet I doubt the definitiveness of the careful distinctions scholars often make between passion, feeling, emotion, affect, and related terms. Influentially, some theorists of emotion have argued for the temporal and ontological "priority" of corporeal feeling over mental life.[24] Thus "affect" comes to name the specificity of a bodily sensation that unseats the sovereignty of the self, while words like "emotion" point instead to consciously inhabited states.[25] Early modern authors are often unconcerned by these distinctions; thinkers as different as the sixteenth-century physician Laurent Joubert and the seventeenth-century clergyman Joseph Glanvill juggle emotion-words as if they were interchangeable.[26] Much more important, even though early modern theories of the passions often define them as corporeal events, many firsthand descriptions of emotional life move freely between cognitive and embodied dimensions of feeling—to say nothing of active and passive, metaphorical and literal, natural and supernatural ones. My demurral to firm distinctions between the mind and matter of emotion follows from theoretical as well as historicist misgivings. Linda Zerilli has argued that the realm of conceptual thought has too often been reduced to the affirmation or denial of fully articulated propositions—the better to exclude affective experience from it.[27] I affirm that there is much more to say about the forms of conceptuality that belong to impassioned states—that are, in some cases, inextricable from them.

My practice is to use emotion-words in nontechnical senses, mimicking common speech—and without any a priori premise about the degree to which they are "thoughtful." Most of the time, I understand them as points of departure for the recounting of experiences. A deliberately loose taxonomy, however, which I have adapted (with some modifications) from Rei Terada, clarifies my terminological habits.[28] "Emotion" foregrounds the cognitive content of a state of feeling, while "affect" and "passion" draw attention instead to physical sensation. "Feeling" usefully blurs that distinction (once again, it is not always easy to decide the extent to which I am "in my head" or "in my body"), while "mood"—departing now from Terada's glossary—suggests nonimperative gentleness. For my purposes, then, emotion-words are more like colors on a painter's palette than technical categories. I maintain a principled vagueness about the nature of emotion (I am again declining to make ontological claims) in order to pursue exactness about a different question: what it feels like to inhabit the series of interconnected literary and philosophical scenes this book investigates. If I were to retain schematic distinctions between emotion-words, I would end up discerning within a single integrated experience a set of related qualities that seem misleadingly to belong to different categories—which would serve not as evidence of the unprecedentedness of the case but rather of the limited usefulness of our classificatory norms. Staying close to what my sources say, I find that "mood" is often the most instructive term for the feeling of receptivity. Yet my ability to approximate the sensory and cognitive richness of the scientific imagination depends on my willingness to exploit the full run of our emotional vocabulary.

One point of convergence between my project and other philosophical and literary approaches to emotion is my interest in evaluative experience.[29] The slight but precious pleasure of carelessness is both a premise and an ongoing discovery. The observational mood implies an awareness of the perceptual and cognitive advantages it affords. From the perspective of my sources, mental drift ultimately (if not presently) produces facts for the "rich storehouse" of human knowledge—and minimal but palpable cognizance of that happy outcome brightens the tone of the experience.[30] Such background pleasure is then replenished when the inquirer makes contact with some undercomprehended feature of the world. There's a sense of exultancy (of a pleasure already enjoyed) in drinking in the world's sights and sounds— and yet, *as* background experience, the observational mood is more like the enjoyment of air in the lungs than the satisfaction of slaking thirst. Often, it barely knows about or takes stock of itself; it does not possess or seize hold of anything so much as take gratified note of passing contact. The pleasure in inquiry is its underlying premise (a vague sense of assurance) as well as its byproduct, the thing assumed to inhere in investigation and the variegated thing generated by the distinct phenomena that compose the experience. This state of feeling, for which Bacon provides the template in chapter 1, is at the very center of this book: an alternation between a minimal expectation that doesn't last long enough to feel like desire and a gentle satisfaction that always arrives without delay because its sources are as manifold as the world.

If the scene of investigation is foreground to the self's background, then the observer is distinctly unselfconscious. Often, indeed, the occasion for the observational mood (this is not to suggest a single determining cause) is the drifting of attention from the self (and whatever it feels) to external sights and sounds.[31] As a quietly incidental experience, carelessness remains at the edges of vision: it's not an attitude on which one is likely to lavish attention. Mood's soft focus mediates access to whatever lies beyond the province of the self: whatever actually does attract interest. To be sure, Baconians sometimes explore aspects of personal experience (one of my interests is the delight they take in what they observe), but this is not the self in the sense of the unified persona with whom the thinking mind is thought to identify. As an experience of wanting something, but not anything in particular, from the field of perceptible objects, the observational mood implies the lingering of attention somewhere "out there" in the world. The vagueness of the wish ensures the mildness of the mood—not because all vague wishes are mild (it's never clear what in particular Marlowe's Faustus wants with the magic power for which he trades his soul), but because this particular one amounts to an ill-defined interest that more closely resembles open-ended waiting than suspenseful anticipation. The constant but partial fulfillment of the observational mood's sliver of desire preserves its ongoing gentleness.[32]

Reading beyond Selves

Unselfconsciousness is a theme with special difficulties. Indeed, literary-critical norms make it difficult to apprehend. In order to complete my anatomy of the observational mood, I need to examine the interpretive patterns that encourage the perception of self-consciousness even when someone drops it altogether—as in my paradigmatic case, when the inquirer comes face-to-face with the perceptual riches of the world. In early modernity, one familiar paradigm of unselfconsciousness is *sprezzatura* or *sprezzata disinvoltura*, the quality of artlessness Baldassare Castiglione attributes to the successful courtier in his *Libro del cortegiano* (*The Book of the Courtier*, 1528). Yet the term, at least as Castiglione uses it, suggests a strategic performance that couldn't therefore be more self-conscious. Scholars such as Frank Whigham and Harry Berger Jr. have made brilliant use of this concept as a point of departure for the interpretation of culture.[33] Like the accusation that someone is "trying to be cool," the motif of sprezzatura casts apparent easygoingness as hidden ambition. Yet Castiglione himself indicates that effortless grace (*grazia*) might be exactly the thing it appears to be; though it can be counterfeited by the social choreography of sprezzatura, it might just as well come "from the stars" (*dalle stelle*).[34] In their emphasis on the drama of self-fabrication, scholars have perceptively followed Castiglione's lead, but the very attraction of this theme is diagnostic of the privilege granted to self-consciousness—to self-management, self-interrogation, and self-presentation—in the field of early modern studies. If taken for a reality rather than an idealization, grazia might be attributed to fate, fortune, or happenstance rather than the restlessly calculating mind. One of my aims in this book is to demonstrate the value of this humble assertion: that effortlessness is not only a mask behind which a person can hide; it's also a plain fact of emotional life. The literature of Baconian inquiry takes up ordinary states of self-forgetfulness, discovering opportunities for insight in distraction, absorption, and abstractedness.[35]

I want to show that several of the key critical conversations in which this book participates are utterly fascinated by the most labored forms of self-cultivation (including the struggle for sprezzatura), clarifying the necessity of carving out a new conceptual space for reflection on the observational mood. First, however, I demonstrate that self-consciousness is not simply a theme, different from my own, that happens to have magnetized critical interest; sometimes it carries the force of common sense and so entails the reflexive unseeing of its others. Consider the case of Stanley Fish, who, in *How Milton Works* (2001), argues directly for artful self-presentation as an undeniable fact of Adam and Eve's relationship to each other in *Paradise Lost* (1667).[36] I return to Fish's account when I take up Milton's epic in chapter 4, but for the moment I speak only of his peremptory dismissal of the possibility of unselfconsciousness. Responding to Marshall Grossman's account of Adam and Eve's debate in book 9 about how best to accomplish the work of gardening in Paradise, Fish argues for the permanent theatricality of their marriage. Grossman shows that their dialogue takes a turn for the dramatic in the sense that each of them wants increasingly to produce affective responses in the other. For this reason, they lose track of the actual "substance" of their dispute.[37] Fish responds as follows: "The criticism makes sense, however, only if the staging of the self in the theater of their relationship is a new action that can be contrasted negatively with an earlier and alternative state in which the self is not mediated but knows itself directly. But there is no such state."[38] He goes on to enumerate earlier moments in which Milton signals that Adam and Eve understand themselves *as if through each other's eyes*, under the assumption that experiences of this kind rule out Grossman's interpretation. This view only makes sense if self-consciousness is a binary choice. Grossman's premise (like Milton's) is simply that people do not always attend to themselves with the same intensity; self-consciousness admits of degrees. It's as if Fish mistakes a psychological for an ontological distinction. The timing of the scene in question reveals the extremity of Fish's position. This is the last of Adam and Eve's conversations before they part ways and Eve consumes the Forbidden Fruit; what Fish therefore implies is that the entire history of their prelapsarian relationship, which has made room for outward-looking activities such as the cultivation of the garden, was pure performance. If Milton's readers have to choose between utter unselfconsciousness and some measure of self-awareness, they can only choose the latter—but the choice is both unnecessary and misleading.

I have tarried with an old textual dispute in order to cast light on what I take to be both the most obvious and the best reason to hesitate before following my lead: the proximity between the claim to unselfconsciousness and the claim to naturalness. That's what's at issue in the interpretation of *grazia*: following Castiglione, scholars have dwelled on cases in which the experience of effortlessness is not actually artless and uncultivated but is rather the careful simulation of those qualities. Many of the authors I explore in this book assert the naturalness of their behavior, which risks raising objections like Fish's. In what world do we get to be utterly free from an awareness of where we stand in the eyes of others? My response, once again, is to refuse the misconstrual of the question of self-consciousness as a binary choice. We cannot let a healthy suspicion of claims to naturalness metastasize into a blanket unwillingness to think about the *feeling* of naturalness.[39] If we do, an ontological position (the justified rejection of the faux legitimacy of "givens") collapses into a distorted view of psychology (as if nothing were ever *experienced* as "given"). Unselfconsciousness does not imply the absence of a role for the self in perception—only a relative lack of attention to it. The point can be extended to answer another possible objection to my line of inquiry—directed this time at the naïve realism for which, given my attention to receptivity, it might be mistaken. Though the observational mood enables experiences of discovery and insight, it does not necessarily presuppose unmediated contact with the real. (Some of my sources do in fact approximate that claim to directness, but this is neither a feature they all have in common nor a necessary premise of receptivity.) The observational mood need only suggest gratitude for the forms things take in the under-managed sensorium: some minimal feeling that what unexpectedly comes into view counts as advancement. On a Baconian timeline, which affirms trust in the future achievement of understanding rather than the immediate unfolding of the truth of things within the experience of inquiry, what gets discovered might be nothing more than latent possibility: perception offers access not to what is true but rather to what stands a chance of mattering to the making of knowledge.

This experience can be swiftly spirited away by setting the bar for unself-consciousness at an unfair height. How would our perspective change if we rejected that impossible standard? To be sure, the perception of guileless-ness never guarantees its truth; a person in aggressive pursuit of some ambi-tion might very well (indeed, often does) manufacture an artless pose. Yet it would obviously be wrong to conclude on that basis that there is no such thing as unselfconsciousness—or that obliviousness to self-presentation is purely imaginary. Everyone has firsthand knowledge of it. Even the most anxiously self-protective person does not sustain a practice of intense self-scrutiny in every waking moment. "Surely," someone might complain, "it's impossible for a person to free herself entirely from self-awareness; no one is exempt from the ongoing labor of cultivating an image for the world's con-sumption." I wish to stress that unselfconsciousness is not a marvel. Indeed, I refrained from countering the prevailing view of grazia (as an anxiously managed fiction) with this line of argument because it flattens out psycho-logical complexity: "Surely," I might have said, "no experience is fully calcu-lated; surely something (a gesture, an expression, a passing thought) remains unscripted in even the most intensely self-conscious performance of personal identity." Reasoning in this way assumes that no psychological state is cred-ible unless it achieves purity—but perfection cannot be the right criterion for admission to reality. Unadulterated self-consciousness would produce an impossible experience of paralysis; its unblinking vigilance could only terror-ize and immobilize the self, turning it to stone. Pure unselfconsciousness is likewise fantastical: the dissolution of the utterly dispersive self in wisps of smoke. As a scholar of emotion's history, I accept as a first principle that psy-chic states are less like bottled distillations (lucid concepts) than samples from the open sea (situations): saline, sweet, ionized, microbial, warm, cold—but never the absolute quintessence of a single quality. To be endlessly suspicious of unselfconsciousness is to take a quality for a substance and so to obscure the habit of mind that animates many of the most important literary and philosophical experiments of the seventeenth century.

Whether attributed to the author, to the author's persona, to fictional characters, or to the written work itself (insofar as, like a good courtier, it labors to "make the right impression" on its readers), scholarship in literary studies has given special attention to the drama of strategic self-presentation. Though I have suggested that this persistent interest might follow from a theoretical commitment (to denaturalization), and though perhaps it also reflects the likelihood that theories of literature place a premium on artfulness as a synonym for literariness, my goal here is not to establish a motive but, much more simply, to explain how an emphasis on self-consciousness hinders comprehension of the observational mood. Nearly forty years later, the title of Stephen Greenblatt's most famous book, *Renaissance Self-Fashioning* (1980), still encapsulates the ascendant critical practice.[40] Literary critics have gone as far as to adapt a concern with achievements and failures of self-cultivation to the inner workings of the human body, recasting affective experience as a problem of physiological self-management.[41] Routing the passions through Galenic medicine and other corporeal vocabularies, scholars have redescribed self-knowledge as bodily awareness—and self-discipline as physical control. Gail Kern Paster and Michael Schoenfeldt, some of the most skillful and persuasive proponents of this approach, offer contrasting accounts of affective embodiment that nonetheless converge in a conception of the corporeal self as a problem of mastery. In Paster's classic *The Body Embarrassed* (1993), she presents us with an image of the early modern body as "a semipermeable, irrigated container in which humors moved" and draws our attention to the subject's leaky unmanageability.[42] In *Bodies and Selves in Early Modern England* (1999), Schoenfeldt rejects the carnivalesque disarray of Paster's vision in favor of the metaphor of the "internal kingdom," arguing that self-discipline has a more decisive influence on early modern culture than she acknowledges.[43] These accounts lose none of their force if I observe that they haven't made much room for states of feeling in which the self *forgets all about* the problem of holding itself together. The observational mood is neither the achievement of self-mastery nor the abjection of the subject who loses control of her bodily functions. For it to come unbidden—for it to be experienced as a gift—it cannot be the success or failure of a disciplinary project.[44]

If it seems that I have borrowed the theme of observation from the history of science in order to suggest a new path for literary criticism, I now suggest that the history of science, at least that portion of the field that bears on my theme, has likewise declined to countenance unselfconsciousness. In this respect, indeed, historians of science have been thinking very much like literary critics. I am thinking especially of Steven Shapin, one of the most influential interpreters of Baconianism, who gives pride of place to the careful performance of personal identity. (Although *Leviathan and the Air-Pump* [1985], coauthored with Simon Schaffer, has earned a wide readership outside the historical field, I understand Shapin's *Social History of Truth* [1994] as a particularly useful crystallization of the sociological perspective I wish to describe.[45]) When Shapin explains that "scientific discourse was a species of *sprezzatura*," he envisions science as a social world in which everyone faces the challenge of making the right impression.[46] I suggest the reverse proposition is equally persuasive: "grazia is a species of science"—in the sense that the natural philosopher's unselfconsciousness is an affective experience with epistemological consequences. If practicing science means putting on a show, a claim I can hardly deny, what's most interesting about it, from where I stand, is how much *less* theatrical it is than almost any other social practice—or, to avoid so sweeping a claim, how conspicuously it distinguishes itself from other discourses by placing a special value on antitheatricality. The experience of undermotivated *looking* at the center of Baconian science is not to be confused with either the stage actor's experience of *being looked at* (thoroughgoing self-consciousness) or the theatrical audience's experience of *looking forward* (expectantly) to being entertained.[47] To be sure, some experiments are conducted in public (before, for instance, the Fellows of the Royal Society), but my point is not that science literally lacks an audience. Although the scientist sometimes faces the pressures of making arguments, producing marvels, and garnering patronage, the observational mood is nothing other than the crucial if necessarily temporary forgetting of those concerns. Philosophers of science draw a contrast between the context of discovery and the context of justification, which can be translated into a literary-critical vocabulary as the difference between the scene of insight and the scene of validation.[48] Although I complicate that distinction below, the observational mood is most easily grasped as a feature of understanding rather than a strategy for being understood.[49] It bids welcome to the world rather than sorting out what should count as bona fide "world."

One cannot simply amend Shapin's view by introducing the observational mood to his picture of scientific practice. The driving force of his account is nothing other than the calculation of personal stature. Unselfconsciousness is not available as an object of analysis; unawareness of other people's personae is similarly absent from his interpretation—much as Fish says it should be with respect to Adam and Eve's relationship in *Paradise Lost*. Shapin narrates the production of factuality as an effect of trustworthiness; the scientist confers credibility on his claims by constructing an aristocratic persona that merits respect. Although he does not put it this way, it makes sense to wonder about the social construction of truth at a time (our own) when objectivity sometimes takes a coercively absolutist form. (Of course, another late modern problem, though not one I can take up in these pages, is the refusal to distinguish between claims that have undergone a good-faith effort at verification and those that have not.) Sometimes, to be sure, the claim to objectivity is an attempt to insulate supposedly incontestable truth-claims from the social world in which they take shape—and thus to foreclose contestation.[50] Shapin's argument offers us resources for answering that concern. However, it does so by sidelining exactly the collective interest that best distinguishes experimental natural philosophers, on their own account, from their predecessors. "The new philosophers of nature and their cultural allies," he acknowledges, "avowed the supremacy of direct individual experience or intuition over trusting the authority of previous writers"—but he swiftly rejects the possibility of dispensing with an attitude that takes things on trust.[51] One does not have to believe that scientific knowledge is uncontaminated by human interest to care about the aspects of Baconian self-understanding this argument excludes. Shapin attributes the force of Baconian natural philosophy's claims to a value it explicitly aims to suspend: reputation. Thus his argument speaks more directly to skepticism about objectivity than it does to the features of scientific culture that interest me: in particular, the contingency of the experimental scene. I am not even sure what it would mean for the scientist to bring personal interests to bear on a thing that remains uncomprehended. To be sure, a person might use some fixed sense of herself to defend against an experience of confusion, but there is no reason to render that case axiomatic. To rule out the professed interest of natural philosophers in the as-yet-unknown is to enforce a stark disidentification. By distrusting experiment, Shapin makes the point that a fact is an artifact of human making rather than a thing that subsists apart from human perception—but in so doing conjures away the Baconian embrace of unselfconsciousness.

I want to be careful here to acknowledge the truth in Shapin's argument. A difference in subject matter need not be a difference in belief, and I am persuaded by his description of the (for instance, Protestant and aristocratic) conventions on which seventeenth-century scientists rely when presenting themselves to the world as deserving of credit. Shapin doesn't take up my question, and he has no obligation to do so; he doesn't argue that the experience of observation is just another style of self-fashioning. Yet there is something dismayingly precise about his omission of my topic or anything like it; his interpretation privileges exactly that aspect of scientific culture dispensed with by the observational mood. Because Shapin's focus is justi- fication, he can argue successfully for the inescapability of the problem of trust; one only casts doubt on a given claim, he points out, by weighing it against claims that aren't subject to the same level of scrutiny. Yet natural philosophy isn't reducible to a set of theses; nor is it a discourse occupied entirely by measuring assertions against each other. Experiences of inquiry, exploration, and speculation are no less important to the New Science. Indeed, Shapin's rejection of the possibility of "thoroughgoing skepticism" is an effect of his selective focus. Because he imagines scientific discourse as a collection of truth-claims, wide-ranging doubt sounds to him like freewheel- ing destructiveness—but it is more than possible, in a spirit of receptivity, to inhabit a world that remains less than solid without collapsing into terror at the groundlessness of things.[52] Indeed, the kind of (Montaignian) skepti- cism that matters most to the authors I examine in these pages is an atti- tude of exploratory openness. (I have already indicated that what I here call "skepticism"—an awareness of the unsteady, provisional quality of knowledge-claims—should not be confused with the philosophical skeptic's therapeutic program.) Shapin quotes Montaigne's "Des menteurs" ("On Liars") on the vital importance of trust to social cohesion. Yet he does not acknowledge that the very same essay attributes the author's credibility *not* to breeding, education, virtue, or any admirable quality of his character, but instead to the sad fact of his bad memory; Montaigne refrains from lying not because he objects to it in principle but because he cannot keep his facts straight.[53] Gentlemanly identity does not explain it; credibility is an effect of cognitive weakness, an accident of circumstance. This is not to say that Mon- taigne is not an aristocrat or that he does not like to remind his readers of it; it is only to cast doubt on the conclusion that his noble status sufficiently explains his professed guilelessness.

Indeed, another benefit of declining Shapin's emphasis on strategic self- fabrication is a newfound awareness of the insufficiency of aristocratic pres- tige as an explanation for philosophical and literary developments, including the observational mood. The ideology of social class has a decisive impact on the culture of science—but one would be hard-pressed to identify any sphere of activity (notwithstanding new possibilities for social mobility in the period) that does not show evidence of the rigid hierarchy early moderns know as a fact of life. There are several other good reasons to have doubts about an overemphasis on aristocratic identity in the seventeenth century's scientific imagination. One is the importance of "maker's knowledge" to the new philosophy: a real investment in the trade-specific know-how of the artisan.[54] Another is the democratizing effect of an experimental program that advocates widespread participation in the reproduction of trials and devalues traditional textual authorities. What is unprecedented about Baconian science is just this claim: that the pudding's proof really is in the eating—that it can't be derived from the cook's reputation. In principle, anyone and everyone can give it a shot. Indeed, Baconianism calls for the distribution of the practice of philosophy into a wider public—out of the cloister or university and into the world. My interest in the affective experience of scientific and parascientific investigation suggests one final reason to strip emotional effortlessness of its aristocratic patina: it appears in these pages as a passing mood so basic to ordinary experience that it would be difficult to imagine life without it. Often, leisure is hard to come by—but the same cannot be said for leisureliness. It is almost too obvious to say: I reject the suggestion that any particular social group has managed to monopolize the experience of mental drift. My dramatis personae come from different stations in life (Izaak Walton and Robert Hooke are not aristocrats), but they all reap the epistemological rewards of observational moods.

I want to conclude this section by reflecting on how unselfconsciousness inflects the practice of literary criticism. By deprivileging emphatic self-stylization ("Look how artless I am!") in favor of the most glancing self-description ("I forget myself in attending to some aspect of the external world"), this book redirects attention from appearance (the question of self-fashioning) to experience (the question of receptivity). Some of the authorial personae and literary characters I examine want less to study their own reflections (in the mirror, if I can put it this way) than to observe (through the window) the sensory riches of the world—an impulse no less deserving of interest than self-scrutiny. I defend this view even about as committed a self-portraitist as Montaigne—for whom, sometimes, the self is less the thing with which the authorial "I" identifies than a strange and unpredictable object of inquiry. Often, the works in question present themselves as both imprinted surfaces (records of wayward perceptions) and inducements to further observation (opportunities for such perceptions on the part of readers). Thus a written display of wandering unselfconsciousness invites readers to participate in the experiment by following an equally unpredictable train of thought. In none of my cases does it concern me whether the author "really is" careless or whether the work "really is" messily unstructured. Nor is my wish to suppress (quixotically) every indication of textual strategy; after all, the perception of design (the direction of an argument, say, or the relative importance of a narrated event) is indispensable to the making of meaning. My practice is to linger with that moment in the experience of interpretation that lets things unfold without any regard for the purposes they serve—not to the exclusion of subsequent moments but in the interest of holding them in tension. The scientific imagination grants value to contingent circumstances and fleeting appearances; to do justice to it as a literary critic is to inhabit a perspective that takes seriously the "unfinishedness" of the artwork, casting a solitary beam of light into its depths (without the expectation of discovering its edges) rather than mapping it with sonar (as a fixed terrain that can be definitively captured). To accept this experience of uncertainty is to gain access to the strange beauty and philosophical daring of a culture-spanning intellectual endeavor that savors disorientation.

Asymmetries: Action without Passion

Having anatomized the observational mood, including the quality of unselfconsciousness that least accommodates the prevailing habits of literary scholarship on the early modern period, I am now prepared to reflect on the intellectual-historical stakes of my argument. To show why and how carelessness matters to extant narratives of modernization, I begin with the influential work of Hans Blumenberg. With nothing other than the metaphor of "labor," he drives a wedge between medieval and modern conceptions of knowledge. Against a contemplative tradition that envisions truth as a "mighty" force that overtakes the mind and compels assent, he argues, the new philosophy locates understanding on the far side of strenuous toil: "Truthful appearance is suspected of being only apparently true. All truth is earned and no longer freely given; from now on, knowledge assumes the character of *labor*."[55] For Blumenberg, then, early modernity at last sees the identification of knowledge as a distinctly human responsibility: not a miraculous gift but a product of determined striving. He grants to experimental science the prophetic role of cementing the terms of a metaphor that remains a defining feature of modern philosophical self-understanding. "In Bacon," Blumenberg writes, "the concepts of *labor* and *truth* appear together for the first time in the course of a *single* sentence. After more than two thousand years, truth's *splendid isolation* from any hint of strain is brought to an end."[56] The rise of science consolidates recognition of the artifactuality of knowledge. However reflective of external reality, a fact is exactly what etymology suggests: a "thing made" in the furnace of human creativity.[57]

I accept Blumenberg's thesis that self-reliance is the self-conscious predicament of the new philosophy. In modernity, the philosopher does not wait for understanding to come; he or she rises to the challenge of achieving it.[58] Yet the pervasive appeal of the observational mood in seventeenth-century science suggests the incompleteness of Blumenberg's formulation. He disregards the misalignment between self-assertion as *feeling* and as *behavior*, which I suggest is a fundamental problem for intellectual history. It's easy to take for granted that Attitude X attends Action Y—that anxious intensity must be the emotional atmosphere of experimental labor—but the truth is different. The expectation that human psychology is uniform and predictable in every time and place distorts the interpretation of historical descriptions of emotional life. In Blumenberg, the metaphor of "strenuous exertion" refers most directly to the claim that knowledge is manmade, but it also implies that the philosopher is emotionally overwrought when confronted with the prospect of unassisted inquiry.[59] Blumenberg actually depicts the beginning of the "modern age" as an explosion of repressed feeling. In *Die Legitimität der Neuzeit* (*The Legitimacy of the Modern Age*, 1966), he argues for the reappraisal of curiosity as a distinctive achievement of the Renaissance (anticipating a claim I already noted in Daston and Park), and pictures the reinvented concept as the open floodgate through which long-constrained passions finally flow. "Pent-up energy," he writes, "[had] to be let out through curiosity once it was rehabilitated, a kind of energy that deprived the ancient ideal's contemplative repose of the qualities precisely of repose and calm."[60] By posing the riddle of repose without repose, he unknowingly requests the very answer this book gives: an experience of carelessness that pervades and interrupts even the most difficult (or apparently difficult) tasks. When he pits Bacon's style against the explicit aims of "instauration," he again implies, but doesn't explain, an asymmetry between feeling and action—which only looks like a problem, I suggest, under the pressure of an unwarranted expectation. He proposes that modern science sets out to solve the problem of "the human spirit's historical indolence," but he also observes that "Bacon defends theoretical curiosity . . . more indolently [than Giordano Bruno], with juristic tricks and shrewd twists of hallowed arguments."[61] Though Blumenberg echoes the old line about Bacon writing philosophy "like a Lord Chancellor," making an observation about cunning calculation that points away from my concerns, he interestingly imbues the scene with a mischievous leisureliness that seems to confute his description of the philosopher's zeal and the enthusiastic temper of the historical moment.[62] The near-paradox is more meaningful than he knows. England's scientific revolution gathers momentum exactly as he says: from experiences of languorous drift.

This book's protagonists take up aesthetic and scientific practices that require intellectual precision, manual dexterity, and bodily exertion—but nonetheless describe their efforts as emotionally undemanding. They conduct experiments, make conjectures, speculate, and argue—but carry it all off, so they say, without breaking a sweat (their perspiration is literal but not metaphorical). One wager of this book is that we have more to learn from taking these self-characterizations seriously than we do from explaining them away. My focus on emotional life rather than the object-world might already have suggested some distance from the work of Bruno Latour, perhaps the philosopher of science who has most powerfully influenced scholarship on literature and science—but on this question of asymmetry he offers valuable guidance. Relocating Latour's dictum that "nothing is reducible to anything else" to the field of affective experience, it helps explain my reluctance to understand emotion as the predictable expression of behavior.[63] I do not imply that mood remains unaffected by circumstances or that it reciprocally leaves them unchanged. What I do mean is that those relationships should not be taken for granted; each case deserves a look. By accepting the disjunction between feeling and procedure (an a priori mismatch in the sense that no affective experience follows naturally or automatically from any course of action), I can affirm the possibility of an experience of carelessness that survives the challenge of technical difficulty. In the seventeenth century, contemplative serenity accommodates, but also disrupts, muscular interventionism.[64] Rather than signing on to the familiar claim that early modernity reverses the traditional hierarchy of *otium* over *negotium*, remaking knowledge production as hard work, I suggest their interpenetration.[65] If emotion were only the predictable aftereffect of behavior, the role of carelessness in the culture of endless trial could only be escapist fantasy. Yet the era of idol smashing really is the era of slackness. For the first Baconian intellectuals, restfulness is not an expression of disavowal but rather an enabling condition of amplified awareness.

To be sure, the suspension of assumed continuities between action and emotion does not proscribe the discovery of specific ones in particular historical settings. It forbids the easy extrapolation of laboriousness from labor but not therefore the identification of patterns of experience that link states of feeling to courses of action. This book consistently associates the observational mood with a stable set of behaviors (to say nothing, for the moment, of literary genres and topoi). I'm thinking, for instance, of inattention, sloppiness, and daydreaming—of forgetfulness, speculation, and the lingering delectation of sensuous particulars. Bacon and his successors profess a commitment to meticulous self-control they cannot be counted on to perform: that's the predictable effect of their observational moods. Therein lies my most basic interest in the subject: the most avowedly rigorous of philosophical projects succumbs to the voluptuousness of neglect. Still, there is more to say—on a different scale—about the unpredictability of mood's practical consequences. The relationship between carelessness and exactitude is not always clear. Over the course of this book, several possibilities suggest themselves. At the extreme, the appeal to rigor might be taken as nothing more than an alibi for liberties taken. Perhaps, however, it authorizes playful digression by promising the counterweight of discipline. The delegation of the duty to regulate experience to a regimented sequence of steps would then be understood as an attempt to absolve the mind of responsibility for the enforcement of discipline. Yet if method sets the mind free for harmless play, it licenses an experience of luxury that risks losing track of its alibi. The diligent philosopher hands off the baton to a less-than-dutiful version of himself, but one who just barely remembers to run the race—and who's to say for how long? The scope of carelessness is likewise variable—and variably interpretable—across this book's several cases. Throughout the culture of the New Science, disarray remains a feature of experimental findings no less than hasty jottings in notebooks—of postexperimental conclusions no less than speculative flights of fancy. It's for this reason that I do not insist on the difference between discovery and justification—by stipulating that the observational mood belongs only to an exploratory moment that precedes (and therefore leaves untroubled) the logical procedure by which perceptions are converted into hard facts. That distinction goes a long way toward explaining the divergence between my account and dominant histories; the effortlessness of inquiry is less mysterious if it animates a moment of insight ("Eureka!") but finds no opening in the practice of confirming hypotheses. Yet moods have an unpredictable life of their own; they arrive without warning.

If there can be no settled answer to the question of whether carelessness "wins out" over method—though it's obvious that method ultimately "wins out" over carelessness in the history of modern scientific self-understanding—there can likewise be no simple conclusion about whether carelessness is, within the context of Baconian inquiry, constructive. To what extent are the disruptions of the observational mood folded back into the process of knowledge production? There's an important sense in which the observational mood simply has nothing to do with productivity—which is less an answer to this question than a good reason not to answer it. When Baconians are most intent on recuperating the fruits of careless observation—on sifting fact from fiction by making careful judgments about their experiences—such procedures might be taken to serve as ex post facto justifications for thoroughgoing cognitive irresponsibility. Those cases would then seem to invert the drama of sprezzatura: rather than making the difficult look easy, they make the easy look hard. Yet my point here is much simpler than that. Because passing moments of mental idleness are often semiconscious, they need not initially solicit close attention or explicit rationalization. In fact, they might not be understood as serving any systematic purpose at all. Because, by definition, the observational mood is incidental, it can also be understood as an interruption of whatever happens to be taking place (including the meticulous execution of procedure): the emotional or psychological equivalent of a non sequitur. Throughout this book, I explore Baconian attention to rigor as well as its abandonment without assuming the necessity of a theory that establishes their coherent interrelation. What I want as an observer might just not be what I want as a theorist. To put this differently, *what I think I want* from natural-philosophical investigation might have little to do with *what in fact happens* when I surrender to the observational mood.

It's for this reason that the thesis of the seventeenth century's labored epistemology obscures as much as it reveals about the prehistory of Enlightenment. It lends credence to the still-influential argument that modern science is best understood as the instrumentalization and domination of the world. No matter the lost currency of the claim's most exaggerated form, scholars continue to narrate the rise of science as the exploitation of nature, the enforcement of discipline, and the suppression of imagination.[66] Equally important to this moment in intellectual history is the beguilingly simple intuition that the world is different from what anyone expects. From the far side of capitalist modernity, it's easy to equate the inquiries of the first experimental scientists with the bid for mastery implied by objective neutrality. Yet their attitude of susceptibility confounds late modern confidence that they champion a philosophy of force. Bacon's avowal of human weakness is not, as it is for Descartes and will be for Kant, a strategic point of departure for the affirmation of human power—or it isn't only that. In moments, he and his successors capitulate to a wandering interest in nature's plenitude; the dream of future comprehension excuses the present fact of delinquent inattention. Thus I suggest an inversion of Blumenberg's disquieting image of scientific fervor: "Swallowing reality whole," he writes, "so as not to lack even a bit of it is arguably the ultimate metaphor for all realism."[67] "Consciousness," he explains, edging closer to my theme, "is the organ for not devouring the world and still being able to possess and enjoy it."[68] Perhaps a better "metaphor for all realism"—or at least for the version of it that underwrites the observational mood—is "being swallowed whole." Seventeenth-century authors luxuriate in experiences of gratifying surrender that they trust will furnish them with insight. In such moments, the mind is the organ for getting devoured by the world and still being able to enjoy it.[69]

I want to round out my picture of the asymmetry of passion and action in Baconian experiments by looking briefly at the most influential late modern interpretations of them in both critical theory and literary scholarship. Frequently, the "father of experimental science" takes the stage of intellectual history as the prophet of "instrumental reason."[70] The social theorists of the Frankfurt School develop the motif with particular acuity (though with much better success when aimed at the ascendant positivism of their own time), identifying the same threat of sheer operationality I observed in Blumenberg's account. By collapsing an unfolding process (labor) and the outcome it achieves (knowledge), Blumenberg replaces an unpredictable course of action with the efficiency of technique. Granting pride of place to method (to any formalized and thus repeatable procedure) itself encourages an impression of breathless intensity, which poses a special interpretive problem when the apparent laboriousness of experimental knowledge production is exactly the claim most in need of qualification. When Max Horkheimer and Theodor Adorno express anxiety about the reduction of thought to mere instrumentality, they take that transhistorical problem for a historical fact. If understanding is simply a task to be completed (and one for which the procedure is set), it suspends basic questions about whether, how, and for what purpose the philosopher takes up the challenge of knowledge production in the first place. For Horkheimer and Adorno, the fungibility of the goal, as far as reason is concerned, is cause for trepidation. Yet the history they recount disregards the disobedience of Baconians to their methodological commitments.

Horkheimer narrates the transformation of reason from a principle of cosmic order (as it is, say, for Plato) into a "faculty of mind," and he attributes to "English thinkers since the days of John Locke" a central role in the transformation of reason from "objective" principle into "subjective" instrument.[71] Thus England's scientific revolution is the staging ground for the reformulation of reason as a thing that *works*, in the double sense that it *labors* under human masters and *functions* in the way of a machine. In their *Dialektik der Aufklärung* (*Dialectic of Enlightenment*, 1944), he and Adorno name Bacon himself as the exemplar of reason's mercilessness.[72] As an endlessly powerful but dismayingly flexible instrument of will, it knows nothing of what it does except that it aims for success. In this way, they allegorize the social and philosophical impasse of their own time as an event in early modernity, conveying a desire for conceptual resources they might have found in the object of their critique. "Ruthless toward itself," they write, "the

Enlightenment has eradicated the last remnant of its own self-awareness."[73] Acknowledging that something has changed, Horkheimer and Adorno look out into an intellectual landscape in which reason works too well to do anything but work. I suggest it hasn't always, and that the place of reason in history is different from what they suppose.

Although Horkheimer and Adorno are sometimes treated as antagonists of Enlightenment, they actually preserve the promise of reason's power, narrating the horror of its degeneration without placing their faith in the irrational.[74] I share their revulsion at instances of calamity for which reason has been invoked as justification (they speak, for instance, of the depredations of supercharged capitalism), but I do not think late modern political and social maladies are the predictable consequences of Baconian proposals. Rational oversight is a commitment from which some of early modernity's champions of reason are unafraid to stray. When Horkheimer and Adorno speak of reason's blindness, they point out its susceptibility to manipulation. For the seventeenth century's intellectual vanguard, however, unknowing is less an outward prospect (blinkered vision) than the very quality of understanding itself. Experimental practice is not flexible in the manner of an implement that might be used for anything at all; its very heart is pliancy. "Making use of it" just doesn't make sense: one cannot hammer a nail with a silken scarf. When Horkheimer imagines his object of inquiry as a "too frequently sharpened razor blade," he laments both the reduction of reason to mere utility and the eventual ruin of the sad remainder itself: the diminishment of the already diminished.[75] In the observational mood, however, the mind does not fail in the manner of an instrument; it happily loses sight of its purpose. The amenability of such a mood to whatever happens to cross its path is a source of perceptual strength, but it also implies a vision of rational activity as a wide-ranging experience that precludes so simple an operation as "use."

Descending now from the heights of abstracted history, I want to acknowledge once more that laborious intensity is not only a retrospective explanatory rubric for the New Science but also a central metaphor for practitioners themselves. Interestingly, two of the best cultural histories of early modern science affirm the difficulty of experimental labor while arguing for the importance of theological reflection on the Fall of Man to the Baconian project—a theme that directs attention back to the very experience of effortlessness I have taken as my subject. Drawing on Charles Webster's account of mid-century Baconianism as an expression of puritan apocalypticism, a reformatory effort to prepare the ground for the millennium, both Peter Harrison and Joanna Picciotto explore the role of unfallen Adam in an emergent scientific culture as a paradigmatic image of sensory and cognitive strength—of those cherished abilities to which human beings once had (but no longer have) ready access.[76] For both Harrison and Picciotto, our distance from our original perceptual powers (the ones enjoyed in Paradise) explains the need for intense self-discipline: efforts at self-correction get us closer to the innocent perspective we no longer enjoy by default. By raising the question of anthropology "in the broadest sense," however (the phrase is Harrison's), they also invite the revisionary account I offer in this book.[77] What distinguishes the period's natural philosophy from its predecessors, Harrison argues, is an Augustinian emphasis on human frailty that ends up raising questions of method in order to make up for or mitigate the disastrous effects of the Fall. I suggest we best understand this "anthropological" focus by lingering with the most detailed representations of human experience Baconians have to offer, many of which are found in literary works. By developing a literary history of Adamic natural philosophy, Picciotto already inhabits this "first personal" perspective—especially as it expands to encompass the first person plural of collective intellectual labor. Yet what both accounts leave to one side is the strangeness of an effort to be like someone whose experience is distinctly nonaspirational (who already is who he wants to be); one goal of this book is show that Milton, notwithstanding his interesting discussion of progress in Paradise, lets the emphasis fall on the pleasure Adam and Eve get to enjoy simply by virtue of their innocence—as well as the satisfaction readers can take in recognizing how much they have in common with them. In this respect, Milton follows a pattern this book identifies in a wide variety of texts: Baconian-influenced authors attend with some regularity to the intellectual affordances of daily experiences of easygoing enjoyment. The surprising leisureliness of scientific aesthetics suggests a less aggressive course of action than the version of the Baconian project that takes shape in Harrison and Picciotto. The idea of effortless labor is not always an object of desire; sometimes, it's descriptive of the present delectation of observational pleasure.[78]

Aims and Methods: Essaying History

Throughout this book, I am guided by fidelity to a concept, but I am no less committed to understanding the historical past. The alternative to the ascendant historicism of literary scholarship, which is especially powerful in earlier fields such as early modern studies, is sometimes understood as too much of an alternative. Whether that sense of a fork in the road follows from poststructuralist resistance to contextualization or from the routineness of the historicist's critique of anachronism, I suggest an alternative to the starkness of the distinction. A research program driven by a conceptual problem need not be ahistorical. The historicist is right to understand vigilance against the scholar's back-projected perceptions as an expression of high regard for the past, but there are other compelling ways of showing such respect. The past is not fully constituted by empirically verifiable facts. Indeed, the extent to which something leaves a legible trace is not a measure of its reality (the ephemeral is no less real than the durable). Some of what is most compelling in the literature of early modernity raises questions and induces reactions for which readers are necessarily partly responsible. Sometimes, the material ground of the inquiry (a poetic verse, say) doesn't rise to the level of ironclad evidence for the historical reality of the experience it generates. In such a case, one of the primary sources that must be consulted is the critic's mind. Much of the richness, complexity, and allure of the historical past depends on the capacity of its artifacts to move us, confuse us, and draw us out—to offend us, repel us, and invite our response. Our willingness to embrace both affective and intellectual responsiveness shows dedication to a past capacious enough to encompass latencies, which are neither purely imaginary nor dependably available to the kind of scrutiny that proves beyond all doubt.

Many of the most influential scholars of early modernity do not share my emphasis on cocreation but are instead focused on a practice of retrieval that seeks to protect features of the past from contamination by the critic's perspective. I want both to distinguish my approach from theirs and to engage them in conversation. As every chapter of this book attests, I am deeply indebted to their line of inquiry, and I am hopeful of reciprocal interest. It's for this reason that I wish to explain in greater detail how I understand this book's commitment to history. What I have to say about the observational mood is dependent on my seventeenth-century sources (I wouldn't know how to say it without them), but they've had to enlist my participation in unfolding their ideas—not because they're guilty of failure or incompleteness but because they don't take my subject as their focus. I find myself involved with them not because I'm interested in my own perspective but because I'm committed to understanding an experience that often remains implicit—that modulates the movement of verse, inflects the unfolding of syntax, or determines the shape of metaphor without necessarily finding direct expression as content. The work of understanding an underdeveloped thought means having to think *with* rather than simply *about* my authors.

I can clarify the practical necessity of my approach by describing two alternative versions of this book that would look more attractive from the perspective of a less deliberately exploratory historicism. Each of them follows from the premise that literary scholarship had better take up questions for which empirical answers are eminently available, but neither reveals very much about the observational mood. I do not deny that each of them would be interesting—only that they speak to my theme. First, I might have written a book about the history of a word: "sprezzatura," "nonchalance," or "carelessness." Yet "nonchalance"—in Montaigne's hands, probably the best early modern term for the observational mood—rarely appears in English-language sources before the eighteenth century.[79] Montaigne knows that Castiglione's "sprezzatura" has been translated into French as "nonchalance" (see chapter 1), yet he does not use his signature term the same way Castiglione uses his. In England, many emotion-words—from "carelessness" to "ease"—can refer to the observational mood, and yet none of them are sure indications of it. (The ones that turn up as translations of "nonchalance" in English versions of Montaigne do not always circulate with the same connotations.) The only way to identify the observational mood in literary history is to find an affective-perceptual pathway through the sources that reveals the distinguishing features of the experience. In other words, the observational mood just doesn't yield itself up to strictly philological inquiry; the inherently conjectural practices of literary interpretation are indispensable. Second, I might have taken up the question of philosophy "as a way of life" (following Pierre Hadot, say)—as personality or "lifestyle," ethos or therapeutics.[80] I understand the appeal of that topic, which would supply me with a pleasingly "round" object of inquiry: a self-fashioning character.[81] Yet questions of self-presentation and self-cultivation direct attention away from the unselfconsciousness that defines the observational mood. Since carelessness diminishes the likelihood of direct self-description, it seldom leaves the highly legible traces a historian would want. This is not to say that my authors do not describe their personae; it's only to say that the observational mood is best understood as an impersonal (outward-looking) experience. Montaigne might seem at first like an exception to this rule; his special status in this book, which I discuss below, owes much to the nuance of the account he gives of himself. Still, the "self" he presents to us takes shape as a sequence of changing prospects on the world: sometimes, his interest is less the self-as-character than the self-as-prism through which all manner of phenomena get refracted. Throughout this book, what matters most to my argument—the unfolding in writing of the experience of the observational mood—undercuts, diminishes, or distracts from the work of self-display.

Having chosen to focus on a concept that often eludes self-conscious reflection, I accept the burden of describing an experience that is not so much underexplained (which again sounds like a failure on the part of my authors) as obliquely described. My success depends therefore on the clarity of a definition that can accommodate a variety of cases—that permits the concept flexibility without allowing it to collapse into a catchall. Were the experiences detailed in this book detached from the description this introduction supplies, they would not all belong to the same set. Were the concept too rigid to accommodate different situations, it would seem an arid artifact of thought. My intention is to subject the concept to ongoing trial—to flesh it out in a variety of scenes and in relation to an array of distinct historiographical problems. Doing so has sometimes meant amplifying the observational mood relative to other aspects of my sources. I certainly don't detect its presence where it can't be found, but I also don't emphasize its subordination to those themes for which my sources are already widely known. Although I seek to offer plausible historical contexts for my interpretations, the observational mood is too often transitory, implicit, or unconscious to benefit from imposing the assumed coherence of the author's apparent will on my critical judgments. I do not argue that all the authors explored in this book intend to foreground my topic as a central theme. A *gain* in clarity about the observational mood is sometimes a *loss* in precision about the extent to which my authors agree with my perception of what they say and what they do. In places, I've gone as far as to inhabit the experience of effortlessness—as several of my sources invite their readers to do. In taking this approach, I've forfeited my claim to the sternest scholarly disinterest, but I've also created an opportunity to pose new questions. How does a poem change under the gentle pressure of the observational mood? When a literary work invites me to breathe its atmosphere, how does my willingness to do so influence my experience of reading?

My ongoing and recursive description of an as-yet-untheorized concept places special pressure on my use of language; there's a sense in which a thoroughgoing account of a multidimensional experience for which we do not already have a vocabulary actually needs to be *made out of language*. It must be rendered in terms at once precise, visceral, and intuitive enough to spark the reader's recognition. My wish is to *animate* the works in question. I do not produce exhibits for the prosecution of a case; I undertake the inherently imaginative task of recreating firsthand experience—a responsibility I share with my authors, who are much more eloquent than I know how to be. By proceeding in this way, I invite a rapprochement between historical and conceptual inquiry. This approach explains distinctive features of the style I adopt in this book. As this introduction already shows, I often write in the first person. In a book that assumes that proto-objectivity has more to do with the ordinary situatedness of the observer than is sometimes thought, it makes sense to mark the location from which I speak. Perhaps more unusually, I also make regular use of the first person plural. When I describe an experience that "we" have together or point out something that happens to "us" as readers, I do not imply that everyone will have the same experience. My first person plural is an extended hand; I want to show you around. On this question, Stanley Cavell is instructive. Though he refers specifically to Wittgenstein's appeal to ordinary language, his argument illuminates other styles of interpretation founded on singular acts of judgment but offered up to a diverse and unpredictable audience.

> When Wittgenstein . . . "says what we say," what he produces is not
> a generalization (though he may later generalize), but a (supposed)
> *instance* of what we say. We may think of it as a sample. The introduction
> of the sample by the words "We say . . ." is an invitation for you to
> see whether you have such a sample, or can accept mine as a sound
> one. One sample does not refute or disconfirm another; if two are in
> disagreement they vie with one another for the same confirmation.
> The only source of confirmation here is ourselves. And each of us is
> fully authoritative in the struggle.[82]

I face the challenge of offering as persuasive an interpretation as possible but without inflating persuasion beyond its modest status as an invitation for others to share my perception—to see how far they are willing to travel with me and how meaningful they find our joint discoveries.

Scholars often make a great effort to be objective about literary works, but less often do they try to be objective about when to be objective. It's not that I deny the value of suspending personal agendas in the interest of fair-minded reading; it's just that I doubt the possibility of suspending all of them—as well as the assumption that every agenda is equally obstructive to understanding. The scholar's attraction to her object should not disqualify her from speaking about it; it may turn out to be the source of her insight. I am wary of the assumption that the only licit scholarly passion is a generalized "desire to get things right." Indeed, emphatic disinterest can have (I do not say "does have") a damagingly neutralizing effect on an otherwise bracing, seductive, repulsive, heartbreaking, or alarming voice from the historical past. The admission of emotional entanglement does not imply capitulation to the merely personal. I'm reminded here of Raymond Williams's well-known description of the problem of understanding history as it presently unfolds—before it "precipitate[s]" out into knowable categories and objects of analysis.[83] For Williams, "structures of feeling" are affective means of apprehending history in this chaotic, unsettled form.[84] They are ways of "understanding" events that otherwise escape comprehension. Yet I wish to observe that we cannot take for granted that things ever really "precipitate" out—or, to put this differently, that it's wise to trust the clear outlines of "precipitated" versions of past events. How are "precipitations" related to whatever it is I originally set out to find? When does the process of "precipitation" amount to rendering something unrecognizable? The anthropologist Kathleen Stewart has described an "improvisatory conceptuality" that responds to the confusions inherent to ordinary experience without flattening them out with inherited categories or a precipitous rush to judgment. "Every day," she writes, emphasizing the salutary disorientation of the experience, "in perfectly ordinary moments, there's an activation of the details of something suddenly somehow at hand."[85] Sometimes, the vertigo of ordinary experience—that sensation of being asked by phenomena to understand them in their very thrown-together-ness—characterizes the scholar's encounter with historical sources as well. We cannot perfectly stabilize the relationship between subject and object in the name of critical sobriety. Our sources draw us in.

The Shape of the Book

My approach in this book is prismatic. I value this metaphor because it suggests a set of distinct views of the same object. Each chapter brings one facet of the scientific imagination closest to the eye, but the larger field of inquiry remains visible. The first chapter is the ground for what follows (the base of the prism), but the remaining three (sides, to extend the metaphor) are adjacent to each other: they are organized by topic rather than chronology (though, after the first chapter, I don't stray far from the middle years and second half of the seventeenth century). Were I not in the habit of referring backward, I would invite the reader to explore those three chapters in any order—but the book's cumulative momentum argues against it. Taken together, they offer a revisionary account of the scientific imagination by refracting it through a set of related themes or motifs: *thought, vision,* and *trial*. With the exception of the last, each chapter develops its theme by way of juxtaposition. In the first three, I stage an encounter between two very different authors—one of whom is usually associated with "science" and the other with "literature." The effect, I hope, is an ongoing sense of the diversity and complexity of the cultural field. In the fourth chapter, I have given priority to my duty as a literary critic and devoted myself to *Paradise Lost.* Yet Milton's multifariousness ends up producing a similar effect to the other chapters: dissimilar perspectives meet within the space of this one poem.

Often, scholars working between literature and science have felt obliged to chart a path from one domain to the other. Sometimes, for instance, they find that science is a source of topical interest for literary writers (Why does Milton allude to Galileo in *Paradise Lost*?), and other times they show that philosophy depends on fiction or metaphor to accomplish its aims (Why does Descartes discover mind-body dualism over six days of fictional meditation?).[86] This binary choice hardly exhausts existing approaches to the topic, but it does illustrate the frequent habit of dropping anchor in one field or the other (notwithstanding the equally frequent habit of reminding readers that no vast disciplinary gulf lies between them in the seventeenth century). My intention in this book is to describe the formal and conceptual resources of different genres without preserving a categorical distinction between them: variegation, then, rather than opposition.[87] I've attributed the observational mood to the "scientific imagination," but attention to the natural world has always been the work of literature as well as science. Just think of *De rerum natura* or the Book of Job. My wish is to keep track of both threads—or, more exactly, to show how a single thread weaves its way through different genres and areas of inquiry. I'm attentive to the observational mood's bearing on Baconian themes not because it's the exclusive property of that philosophical tradition (my insistence on its ordinariness implies that it is to be found everywhere) but rather because it disrupts a consensus view of the New Science.

One weakness of this book, which I hope is also a strength, is its modest sample size. Relative to the work of the historian, literary criticism is often like this; as if by synecdoche, it advances a claim about the historical past by way of just a few examples. I want to specify the trade-offs I have chosen to make, and to explain why I take them for enhancements of, as well as limitations on, my analysis. Because my intention is to read early modern authors with exploratory interest—indeed, to exemplify the interpretive latitude of the observational mood—I have adhered to the norm of selectivity. A single book has only so much space for wide-ranging interpretations. Yet this presents a problem for an argument that aims to revise the prevailing view of the culture of experimental science. I have responded to the problem of synecdoche by paying special attention to those authors who figure prominently in extant intellectual and literary histories. A new interpretation of Boyle has a more immediate impact on our sense of the Baconian moment than one, say, of Nehemiah Grew. The argument depends on the cumulative strength of a small set of carefully chosen interventions. While some scholars seek to revise an established narrative by bringing new characters into it, my strategy here is to show that characters we know are not who we thought they were. Yet this way of putting it understates my attention to literary and philosophical works other than my primary objects of inquiry. In addition to the figures who receive name billing in my table of contents, plenty of other ones—Robert Burton, Thomas Browne, and Margaret Cavendish are three important examples—make appearances.[88]

One historical circumstance to which I repeatedly refer, even if I cannot give it sustained attention, is the event of civil war. Because I discover a strain of irenic thoughtfulness in Bacon's philosophy and then go on to show its importance during the Interregnum and Restoration, I inevitably raise the question of why seventeenth-century Baconians are sensitive to a subtheme (they are often more sensitive than Bacon himself) that later readers would frequently ignore. One answer is the terrifying experience of civil strife. Another is the religious zealotry on which many of the characters in this book place the blame for war. I believe Bacon's observational mood, the feeling of lowered stakes and peaceful accommodation he describes, would have much to recommend it in this context. I also suggest that Montaigne's *Essais* (1572–92), the great theme of which is wartime bloodshed, but which is often explicitly focused on other (often epistemological) questions, sets a good precedent for my approach. I do not go as far as to elevate the question of violence (or, in this epistemologically oriented book, any other moral issue) to my own "great theme," but it does not completely recede from view.

With the exception of Bacon, no one matters more to my argument than Montaigne, and for several reasons other than his response to the French Wars of Religion. Most important, he's the best theorist of the phenomenology of artlessness I defended in this introduction. When Montaigne presents himself as "nonchalant," "lazy" (*oisif*), and "disdainful" (*desdaigneux*) of effort, he nods at Castiglione's paradigm of sprezzatura—but without the emphasis on hidden calculation scholars have taught us to expect. It's Montaigne, moreover, who most explicitly argues for the epistemological value of gently aimless psychic states. Though this book sets its sights on seventeenth-century England, Montaigne is my investigation's guiding light. One might even describe my line of inquiry as an extended answer to the following question: What if the Baconian experiments of the seventeenth century, both in the laboratory and on the page, are much closer to the spirit of Montaigne's casual, digressive essays than has been hitherto acknowledged? For me, this is less a question of genre than of mood—though I discuss, among other Montaignian echoes, Izaak Walton's citations of the *Essais* and Boyle's rather offhanded use of the word "essay" to describe his distinctive writing practice.

Chapter 1, "'Nonchalance' and the Making of Knowledge," advances a fresh interpretation of Bacon's philosophy that foregrounds his interest in Montaigne. Reading between their bodies of work, I discover the conceptual resources on which the remainder of the book depends. Though Bacon's awareness of the *Essais* is no surprise, I explore a different point of contact: the remarkable similarity between Montaigne's description of "nonchalance" and Bacon's conjectures on the importance of mental clarity to scientific discovery. Although I raise the possibility of influence, my argument is not for direct indebtedness. Because mid-century Baconians have more to say about the observational mood than Bacon himself (in which it's a persistent but subtle theme), Montaigne's perspective is useful to us as a kind of magnifying glass: it brings Bacon's carelessness into full relief. Reflecting on the Epicurean tradition as well as the paradigm of artlessness he discovers in the *Cortegiano*, Montaigne departs from those precedents by embracing the virtues of unfeigned and uncultivated dispassion. In the face of civil war, he presents his own propensity for emotional cool as an accidental countermeasure to violence—as well as the very condition of intellectual freedom. I demonstrate the traction of this idea across Bacon's body of work, setting the stage for subsequent chapters that focus on epistemological questions but nonetheless retain an interest in the ethico-political circumstances of interpretation. With special attention to the *Advancement of Learning* (1605), but looking also to other parts of Bacon's oeuvre, I argue that calm flexibility and casual distraction belie the relentless intensity of purpose with which his project is now associated. That surprise prepares us for the interpretation of his literary and philosophical heirs.

Having anatomized the observational mood and explored its role in the *Advancement*, chapters 2 through 4 move ahead to the second half of the seventeenth century, each of them taking up a different theme in order to revise our understanding of Bacon's cultural legacy. The first of these is *thought*—which here goes by the name of "occasional reflection." In chapter 2, "The Angle of Thought," I ask how "carelessness" animates the intellectual life of the chemist Robert Boyle. Juxtaposing his long dialogue on the pleasures of fishing (*Angling Improv'd to Spiritual Uses*, early 1650s) and Walton's popular handbook on the same theme (*The Compleat Angler*, 1653), I argue that they together typify an emergent literary field energized by developments in science. Taking Walton's lead, I propose a new reading of Boyle's career, ranging from laboratory notes to natural theology. Boyle's meditative practice ("occasional reflection") is an experience of easygoing susceptibility to "chance" occurrences that frames not only his devotional writing but his scientific pursuits as well. Though often held up, in his time and ours, as a model of conscientiousness, Boyle is also the very picture of discomposure—and of the mind's propensity to wander. "Carelessness" spans his literary, devotional, and philosophical works as a name for a "way of thinking" (the phrase is Boyle's) that enables sudden and unexpected revelations—whether he happens to be rambling the countryside with fishing companions or testing out the properties of glowing minerals in his laboratory.

Chapter 3, "The Microscope Made Easy," examines the perceptual powers of the period's new optical technologies, putting the poet Andrew Marvell into conversation with the natural philosopher Henry Power. Here, my organizing theme is vision. Both Marvell and Power wield magnifying lenses with flamboyant abandon, savoring the participation of untamed "fancy" in the observation of nature. The microscope, we discover, materializes the fantasy that discovery is effortless. Notwithstanding the actual technical difficulties of manipulating the instrument, the lens offers up to the imagination a perfect picture of a perceptual enhancement from which the inquirer reaps immediate rewards. In *Upon Appleton House* (1651), Marvell compares the sensory sumptuousness of the poet's experience to the amazing intensification of visual power anyone can accomplish with no more arduous a gesture than holding a lens up to the eye. Thus the poem seamlessly integrates Montaignian insouciance and Baconian optimism. In *Experimental Philosophy* (1664), the first book of microscopy in English, Power offers readers a similar experience of vertiginous receptivity as they examine hitherto imperceptible objects. Poetry is especially rich with formal resources for pleasurable torsions of perspective (Marvell puts georgic and pastoral traditions to new use), but Power finds means of his own for a calm but chaotic expedition into the microscopic world. One strange lesson of their joint experiment is that the enhancement of vision is an experience of blurry confusion, but one that nonetheless contributes to the production of knowledge.

In the final chapter, "The Paradise Without," I set aside my practice of tacking between "literary" and "scientific" figures (Montaigne, Walton, and Marvell, on the one hand, and Bacon, Boyle, and Power, on the other) in order to focus on what might be taken for the greatest literary achievement with Baconian ambitions: Milton's *Paradise Lost* (1667). I argue that some of the most powerful interpretations of the role of science in the epic too quickly assimilate Milton's interest in experimental trial (the final term in the thematic triptych of chapters 2 through 4) to the vision of Baconian rigor I have called into question. Reflecting on this theme of trial, I discuss how the feeling of effortlessness can buoy the experience of challenge; in so doing, I bring into full relief the question of easy exertion that has likewise animated the preceding chapters. Focusing on Adam and Eve's debate about the work of gardening (which prompts Fish's insistence on the utter impossibility of unselfconsciousness), I demonstrate the importance of a distinction between labor and laboriousness to Milton's poem. Paradisal gardening perfectly accommodates the observational mood: it's a flexible practice of exploration, inquiry, and conversation that never has to move more quickly than the first laborers wish. I also show that the story of Adam and Eve's departure from Paradise is a relinquishment of carelessness—but that Milton invites readers to inhabit the experience of innocence by reclaiming the observational mood.

The book ends with a brief postscript that reflects on the role the observational mood has played—and the other roles it could play—in the practice of literary criticism.

CHAPTER 1

"Nonchalance" and the Making of Knowledge

Francis Bacon after Michel de Montaigne

Francis Bacon has a bad reputation. To many philosophers and social theorists, the good news of the "invention of modern science" is anything but: it signals the precipitous rise of an impersonal perspective that gazes down imperiously on the earth and its inhabitants, licensing their exploitation. Max Horkheimer and Theodor Adorno crystallize our wary late modern reception of Bacon's vision: "Knowledge, which is power, knows no limits, either in its enslavement of creation or in its deference to worldly masters."[1] I do not refute the chilling précis. The fathers of critical theory have good reason to recoil from one cruel legacy of Enlightenment: the use of reason as an instrument of domination. Indeed, revulsion at Bacon's philosophy illuminates features of his thought we can't explain away. Another influential example is Carolyn Merchant's ecofeminist critique of the New Science, which calls attention to the disquieting language of torture we find in Bacon's descriptions of experimental procedure; her polemic responds to the stark fact of his rhetorical violence.[2] In his *De sapientia veterum* (*Wisdom of the Ancients*, 1609), for instance, Bacon takes Proteus as a figure for "Matter" and recalls that the mythic prophet jealously guards his clairvoyance unless he is captured and forced to share it.[3] Recommending chains and abuse as instruments of interrogation, Bacon remarks that the able interpreter gives Matter no choice but to disclose her secrets by "vexing and urging her with intent and purpose to reduce her to nothing."[4]

Many scholars of early modernity are less averse to Bacon's vision than postfoundational critics of Enlightenment (whether because of professional disinterest or genuine enthusiasm), but they often affirm the premise that his philosophy is best understood as a sustained reflection on the exercise of power. Throughout the scholarly literature, Bacon is less brain than muscle.[5] We can discern the deep entrenchment of the idea, the near-synonymy of Bacon's name with ardent exertion, by observing its importance to even his most adventurous interpreters. I'm thinking, for instance, of Ronald Levao, who borrows the image of "wrestling" (*luctari*)[6] from the *Cogitationes de natura rerum* (*Thoughts on the Nature of Things*, 1624) and speaks of aggressive "athleticism" as he tracks Bacon's alternation between "suppleness" and "severity."[7] Similarly, John C. Briggs sums up his subject's contribution to natural philosophy as "a way of understanding nature and mankind in terms of a code to be broken and exploited through self-abnegation and an assiduously analytic inquiry undertaken for the sake of charity."[8] Though he ultimately presents us with a startling image of Bacon as a *magus* who coaxes secrets from persuadable matter, he repeats the verdict we've come to expect: the philosopher pursues "assiduous" *askesis* and the exultant accumulation of power.

We are too familiar with Bacon the disciplinarian to make out the other roles he plays—including the theorist of the observational mood from whom the first experimental scientists draw inspiration. What's intriguing about this particular alternative is that it's exactly the opposite of what we've come to expect: Bacon the apostle of procedural rigor forgets his commitments and savors the perceptual advantages of absent discipline.[9] This chapter makes this other perspective available to us. In pursuing this goal, I echo an essay by Guido Giglioni that distinguishes itself from the standard view of Bacon's work by exploring his use of the image of the *silva*, or forest, to convey "the need for the human mind to be … lost in experience" and to resist the imposition of order on "natural particulars."[10] Yet my emphasis on states of feeling—and my reluctance to subordinate Bacon's carelessness to his teleological ambition—demands a different line of inquiry. I conduct an experiment in reading, risking underadherence to some of Bacon's most explicit wishes by highlighting an underappreciated aspect of his thought. Though I aim to understand how my theme relates to Bacon's better-known commitments, adopting his hierarchy of values wholesale would return my topic to obscurity. I am emboldened in my approach by the more salient observational moods of Bacon's successors in the second half of the seventeenth century (one of whom, Robert Boyle, is the subject of chapter 2); here, I seek a precedent for the alliance between waywardness and perceptiveness on which the remaining chapters of this book are focused. Like a doctor who administers a radioactive dye, or "tracer," to highlight an area of concern in a patient's body, I adopt a method that privileges what is otherwise a hidden feature of Bacon's philosophy.

My "tracer" has a proper name: Michel de Montaigne. Bacon's familiarity with Montaigne's *Essais* (1572–92) should come as no surprise to scholars in the field; the three-volume experiment in self-portraiture lends a title, if not exactly a model, to Bacon's less prolix *Essayes or Counsels, Civill and Morall* (1597, 1612, 1625).[11] Yet the connection I propose is a new one—perhaps, as it has nothing to do with Montaigne's representation of nature or his engagement with natural philosophy, an unexpected one. One through-line of Montaigne's book is exactly the sort of unlabored indifference that intermittently animates Bacon's writings. Montaigne's word for it is "nonchalance," and he delineates a coherent perspective on the epistemological strides it effortlessly propels, serving as the most insightful (though proleptic) expositor of Bacon's observational mood. When I eventually take up their similar responses to the Epicurean concept of *ataraxia*, or "imperturbability," I raise the possibility of direct influence, but I don't press the point.[12] I take Montaigne as a guide to Bacon strictly on the basis of formal and conceptual affinity. Another reason I've chosen to begin with Montaigne is that I wish to foreground what we *lose* in passing from his perspective to Bacon's: in particular, an emphasis on the moral danger inherent to the observational mood. Though, in subsequent chapters, I linger with the optimism of the scientific imagination, Montaigne presents us with an initial formulation of our theme that denies us the luxury of forgetting its costs. In this too he is a great anticipatory analyst of Bacon's intervention in natural philosophy. Montaigne lingers with what is genuinely risky about the experience of effortlessness this book explores, developing an early, subtle, and strangely self-indicting version of the critique we have seen taken up by Bacon's late modern interpreters. Notwithstanding my emphasis on what is Montaignian in Bacon, our awareness of disjunction should inflect our understanding of this book's remaining cases—as an early indication of the unruly multiplicity of literary and philosophical projects that achieve moments of harmony without falling into perfect alignment.

The first half of this chapter explores Montaigne's conjectures on "non-chalance"; the second shows how powerfully they illuminate Bacon's celebration of scientific progress. Though I range over Montaigne's *Essais*, I give special attention to just one of them, "De l'utile et de l'honneste" ("Of the Useful and the Honorable"), where he conveys most clearly the moral stakes of his ruminations on the observational mood. At the center of the chapter, I offer a brief reflection on "De l'experience," the climactic final entry in the *Essais*, which serves as the hinge around which my argument turns, tracing the path of "nonchalance" from ethical to epistemological questions. Taking up Bacon's body of work, I focus on the *Advancement of Learning* (1605), but without losing sight of the total project he calls "instauration." Though the *Novum Organum* (1620) plays an important role in familiar accounts of Bacon's legacy, I make a case for charting a different course. Pierre Villey identifies the *Novum Organum* as the single text in Bacon's oeuvre that clearly bears the imprint of Montaigne's influence, but the argument holds in only one respect.[13] Villey's interest is the transmission of an idea. In the *Novum Organum*, he argues, Bacon adopts Montaigne's thesis that the mind is untrustworthy—and thus concludes that we would do well to bridle it. Yet it's the Bacon of the *Advancement* who comes closest to Montaigne's observational mood—as reflected in his literary style no less than his thematic focus.[14] Bacon describes a state of calm satisfaction that animates understanding, and he finds distinctive means of conveying it in prose. Bacon's *Advancement* is more "essayistic" than either the *Novum Organum* or even his *Essayes*; it shows a willingness to digress. We should not be surprised by the difference between the *Advancement* and the *Novum Organum*; in the *De augmentis scientiarum* (1623), the significantly expanded and Latinized version of the *Advancement*, Bacon underlines what the earlier version already tells us: that here he "say[s] nothing, nor . . . give[s] any taste" of the method expounded upon in his other works.[15] Bacon tells us where to find him at his least methodical. In this chapter, I go looking for him there.

The Passions of Strife

Like his beloved Socrates, Montaigne thinks any ordinary thing a worthy candidate for philosophical reflection.[16] When, for instance, he explains his aversion to playing chess, a trifling matter of personal preference competes for our attention with no less exalted a subject than "the human condition" (*l'humaine condition*).[17] His purpose in this essay, "Of Democritus and Heraclitus," is to "weigh" or "judge" (I translate from Montaigne's sense of *essayer*) the contrasting perspectives on human folly signified by the names in the title. "Democritus and Heraclitus," Montaigne writes, "were two philosophers, of whom the first, finding the condition of man vain and ridiculous, never went out in public but with a mocking and laughing face [un visage moqueur et riant]; whereas Heraclitus, having pity and compassion on this same condition of ours, wore a face perpetually sad, and eyes filled with tears" (220, 1:359). Montaigne prefers ridicule to pity because "it seems to me we can never be despised as much as we deserve. Pity and commiseration are mingled with some esteem for the thing we pity; the things we laugh at we consider worthless" (221). Montaigne claims to trivialize, but he refrains from doing exactly that when it comes to the supremely trivial matter of playing chess, betraying a much straighter face than the "visage moqueur" he apparently wears. "It is not enough a game," he writes, "and too serious an amusement. I am ashamed to devote to it the attention that would suffice to accomplish something good" (220). Thus Montaigne the mocker wishes for a world of justly apportioned seriousness—one in which things matter exactly as much as they deserve. He even asks us to notice the disjunction, remarking that he "hates" chess ("Je le hay et fuy") before excluding "hate" from his Democritean perspective (220, 1:358). "For what we hate we take seriously" (*car ce qu'on hait, on le prend à coeur*), he explains, inclining instead to observe calamitous error without the misfortune of "taking" it "to heart" (221, 1:359)

Why does Montaigne suggest that one thing in particular is unworthy of affective investment when he claims to withhold "esteem" from the manifold world? We discover an answer to the question if we observe that disconnection between successive modes of feeling is central to Montaigne's self-understanding: he presents us with a disposition that slips away from the paradigms it entertains, including the very atmosphere of amenability that characterizes so many of the essays. Though he tends to inhabit an attitude of gentle irony, the very lightness of the disposition permits surprising affective transitions, as illustrated by the uncharacteristic humorlessness of the admonition: Chess players had better put their intellectual energies to better use! Montaigne has been celebrated for a "pliant goodness" that counters the stern determination of classical virtue; my point here is the simple but perhaps surprising one that such flexibility lives up to its name when it hardens into frigid disapproval.[18] The defining feature of Montaigne's unseriousness is freedom from labor, including the effort of adhering to principle. Even as sensibly minimal a practice as steering clear of hate, which seems both guiltless of solemnity and unmistakable for virtue's steadfastness, is too programmatic to suit Montaigne's temperament. When the stakes are considerably higher, he likewise narrates affective disunity instead of ensuring the coherence of a delimited portrait. In "De la cruauté," for instance, he doesn't make a straightforward bid for merciful kindness but explains instead that "cruel" is exactly the right word to describe the manner in which he "hate[s]" the vice in question: "I cruelly hate cruelty" ("Je hay, entre autres vices, *cruellement la cruauté*"; 313, 2:98; emphasis added).[19] The seeming paradox of the phrase illustrates my point about the degree of flexibility to which he aspires: loosening cruelty's hold means refraining from fending it off.

Montaigne's favorite word for easy accommodation is "nonchalance," exactly the term he finds standing in for *sprezzatura*, along with "mesprizon" (contempt), in French translations of the *Cortegiano* (though a good Italianist, Montaigne alludes specifically to these editions).[20] We hear echoes of this language when he observes that humankind can never be "assez mesprisez selon nostre merite" (despised as much as we deserve), just as we do when he praises the Democritean "humeur" for being fittingly "desdaigneuse" (disdainful)—a term that likewise belongs to the courtier's lexicon.[21] It comes as no surprise, then, that scholars often take sprezzatura as a rubric for reading the *Essais* (and not without considerable rewards), but these assessments, focused as they are on hidden artfulness, can only obscure the question of atmosphere.[22] I've already suggested that sprezzatura, in both early and late modernity, is usually taken to imply the feint of a clever illusionist whose calculating ambition busily spins wheels behind a deceptively placid face. Montaigne's very different habit is to treat his observational moods as affective accidents, and ones that can't be preserved by feats of deliberateness.[23] He describes a disposition defined by constitutional weakness, originating in circumstances beyond his control (heredity, nature) and dissolving unpredictably at the slightest pressure. One meaning of "nonchalance" is unresisting readiness, a facility for yielding not only to powerful emotions like anger and joy but even to the very thing it isn't: unyielding intensity of purpose.

My point here isn't that we could reasonably exclude "art" from the *Essais*, but rather that the claim of artlessness only rings false (demanding redescription as strategic performance) if held to the impossible standard of utter self-abandonment (the ecstatic death of the subject) or sheer disorganization (the essay as mess). One negative consequence of the otherwise salutary emphasis on the constructedness of things that pervades most quarters of literary scholarship is the premise that fabrication—usually a predicate of society or one of its constituents—wholly saturates selves and the artifacts they make, leaving nothing slack. By contrast, Montaigne's "nonchalance" assumes underdetermination—a gap between self and self-management that takes the form of affective drift. As if in anticipation of late modern incredulity, Montaigne writes: "Those who commonly contradict what I profess, saying that what I call frankness, simplicity, and naturalness in my conduct is art and subtlety, and rather prudence than goodness, artifice than nature, good sense rather than good luck, do me more honor than they take away from me. But surely they make my subtlety too subtle [ils font ma finesse trop fine]" (603, 3:10). Montaigne professes respect for virtue but distinguishes it from undeserved "naturalness." He simply hasn't the capacity for such "fine" workmanship.

We begin to sense the deep strangeness of Montaigne's observational mood when we observe that witty self-description and flirtatious indirection serve no less straight-faced a purpose than parrying the violence of the *guerres civiles*, the essayist's preeminent concern. By adhering to the thesis of Montaigne's preoccupation with violence, I write in sympathy with David Quint, who offers a sustained interpretation of the *Essais* as a response to the French Wars of Religion.[24] Yet my interest in the thoroughgoingness of Montaigne's malleability (rather than, say, a "flexible" attachment to principle that would only be "soft" in the relative sense that other attachments are more insistent) sets me apart from critics who share my point of departure in ethical considerations. Quint, for instance, argues that Montaigne views "natural goodness" as a better premise for ethics than principled commitment, since the latter would seem much more likely to encourage violence than interrupt it (especially when considered from the vantage of war-torn France, where combatants appeal to God's cause and aristocratic honor as they lay the country waste).[25] Although Quint notes the shortcomings of an ethics predicated on "natural goodness" when "the cruelty [Montaigne] deplores is equally natural to human beings," he nonetheless speaks approvingly, if with an air of disillusioned realism, of that position: "live and let live may be as good a start as we shall ever find in thinking about morality."[26] I think we miss one of the challenges of Montaigne's moral reflections if we understand his acknowledgment that "natural behavior" isn't uniform as a realist concession to the necessary imperfection of social harmony. The force of the assertion that "cruelty" is "natural" is not that we should expect violence to persist no matter how much we encourage kindness (the point is true enough). "Cruelty," the bête noire of the *Essais*, is "natural" even to our gentle author himself, and his predilection for acting "naturally" (for acting on the basis of predilection) conveys the terrible delicacy of his own moral standing: not knowing in advance how he will behave or whether it will be kind. The claim sounds unlikely for an author who has proved himself endlessly likable (I refer to generations of admiring scholarship), but one task of my interpretation is to show how genuinely frightening Montaigne can be—how alarming he is willing to be when he lingers with the problem of characterological drift. The sovereignty of chance ensures the possibility of aggression (or even straight malevolence) just as resolutely as it sustains an affective flexibility that tends, or has tended so far, to discourage it.

Another way to put this is to say that Montaigne is less a practical ethicist who comes up short than a perceptive observer of his times who finds good reason to refrain from dispensing moral counsel. He exposes the limits of ethical tenacity without wishing to reproduce its violence; handing down a new moral program—even one predicated on something like "natural goodness"—might duplicate the doubt-free adherence to principle from which he takes his distance. Montaigne's "naturalness" is a style of skepticism; often, "natural" reads as "effortless," describing both the manner and the mood in which Montaigne wanders away from orthodoxy.[27] He thus dependably disarms late modern misgivings about the norms smuggled in by such claims; Montaigne's "nature" doesn't imply self-evident standards of behavior.[28] It's true that the observational mood guarantees nothing (indeed, it leaves the door open to the most perilous passions, including the ones Montaigne "hates"), but risk follows from affective looseness rather than a foregone conclusion about what counts as possible. Montaigne is much less an advocate for anything in particular than a horrified witness to the enlistment of reason in justifying bloodshed.[29]

Indeed, he most often describes violence as a matter of drive rather than decision. In "De l'utile et de l'honneste," he exposes the affective determinants of rationalized acts of injury, discovering vectors of passion where virtue supposedly reigns. ("De la cruauté" is an equally important statement of Montaigne's moral vision, but it pays less attention to Montaigne's dispositional "nonchalance.") The cruelties of war, though cloaked in conviction, find their origin in mere "malignity"—whatever that means. Of his bloodthirsty countrymen, Montaigne has the following to say: "Their propensity to malignity [propension vers la malignité] and violence they call zeal. It is not the cause that inflames them [ce n'est pas la cause qui les eschauffe], it is their self-interest [interest]. They kindle [attisent] war not because it is just, but because it is war" (602, 3:9). Montaigne says nothing about the specific claims with which one justifies violence (he takes no position on this or that ideology), suggesting only that assertions of this kind have little to do with motivation. One detects in Montaigne's remark a fear of exponential intensification; it's not just that people deceive themselves and others about why they wage war but also that fending off doubt by clinging to ironclad reasons has the effect of escalation. For Montaigne, "zeal" is shorthand for passion hardened by principle and thus made resistant to softening influence. "They kindle war . . . because it is war" formulates such willful adherence as tautology. We might describe this process as the reification of passion as allegiance. The interpretive challenge of Montaigne's perspective is that it offers one style of nonexplanation in place of another—the bluntness of saying little over the delusion of self-justification. He affirms sheer errancy rather than suppressing it by stiffening his neck. If rationalization only stokes the flames of war, perhaps the affirmation of untamed emotion, however minimal a gesture, is a bid for peace. Montaigne's diagnosis rests vaguely on the vocabulary of disposition:

"propensity," "interest," and verbs suggesting "heat" (*eschauffer* and *attiser*). Something wells up (or heats up) within us, and that's as much of a "cause" for violence as Montaigne is willing to name.

If the passions fail to explain everything, they succeed at least at explaining something away (they expose ideology as pretext). Such is the elusiveness of Montaigne's moral vision, which predicates comportment on passion, an unpredictable engine for action. Since his irenic disposition isn't simply fabricated, it's hard to square with the imperative mood of moral reflection. Indeed, one of his fundamental premises—and on this score I'm again indebted to Quint—is that goodness isn't rectitude. The observational mood animates the essay as an involuntary ethics, a conspicuously unreliable alternative to the principled violence of the Roman Catholic or Huguenot partisan. "Nonchalance" is not alone among the motifs of the *Essais* in suggesting that ethical engagement precludes unimpeded self-determination. As I mentioned in the introduction, Montaigne describes himself in "Des menteurs" ("Of Liars") as a rare specimen of absolute honesty (a person who can truly be said never to tell a lie) but claims that he owes this admirable quality to something like congenital defect. Good liars need good memories to carry off their fictions, and Montaigne's memory is singular in its weakness. "There is no man who has less business talking about memory," he writes. "For I recognize almost no trace of it in me, and I do not think there is another one in the world so monstrously deficient [une autre si monstrueuse en defaillance]" (21, 1:71). No matter how charming its effect on readers, "nonchalance" can be understood as a form of "monstrosity"—an aberrant quality for which Montaigne bears little responsibility.[30] It's an attitude he can't quite recommend, since he discovers unsought weakness (I'm incapable of lying, I can't sustain a passion) where others see ethical triumph (I refuse to lie, I refuse to lose myself in anger).

Though difficult to explain (in, say, the Baconian sense of locating the mechanism by which it's produced), "nonchalance" is easy enough to locate. "De l'utile et de l'honneste" begins with a description of disposition as atmosphere and style; for Montaigne, it's both an object of interest (he expounds upon it) and the mood in which his sentences unwind (he displays it). When explicit, "nonchalance" is most often a manner of speaking or writing, but language is less the special province of "nonchalance" than the case Montaigne finds most interesting (unsurprisingly for someone whose métier is composing sentences). "No one is exempt from saying silly things," he writes, by way of introduction. "The misfortune is to say them with earnest effort . . . Mine escape me as nonchalantly [nonchallamment] as they deserve . . . I would part with them promptly for the little they are worth" (599, 3:5). The passage describes a practice of shrugging improvisation, a habit of passing things by (including the things we say) without growing too fond of them. It goes on to suggest that the remainder of the essay serves as evidence of the disposition in question: "I speak to my paper as I speak to the first man I meet [au premier que je rencontre]. That this is true, here is proof" (599, 3:5). The offhandedness of the accidental encounter, Montaigne promises us, is perceptible on every page.

Such easygoing self-description becomes a dependable refrain across the book's three volumes. Montaigne's "Democritean" perspective, for instance (which now reveals itself as a rough synonym for "nonchalance"), encourages the habit of "part[ing] . . . promptly" from whatever he happens to say: "Scattering a word here, there another, samples separated from their context, dispersed, without a plan and without a promise, I am not bound [tenu] to make something of them or to adhere to them myself [n'y de m'y tenir moy mesme] without varying when I please and giving myself up to doubt and uncertainty" (219, 1:357). The difference in "De l'utile et de l'honneste" is that Montaigne presents his aversion to "promise[s]" and statements of purpose, his preference not to be "held" (tenu) by what he says, as a counterweight to bloodshed. He extends his account of uncommitted style to the wider realm of action: "I aspire to no other fruit in acting than to act, and do not attach to it long consequences and purposes. Each action plays its game [jeu] individually: let it strike home if it can" (601, 3:7). Though Montaigne meditates on no less weighty a subject than the urgent imperative to "act" (agir) in the midst of factional strife, he reaches nonetheless for the language of play. Intervention in a crisis is "game" (jeu) enough to count as pleasurably idle experiment—unlike, remember, the "game" (jeu) of chess, with its atmosphere of earnest concentration (220, 1:358).

Montaigne proposes that the observational mood encourages exactly those styles of comportment for which one would most desperately hope in the face of spiraling violence. Writing sentences, for instance, seems here to evade the potential friction of the face-to-face encounter. "We have no need to harden [durcir] our hearts with these plates of steel," he writes. "It is enough [c'est assez] to harden our shoulders; it is enough to dip our pens in ink without dipping them in blood" (609, 3:18). Square shoulders do the work of an armored "heart" with less potential for violence, and softhearted affection seizes the advantage from stony relentlessness. An affirmation of the body's strength is more desirable than clothing the body in "steel," a preference that draws our attention to the nuance of Montaigne's position, and of his disposition. He describes a tendency (toward peace) rather than a stance (pacifism), which is to say that he can't quite rule severity out. The rhetoric of adequacy ("assez") treats peacemaking as if it were already under way, yet our assurance in this regard might not last more than a moment.

When Montaigne turns to his own experience of political engagement (his focus here is diplomatic service), he deepens his reflection on a pacific predilection for which he deserves little credit. In the midst of his account of the "feverishness" (fiévre) of war, he portrays himself as a cool, contented, and entirely ingenuous go-between (601, 3:7). "I have an open and easy way," he writes, "that easily insinuates [s'insinuer] itself and gains credit on first acquaintance. Pure naturalness and truth, in whatever age, still find their time and their place . . . My freedom has also easily [aiséement] freed me from any suspicion of dissimulation by its vigor—since I do not refrain from saying anything, however grave or burning [cuisant], I could not have said worse behind their backs—and by its obvious simplicity and impartiality [nonchalance]" (600–601, 3:7). One sees here why Catherine Belsey has located in Montaigne and his English imitators the precedent for "Honest Iago," whose "naturalness" so effectively "insinuates itself" in the susceptible mind of Othello.[31] Iago's strategy succeeds with a completeness that ensures Othello's blindness to deception even as he provides a perfect description of it.

> I know thou'rt full of love and honesty
> And weigh'st thy words before thou giv'st them breath,
> Therefore these stops of thine fright me the more.
> For such things in a false disloyal knave
> Are tricks of custom, but in a man that's just
> They're close dilations, working from the heart
> That passion cannot rule.[32]

Unlike the tragic protagonist, the audience knows Iago's seemingly accidental pauses epitomize the strategic "knave[ry]" Othello conjures forth in his imagination. Understanding Montaigne, I suggest, means risking Othello's credulity (even if we know very well that Montaigne is no Iago): entertaining the thought, however much it may embarrass us, that Montaigne is just as irrepressible as he says (speaking from a "heart / that passion cannot rule"). He tells us he doesn't hide discomfiting facts and opinions, just as he foregoes flattery and overdelicacy. We might say simply that Montaigne speaks the truth, as long as we hear in this weighty word only an attunement to relevance (encouraging the disclosure of "anything, however grave or burning," that seems to matter) rather than the undeniability of the given.[33]

The claim is elusive, since utter freedom from calculation would mean capitulation to infantile burbling (or some other state of psychic anarchy), but it's no sophism to say that his characteristic mood need not entirely abandon care. Montaigne invites us to accept "nonchalance" as an atmosphere in which all manner of thinking and speaking unfold. It takes shape as a feeling rather than a quality of action, a less-than-calculative attitude rather than a prohibition on calculation. Extending his thoughts on diplomacy, Montaigne writes, "I say nothing to one [party] that I cannot say to the other, at his own time, with only the accent [l'accent seulement] a little changed; and I report only the things that are indifferent or known, or which serve both in common" (602, 3:9). Passing a message virtually unchanged minimizes self-assertion and exemplifies the attitude Montaigne the "nonchalant" political agent seems to favor: not having a stake in the outcome. Yet the suggestion of deliberate withholding (of, say, information that would give one party an advantage) confirms that "nonchalance" is unfeigned without having to be passive; it doesn't rule out the minimal calculation of deciding what one had better not say. My reader might object that Montaigne here plays against type by narrating unambiguously disciplined behavior—to which my response is twofold. First, I am happy to concede the point; there have to be limits to Montaigne's "nonchalance," and this might be one of them. Second, effortlessness can only be relative; perhaps Montaigne remains "nonchalant" in comparison to the diplomatic agent who does have a stake in the outcome. In that case, Montaigne would present us with an image of ordinary affective evenness—not the strategic brilliance of an Iago or the aesthetic mastery of the successful courtier but simply the everyday thoughtfulness of choosing words wisely.

Casual Malice

I've said that Montaigne's observational mood is risky, but I haven't yet demonstrated how wide a space it leaves open for danger. Because, in just a few pages, this book redirects its gaze at scientific optimism, I want to be as clear as possible about what the dream of Baconian progress fails to countenance. The observational mood is a hole at the center of Montaigne's ethics. We're often told that he chooses the honorable over the useful, Roman virtue over Machiavellian realpolitik—but the essay's title names a conundrum rather than a simple choice.[34] Whereas Cicero, in *De officiis* (*On Duties*, 44 BCE), subordinates the pragmatic to the honorable (a dishonorable pragmatism would harm the perpetrator no less than the victim, and so it fails to count as pragmatism), Montaigne casts doubt on the voluntarism that Cicero's moral imperative assumes.[35] He holds on to Cicero's distinction and even repeats his prescription, but the basic choice at the center of the essay is disabled by the blur of "nonchalance." It's no surprise that an ethics of weakness has its disappointments, but the point would be lost without Montaigne's disarming forthrightness about the vulnerability of his position.

No private utility is worthy of our doing this violence to our conscience [cet effort a nostre conscience]; the public utility yes, when it is very apparent and very important. (607, 3:15)

If there should be a prince with so tender a conscience that no cure seemed to him worth so onerous a remedy [un si poisant remede], I would not esteem him the less. (607, 3:15)

Montaigne doesn't like expediency when it means abandoning virtue, and yet the common good justifies vice (even if his language of "effort" reminds us that the latter position could not be less amenable to his default way of looking at things). Montaigne also tells us that anyone who chooses not to bear the "weight" (*poids*) of such self-compromise is blameless. On the very question for which the essay is named, then, Montaigne simply doesn't prescribe a course of action. Nor are two behaviors suspended on either side of a scale, each having its merits. It's clear Montaigne prefers gentleness, but he's prepared for it to yield to grim determination. The specific attitude in question (peaceful "nonchalance") is far from deliberate, and is therefore liable to lose track of itself—even, at the limit, in severity.

Montaigne's predilection to lose touch with his ordinary disposition means engaging in unexpected behaviors (without losing his cool) and giving way to passion (which does mean losing it). We perhaps saw a muted version of the first case in the diplomat's serene meticulousness, and the second we glimpsed in the overseriousness of Montaigne's hatred for chess. These bare indications of imprecision (of a state of feeling defined by drift) have nothing on Montaigne's most shocking examples, which present a challenge to his too-frequent enlistment in an ethics of cultivated good-heartedness (which risks collapsing into complacency about what constitutes virtue).[36] The most disturbing of such examples of the disjunction between mood and comportment (the insufficiency of the former as a check on the latter) is Montaigne's occasional embrace of the martial heroics at the foundation of the French aristocracy's self-understanding (the very ideological ground of its wars). Epaminondas of Thebes, a great hero of Montaigne's (and a sometime stand-in for his friend Étienne de la Boétie), exemplifies the role of an easy disposition in the "innocent" conduct of war.[37] He is not alone among figures from antiquity in his confounding capacity to intermingle gentleness and violence. In "De l'experience," Montaigne celebrates Scipio Africanus, whose fame follows (like his name) from his defeat of Hannibal's army, for his habit of "playing nonchalantly and childishly [nonchalamment et puerilement] at picking up and selecting shells and running potato races by the sea"; Scipio even "[attends] lectures on philosophy," which Montaigne describes as a leisurely pastime, while rehearsing wartime exploits in his mind (851, 3:321).[38] Montaigne pairs Scipio with the example of Socrates, who, in his final years, turns to "dancing and playing instruments" with an attitude of childlike innocence—and yet, Montaigne reminds us, Socrates was "first among so many valiant men of the army" to throw himself in front of the sword that would have slain Alcibiades (852). It's Epaminondas, however, who captures most completely how affective ease accommodates an inversion of values.[39]

> To the roughest and most violent of human actions he wedded good-
> ness and humanity, indeed the most delicate that can be found in the
> school of philosophy. That heart, so great, full, and obstinate [obstiné]
> against pain, death, and poverty—was it nature or art that had made
> it tender to the point of such an extreme gentleness and goodness in
> disposition [une si extreme douceur et debonnaireté de complexion]?
> Terrible with blood and iron, he goes breaking and shattering a nation
> invincible against anyone but himself, and turns aside in the middle of
> such a melee on meeting his host and his friend. Truly that man was
> in command of war itself, who made it endure the curb of benignity
> at the point of its greatest heat [chaleur], all inflamed as it was and
> foaming with frenzy and slaughter. It is a miracle to be able to mingle
> some semblance of justice with such actions; but it belongs only to the
> strength of Epaminondas to be able to mingle with them the sweetness
> and ease of the gentlest ways [la douceur et la facilité des meurs les plus
> molles], and pure innocence. (609, 3:17)[40]

Here, the disposition Montaigne teaches us to associate with peace permits the most extreme violence ("breaking and shattering a nation"), which
approximates the Stoic form from which Montaigne often distinguishes his
moral outlook (Epaminondas is "obstinate against pain, death, and poverty"). Indeed, throughout the *Essais*, "obstinacy," like "cruelty" (though
often, unlike in this case, signaled by the word "opiniastreté"), is a mark of
ethical failure.[41] Montaigne takes leave of his irenic stance to celebrate the
power of martial might. Even a moment of withdrawal from battle takes
metaphorical form as the violent exercise of power, recalling Montaigne's
slippery observation that he "cruelly hates cruelty": "Truly that man was
in command of war itself, who made it endure the curb of benignity." Yet
amid the thrill of bloody conquest ("terrible with blood and iron"), we're
reminded of an underlying "gentleness" or "sweetness" (both good translations of *douceur*) that seems less an achievement on the model of Stoic
apatheia than an unexplained disposition or habit of feeling. Answering Montaigne's question about the origin of douceur ("Was it nature or art . . . ?"),

we can only repeat his motto, which is perhaps exactly what he asks of us: "What do I know?" (*Que sais-je?*).

The moral precariousness of "nonchalance" is no vague theme. Indeed, the example of Epaminondas makes a conspicuous display of the stakes of suppleness. Like the fiery vocabulary that describes the intensity of violent passion (*eschauffer* and *attiser*), the image of Epaminondas links violence to "heat" (*chaleur*). Etymologically speaking, of course, "nonchalance" is an absence of exactly that. Montaigne doesn't use the word for Epaminondas, but we are asked to imagine that the hero charges into the heat of battle with something like the affective cool of his eulogist. However, we might decide instead that his "obstinacy" shows a significant affective transformation (in the direction of cold severity) rather than psychic stability. Whether "nonchalance" persists as it accommodates violent action or gives way to the soldier's stony intensity, we can hardly escape the lesson that we have no good reason to count on it.

Of all the illustrations Montaigne offers us for the unreliability of disposition and mood, he borrows his most instructive from Lucretius. Near the beginning of the essay, Montaigne inhabits a scene in which the emotional peace of ataraxia looks like "nonchalance" but betrays the sort of cruelty from which he everywhere recoils. He then goes on to make the image one of the essay's organizing figures, a persistent reminder that the atmosphere he celebrates is difficult to distinguish from its contraries. Montaigne offers the following quotation, the famous opening lines from book 2 of *De rerum natura*, a poem itself written in the context of civil war, as an emblem of the proximity of serenity and cruelty:

> Pleasant it is, when on the great sea the winds trouble the waters, to
>   gaze from shore upon another's great tribulation . . .
> Suave, mari magno, turbantibus aequora ventis,
> E terra magnum alterius spectare laborem . . .[42]

Lucretius offers a metaphor for the serenity of wisdom, which protects its bearer from the worldly cares materialized here as the struggle to survive in choppy waters. The beneficiary of philosophy savors freedom from needless anxiety. Yet Montaigne's interest in what might be understood as Lucretian *Schadenfreude* (enjoying the suffering of another) has less to do with a simple contrast between pleasure and pain than with a specific sense of gratitude at having been spared exertion.[43] (Lucretius's formulation itself uses the vocabulary of "labor" [*laborem*] to describe by contrast the wise man's tranquility.) It's in this sense that we can read the Lucretian spectator as an image of "nonchalance" or its approximation. Though Lucretius specifies that "man's troubles" are not a "delectable joy" but that freedom from harm, an altogether different matter, is the source of the spectator's pleasure, we need not take for granted that Montaigne follows his line of reasoning.[44] I suggest we have good reason to worry that Montaigne's irenic mood resembles the disconcerting pleasure of observing the panicky struggle of the soon-to-be-drowned.

The image translates the problem of "usefulness" invoked by the title into a question about emotional life. At this early moment in the essay, Montaigne has said almost nothing beyond his introductory self-characterization as "nonchalant": "There is nothing useless in nature, not even uselessness itself. Nothing has made its way into this universe that does not hold a proper place in it. Our being is cemented with sickly qualities; ambition, jealousy, envy, vengeance, superstition, despair, dwell in us with a possession so natural [une si naturelle possession] that we recognize their image also in the beasts—indeed, even cruelty, so unnatural [denaturé] a vice" (599, 3:5–6). Thus Montaigne slides from the abstract notion of "uselessness" to a list of vicious passions, which we can only take to be instances of unsavory but nonetheless "proper . . . qualities" of "being." The next sentence introduces the quotation from Lucretius (not reproduced here): "For in the midst of compassion we feel within us I know not what bittersweet pricking of malicious pleasure [aigre-douce poincte de volupté maligne] in seeing others suffer; even children feel it" (599, 3:6). Montaigne suggests that "even" as "unnatural" a sensation as "cruel" pleasure is in fact "natural," confirming that his "natural[ness]" doesn't convey an ethical norm but rather an avowedly risky preference. When he speaks of "volupté maligne," he seems not only to avow but also to savor cruelty, the cardinal sin of the moral universe he elaborates and defends. Recall that "malignité" is exactly the word Montaigne uses in this very essay to characterize his warlike countrymen, who disguise their "propensity" for violence as "zeal" (602, 3:9). He later speaks of the "malignity" (malignité) of the "heart" we discern in those who perpetrate violence, even when they helplessly follow orders (606, 3:14). An essay that values affective naturalness reminds us right away that unnatural passions like cruelty are no less natural than sympathetic kindness, using the vocabulary it elsewhere reserves for those who compromise themselves in unforgivable acts of cruelty. Montaigne's observational mood is not an ethical wager in the weak sense that there are exceptions to its irenic effects; exposure to contingency—to the most catastrophic turns of events—is exactly what it is.

Montaigne ensures that we keep the Lucretian image in mind as we proceed. The stormy sea remains in view. Montaigne displays a habit of pitting his tranquil protagonists against watery chaos: Atticus "escape[s]" the "universal shipwreck of the world [cet universal naufrage du monde]" thanks only to "moderation"; the wise ambassador is smart enough to "glide in troubled waters [couler en eau trouble], without wanting to fish in them" (601, 3:8; 602, 3:9). We also learn that "those who espouse a cause completely can do so with such order and moderation that the storm will be bound to pass over their heads without harm [l'orage devra couler par dessus leur teste sans offence]" (601–2, 3:8). These scenarios look at first like images of safety and steadfastness, and they can certainly be read that way—but Montaigne shows that the benefits of moderation are relative, vulnerable, perhaps dismayingly temporary. Atticus "escape[s] . . . shipwreck," but he cleaves to the "losing side," giving us an image of bare survival rather than triumph. Montaigne underlines the delicacy of the diplomat's position by immediately giving up the goal of impartiality and opting instead for less-than-total investment: "Conduct yourself in this case with an affection, if not equal (for it may allow of different measures), at least temperate" (602). Finally, those who take up a "cause" without compromising themselves are only "*likely* to remain on their feet" (*qu'ils seront pour demeurer debout*) when the storm passes overhead; they might very well get thrown to the ground (602, 3:8, emphasis added).[45] The sheer pervasiveness of bad weather conveys the danger of Montaigne's world and signals the ongoing susceptibility of his exemplary characters. We should credit Montaigne with the insight Hans Blumenberg, who reconstructs the intellectual history of the Lucretian shipwreck, attributes instead to Blaise Pascal, that overanxious reader of Montaigne—that "we are already at sea," in the sense that thinking takes place in the midst of chaos rather than from the safe distance of shore.[46]

Montaigne wishes not only to talk about affective vulnerability but also to place the reader in a situation where she experiences it firsthand. My view is that the essay's sequence of graphic descriptions of violence makes an experiment of the reader much like the one Montaigne makes of himself in his opening lines. Do we find ourselves divested of the gentle sympathy we hope is our birthright? Are these occasions for horror, pity, or the thrill of the spectacular? Montaigne's tone is mostly neutral and matter-of-fact, but the descriptions are vivid in their violence: the reader watches the ship founder and sink without rhetorical cues that would elicit any particular response. "But let us continue our examples of treachery," Montaigne writes, signaling that the essay is about to take a turn for the worse (604). He narrates a sequence of episodes in which people are horribly punished for dishonorable behavior. In many cases, they perform acts of violence on behalf of the prince. Often, hapless villains doubly suffer when they are forced to betray their honor; they perpetrate heinous acts and are heinously punished for doing so. Because these episodes are imported from the past, they are conspicuously distant from the present-day experience of warfare. Yet their provenance does not diminish the stark experience of horror they induce.

> He began to feel such remorse and revulsion that he had his agent's eyes put out and his tongue and private parts cut off. (605)

> He was thrown headlong from the Tarpeian rock. (606)

> They have them hanged with the purse of their payment around their neck. (606)

> She, in his presence, opened up the murderer's stomach, and while it was warm, reaching with her hands for his heart and tearing it out, she threw it to the dogs to eat. (606)

> Since Sejanus' daughter could not be punished with death in a certain type of judgment at Rome because she was a virgin, she was, to give way to the laws, violated by the executioner before he strangled her. (606)

One of Montaigne's achievements is his willingness to gaze into the abyss of naturalness. Having witnessed the horror of civil war, and having grown skeptical of principled commitment as an exit strategy, Montaigne explores the ethical advantages of doing what comes naturally. Here, however, as we take in the view from the brink of calamity, we cannot help but wonder at our moral precariousness. Surely one possible response to these images is mounting horror at the depravity of which human beings are capable, especially under the delusion of duty. Indeed, my best guess is that Montaigne hopes for exactly this response. Yet Montaigne's attention to the minimal transition whereby something like ataraxia unfolds into cruelty disabuses us of faith in our goodness. When we observe calamity, we haven't experienced the encounter ahead of time, at least not as we do in the moment. Montaigne wants us to recoil from violence, but the wager he makes is a real one. His disarming thesis is that "nonchalance" mitigates cruelty more successfully than well-intentioned self-discipline—except, alas, when it doesn't. What is most astonishing about the essay is Montaigne's attention to the gravity of this concession.

From Ethics to Epistemology

We often consign Montaigne to the realm of hopeless Pyrrhonian skepticism while crediting Bacon with modern scientific confidence. Yet it's easy enough to identify a large repertoire of shared premises: respect for firsthand observation, qualified by knowledge of human understanding's susceptibility to error; suspicion of established repositories of wisdom; wariness of technical languages, insofar as they exchange clarity for sham authority (what we might call "authoritativeness"); and a reluctance to impose systematic coherence on accumulating observations. I suggest we add an item to this list: their joint assessment of the relationship between knowledge and feeling. As I pass from ethics to epistemology (construed broadly to encompass general problems of understanding), it's worth pointing out that this book has already arrived at a thematic threshold. Rather than make an ongoing attempt to triangulate the observational mood with its moral and epistemological consequences, I've prioritized the latter. Doing justice to my scientific through-line has meant privileging the question of feeling's bearing on understanding. In this respect, I follow the lead of my sources: there's an important sense in which scientific observation really does lose touch with questions of moral obligation—to say nothing of political commitment. Yet it's also the case that much more can be said about the moral background, justification, and evaluation of scientific practices. My hope is that my reliable but typically abbreviated return in subsequent chapters to the problem of moral evaluation (including the challenge of responding to war) is enriched by my detailed assessment of it in Montaigne's paradigmatic case.

In the *Essais*, epistemology is a corollary of disposition, which is the basis of moral comportment. "Nonchalance" is the (shifting) ground of both ethical engagement and interpretive practice, and gentleness is no less a style of understanding than an atmosphere for action.[47] Bacon follows this precedent by coordinating affective minimalism and the "advancement of learning," inhabiting an irenic atmosphere that seems at once to keep the peace and to draw Nature's secrets from their hiding places. In order to observe that continuity, we need first to understand how seamlessly Montaigne integrates knowledge and morality—suggesting, indeed, that nothing ever stood between them. In "De l'experience," the final chapter of the *Essais*, the tracing of that connection bespeaks a soft reorientation. Though conceptually joined, the true receives more attention than the good. Montaigne concludes his final volume of essays by making the same transition this book presently undergoes. In so doing, he brings into focus one of my basic premises. Though we often think of Montaigne as preoccupied with the problem of the self (with self-reflection), the disintegration of intimate experience into particles (of divergent if not contradictory perceptions, convictions, and interests) reveals the impersonality of apparently personal matters. The self takes shape as a receptive surface; it needs to be understood less for its own sake than for its role as a participant in scenes of understanding. I make this point not to weigh in on ongoing debates about the nature of Montaignian selfhood (though I have incidentally done so), but to explain that it makes better sense than one might assume to speak of unselfconscious self-portraiture.[48] Though, with Montaigne's help, I've already theorized morality as a question of receptivity to the variegated world, his epistemological musings bring this theme into starker relief.

Assimilating epistemology to naturalness, the Montaigne of "De l'experience" refashions motifs we recognize from "De l'utile et de l'honneste" as figures of understanding. The essay's first sentence, which echoes the opening of Aristotle's *Metaphysics* (fourth century BC), clues us in: "There is no desire more natural than the desire for knowledge" (815).[49] Seeking out the truth means doing what comes naturally, and the philosopher's pursuit of knowledge is only one manifestation of an all-too-ordinary predilection. The rhetoric of unmethodical lightness in "De l'experience" should by now be familiar—but notice how spectacularly the following passage breaks on the shores of the not-yet-comprehended world:

> The scholars distinguish and mark off their ideas more specifically and in detail. I, who cannot see beyond what I have learned from experience, without any system, present my ideas in a general way, and tentatively [à tastons]. As in this: I speak my meaning in disjointed parts, as something that cannot be said all at once and in a lump. . . . I leave it to artists, and I do not know if they will achieve it in a matter so complex, minute, and accidental, to arrange into bands this infinite diversity of aspects [cette infinie diversité de visages], to check our inconsistency and set it down in order. Not only do I find it hard to link our actions with one another, but each one separately I find hard to designate properly by some principal characteristic, so two-sided and motley do they seem in different lights [doubles et bigarrées à divers lustres]. (824–25, 3:287)

In "De l'utile et de l'honneste" (and elsewhere), Montaigne presents formal symptoms of "tentative[ness]" as evidence of "nonchalance." Here, the very same features of his style answer the world's unyielding resistance to understanding. He knows what he sees, but only in the relative sense that he knows his own experience better than anyone else's. Montaigne's phrases convey open-ended interest in the endlessly multiplying "faces" that render things elusive ("cette infinie diversité de visages"), but this is not wonder in Aristotle's sense: the affective register of a question we can answer. Abandoning questions of cause and effect, Montaigne refrains even from choosing an object of inquiry.[50] Things dissolve into multicolored confusion, and he rests content with findings that can only be preliminary. Anticipating this book's journey into the second half of the seventeenth century, we should not be surprised, when we turn to the same passage in Charles Cotton's 1686 translation, that he converts Montaigne's "à tastons" into "as an Inquirer," which William Carew Hazlitt later emends to "experimentally."[51]

Before "De l'experience" unravels into threads of unsynthesized firsthand observation, it explores the coercive pressure of codified authority on human judgment. The critique legitimates the subsequent practice of disjointed commentary, which encompasses both the general and the personal. Rigid legal interpretation is a good example of distortive authority, Montaigne argues, impeding ethical action just as quickly as it misconstrues fact. (Montaigne has much to say as well of medical malpractice; he insists that he knows his own body better than any doctor.) We might observe that a court of law is one place where *being* wrong (incorrect) means *doing* wrong (being unjust). The essay's early Ovidian fantasy invites us to inhabit a scenario in which sheer naturalness skirts the obstacles of labored legal reasoning: "Nature always gives us happier laws than those we give ourselves. Witness the picture of the Golden Age of the poets, and the state in which we see nations live which have no other laws. Here are some who employ as the only judge in their quarrels, the first traveler passing through their mountains [le premier passant qui voyage le long de leurs montaignes]" (816, 3:276). The newcomer's freshness of perception trumps the care of juridical reason. "The first traveler passing through" resembles "the first man I meet" in "De l'utile et de l'honneste" ("le premier passant" and "le premier que je rencontre"), both of them selected at random (and designated by that randomness) as emissaries from the realm of human artlessness—a judge with no special knowledge beyond the contingencies of his experience and an interlocutor whose everydayness ensures his tolerance for the infelicities of Montaigne's observations. When, later in "De l'experience," Montaigne famously recounts the idiosyncratic details of his life (how he takes his dinner, when he relieves his bowels), he performs this commitment to the wisdom of the quotidian, the habitual, and the accidental. Perhaps the apparent pun is accidental too, but Montaigne would surely have appreciated the insertion of his name ("le long de leurs montaignes") in the landscape of unconstrained because unlabored judgment.

In "De l'utile et de l'honneste," we ruin ourselves when we succumb to the violent passions that generate conflict. In "De l'experience," Montaigne describes recourse to legal precedent as exactly the same sort of error. Recall his critique of those who foment civil war ("They kindle war not because it is just, but because it is war"), and consider now his account of legal authority: "Now laws remain in credit not because they are just, but because they are laws. That is the mystic foundation of their authority" (602, 821). Not "just[ice]" but mere tautology, self-justifying violence and self-authorizing judgments are instances of the same groundless but ferocious "mystic[ism]." Montaigne materializes tautology by picturing the infelicitous interpreter as a silkworm, endlessly working over the same sticky stuff. He offers the image in immediate apposition to the critique of legal reasoning, as if drawing out the larger implications of the specific case of law: "Men do not know the natural infirmity of their mind: it does nothing but ferret and quest [fureter et quester], and keeps incessantly whirling around [va sans cesse tournoiant], building up and becoming entangled in its own work, like our silkworms, and is suffocated in it" (817, 3:278). The passionate drive to make something of nothing (of just this one thing) ends in disaster.[52] The world is large, and an interpretation made only of "mind" never touches it. The jurist or judge endlessly spins out the logic of doctrine, entangling the interpreter in mental arabesques when he might simply have opened the window and looked out on the case. Bad interpreters claw madly for a world they never touch.

Montaigne's alternative is easygoing awareness, which again presents the slippery solution of desirable but unreliable consequences (though here it's a matter of comprehension rather than kindness). Provisional lessons from experience, he suggests, tell us more than codified authority. Thus he proposes an "epistemology of convenience," a phrase I use to suggest both ready availability (effortlessness) and the etymological sense of things coming together (*convenire*).[53] What counts for Montaigne is the matter at hand—not actual immediacy but the feeling we associate with self-evidence. Whether he's allowing his mind to wander during a church sermon or enjoying the pleasures of eating melons, he does not actively contend with obstructions to understanding (848, 846). The same is true of his recollection of lines from Virgil (825). Montaigne's critique falls not on any particular source of knowledge but rather on an attitude of zealous adherence—to precedent, doctrine, cause, or any chosen object of attention. Montaigne looks to tradition all the time, but without the premise that it delivers a verdict. He suspends the assumption that ancient sources have more to tell us than we can tell ourselves, even about the very messages he calls on them to transmit.

The effortlessness of observation, reflection, and citation is a promise, if not a guarantee, of felicity, in both affective and practical senses (happiness and aptness). It isn't, however, an experience of pristine contact with the real. As I turn to Bacon's bid for the reinvention of scientific inquiry, and thus to the problem of cumulative empirical knowledge, I want to underline the point that Montaigne's observational mood is no flight from mediation. Whatever comes easily to mind need not be literally at hand; it only has to feel that way. In the following passage, he immerses himself in the immediacy of a situation, but he also makes sure we notice that immersion is no enemy to absent-mindedness. It's a style of engagement rather than a commitment to fixed attention on literally proximate or adjacent objects: "When I dance, I dance; when I sleep, I sleep; yes, and when I walk alone in a beautiful orchard, if my thoughts have been dwelling on extraneous incidents [occurences estrangieres] for some part of the time, for some other part I bring them back to the walk, to the orchard, to the sweetness of this solitude, and to me [à la promenade, au vergier, à la douceur de cette solitude et à moy]" (850, 3:319). What I love about this moment of self-reflection is how close it comes to the very language of tautology on which we have learned to cast suspicion (war's pseudorationality, mindless devotion to authority). The gift of "experience" is not self-evident fact but a feeling of naturalness that attends reflection. Montaigne emphasizes that knowledge acquired in this way is incomplete by picturing a world that is never quite there for the taking. Montaigne might easily have completed the passage's pattern of repetition by saying, "When I walk alone in a beautiful orchard, I walk alone in a beautiful orchard" (Quand je me promeine solitairement en un beau vergier, je me promeine solitairement en un beau vergier). Instead, he departs from himself and visits "extraneous incidents." When he reigns himself in ("nonchalance" giving way to self-management), the place to which he returns is disjointed: "I bring them [my thoughts] back to the walk, to the orchard, to the sweetness of this solitude, and to me." We can read this list as a collection of attributes belonging to a unified experience, but Montaigne does his best to spread them out. The activity, the place, the sensation, the subject as receptive surface for the whole sequence— the here and now, elongated by parataxis, is "hard to designate properly by some principal characteristic." Things make themselves known with both a "douceur" that recalls the "suave" of Lucretian ataraxia and a delectable confusion of glittering facets.

Passion's Pace

Bacon's affinity with Montaigne casts doubt on the blunt charge of teleology with which his program of scientific advancement is frequently saddled. Though Bacon indulges in a fantasy of total comprehension, such utopianism need not be taken as the limiting horizon of his thought. His thinking is capacious, even cornucopian; he says much more than what we remember him saying. Most importantly, for my purposes, he describes and encourages exploratory immersion in what he calls, simply, "learning," referring broadly to the experience of coming to know the world. It's within that condition of outward-facing awareness that we discover a cast of mind that differs from the one Bacon's late modern critics have taught us to expect. If we choose to keep step with Bacon's Montaignian amble, we begin to wonder whether his stride is as purposeful as it seems. He reminds us of the provisional and preliminary status of his claims with an earnest confidence he sometimes fails to muster when he imagines eventual epistemological victory. We can best understand Bacon's observational mood and the bearing it has on the long-term progress of the sciences by attending to his account of "learning" as an experience of temporal disorientation. The momentum of "advancement" depends on a state of easygoingness Bacon formulates as freedom from (if I can put it this way) "the time of desire." He shares Montaigne's habit of coordinating attenuated passion and perceptual power, and he offers a surprising description of that experience as a matter of pace and rhythm rather than cumulative or end-driven development. In the *Advancement*, Bacon narrates the observational mood as an unfolding process—enabling our understanding of the cases I take up in this book's subsequent chapters.

The intimacy of knowledge and feeling is no minor or incidental theme in the *Advancement*. Bacon singles it out for discussion at the beginning of book 1. The very first objection to learning to which Bacon responds is the charge that it makes us suffer. Bacon's interest in clearing the search for knowledge of the charge of "contristation" might remind us of Montaigne's desire to distinguish his own observational practice from the mad zeal of the "quest[ing]" silkworm, tangled up in thread.[54] Don't worry, both men seem to say, the pursuit of understanding is no cause for frustration unless we contaminate it with avarice, ambition, or small-mindedness. Bacon describes the schoolmen as "fierce with dark keeping"—as if they had devolved into ill-tempered beasts, living out their days in a hermetic darkness unalleviated by the light of experience (278). The acted-on ferocity Montaigne evades in his *Essais* takes shape in Bacon's hands as a quality of feeling—not unrelated to the perpetration of violence but distant from literal scenes of bloodshed. For Bacon no less than Montaigne, affective vehemence and epistemological failure are predictably conjoined.

Like the Montaigne of "De l'utile et de l'honneste," Bacon offers one of his most searching reflections on affective buoyancy while eyeing Epicurean ataraxia. Indeed, he takes up the same (well-known) image from Lucretius, and like Montaigne finds a conspicuously oblique way to relate himself to it. Bacon quotes the passage from *De rerum natura* virtually without comment, but it appears at the end of his most instructive description of the atmosphere of "learning." As with Montaigne's comparison of "nonchalance" to Epicurean tranquility, we're left with a sense of vague resemblance rather than sharp distinction or exact equivalence.

> For the pleasure and delight of knowledge and learning, it far surpasseth all other in nature; for shall the pleasures of the affections so exceed the pleasures of the senses, as much as the obtaining of desire or victory exceedeth a song or a dinner; and must not of consequence the pleasures of the intellect or understanding exceed the pleasures of the affections? We see in all other pleasures there is satiety, and after they be used, their verdure departeth; which sheweth well they be but deceits of pleasure, and not pleasures; and that it was the novelty which pleased, and not the quality. And therefore we see that voluptuous men turn friars, and ambitious princes turn melancholy. But of knowledge there is no satiety, but satisfaction and appetite are perpetually interchangeable; and therefore appeareth to be good in itself simply, without fallacy or accident. Neither is that pleasure of small efficacy and contentment to the mind of man, which the poet Lucretius describeth elegantly,

> *Suave mari magno, turbantibus aequora ventis, &c.*

> "It is a view of delight" (saith he) "to stand or walk upon the shore side, and to see a ship tossed with tempest upon the sea; or to be in a fortified tower, and to see two battles join upon a plain. But it is a pleasure incomparable, for the mind of man to be settled, landed, and fortified in the certainty of truth; and from thence to descry and behold the errors, perturbations, labours, and wanderings up and down of other men." (3:317–18)

Bacon seems to imply that "learning" affords something like the pleasure of Epicurean security (or sometimes does), but it's not quite the same. By introducing the Lucretian spectator in a supplementary but discontinuous thought ("Neither is [it] of small efficacy and contentment to the mind of man"), he repeats Montaigne's gesture of simply placing the image alongside a searching account of pleasurable freedom from intensity. I am not the first to suggest that Montaigne and Lucretius occupy the same region in Bacon's mind. Gerard Passannante has observed that "a quotation from Lucretius had become a kind of calling card for Montaigne," who quotes lines from *De rerum natura* "more than any other Renaissance writer not making a commentary on the poem."[55] Kenneth Alan Hovey points out that Bacon, in "Of Truth" (the first of his *Essayes* in the expanded and final edition of 1625), attributes the same deceptive elegance to both Montaigne and Lucretius: he quotes them for the beauty of their words but not for the content of their utterances—about which he expresses reservations.[56] Thus Bacon praises Montaigne for saying something "prettily" (if not persuasively) and describes Lucretius as "the Poet that beautified the sect that was otherwise inferior to the rest."[57] Here, in the *Advancement*, we likewise observe an oblique invocation of Lucretius, which raises the possibility (and it's only a possibility) that he nods at Montaigne as well.

Bacon's perspective on the Lucretian prospect is inconstant. Elsewhere, he suggests that the pleasure of looking on from the safety of shore is in fact a "vain chimera," and in yet another place he judiciously amends the quotation: "*No pleasure is comparable to the standing upon the vantage ground of Truth,* (a hill not to be commanded, and where the air is always clear and serene,) *and to see the errors, and wanderings, and mists, and tempests, in the vale below:* so always that this prospect be with pity, and not with swelling or pride."[58] The palpable wariness of these examples is an important reminder that the exercise of sheer mastery is not Bacon's theme (his concern for "pity" and affirmation of the impossibility of perfect "command" undermine Horkheimer and Adorno's appraisal); the reference in the *Advancement* likewise softens the distinction between safety and danger, ease and worry. Our first response to Bacon's image of Lucretian spectatorship is the vague impression that his developing account of the experience of "learning" reminds him of it. On closer inspection, we notice that Bacon has again encoded his partial demurral in the quotation itself, offering a moderating paraphrase that pictures a ship threatened by waves rather than dashed to pieces by them, which is what readers of the original often make of the watery struggle for survival.[59] Indeed, in the *De augmentis*, Bacon foregoes the original quotation and instead translates his gentler English paraphrase into Latin.[60] One explanation for that choice is that it lessens the contrast between "learning" and quotidian (rather than extreme) experiences: safety from mere turbulence versus safety from imminent peril. The image points in a different direction from Bacon's idealizing comparison of "learning" to "the obtaining of desire or victory," which he tells us is much better than the lowly enjoyment of "a song or a dinner." We might notice as well that Bacon abstains from Epicurean pedagogy: the tranquility of the spectator in *De rerum natura* is an effect of lessons successfully learned, but here the experience of "learning" seems not to be contingent on knowledge. What is this pleasure that "exceedeth" so many others—but in terms of neither intensity nor rarity? What are we to make of an experience at once superlative and ordinary?

Our interpretation hinges on how we understand the riddling proposal at the center of the passage: "But of knowledge there is no satiety," Bacon writes, "but satisfaction and appetite are perpetually interchangeable; and therefore appeareth to be good in itself simply, without fallacy or accident" (3:317). The best account we have of Bacon's remark is Lorraine Daston and Katherine Park's, but they see endlessly propulsive pleasure where I see an experience of mild enjoyment set altogether free from sharpness. For Bacon, they explain, "the rhythms of curiosity were those of addiction or consumption for its own sake, cut loose from need and satisfaction."[61] Bacon surely dispenses with "need," but my interest here is the conservation of "appetite" as the diluted double of that absent term. "Appetite" differs from "need" *because* it's "interchangeable" with "satisfaction"; the ever-present possibility of substitution rules out the intensities of both longing and delight. Slipping between losses and gains, anyone engaged in the process of "learning" eludes both the "satiety" Bacon names and the destitution that would be its opposite. If, speaking of "addiction," Daston and Park have the late modern discourse of chemical dependency in mind, as their vocabulary of wild "consumption" might suggest, I propose the inverse: not the momentum of rapacity but the delicacy of a "simple" pleasure that can be trusted to surface as soon as it sinks.[62] If they intend "addiction" in the early modern sense of mere attachment or devotion, their description nonetheless underplays the extent to which "learning" is envisioned as relief rather than compulsion.[63]

It's not that Daston and Park's interpretation misses the mark but rather that it attends to only one of two possibilities; it powerfully explains Baconian thought in some—but only some—of its forms. My alternative illuminates features of the scientific imagination about which scholars have so far had little to say. In Bacon's alternation between "satisfaction" and "appetite," he offers us a narrative schema for the observational mood, which materializes as an experience of affective change (of wanting and obtaining) like any other—but one distinguished by the relative softness of its highs and lows. By giving a temporal shape to "learning," Bacon presents us with a resource for understanding several of this book's subsequent cases. On Bacon's account, the story of knowledge acquisition is strangely self-canceling; it brings different moments into such proximity that, in the end, we may not wish to speak of narrative after all. Our plot moves gently within a narrow range of possible sensations; they come so close to each other that we have trouble holding them apart as distinct states of feeling.[64]

In describing an experience of ongoing satisfaction that hovers dependably in the middle altitude between extremes, Bacon's interest is less the reliability of enjoyment than the distinction between the enjoyment of "learning" and pleasure's surfeit.[65] It's not the first time he affirms the value of freedom from fullness. Indeed, at the beginning of book 1, he points out that the "swell[ing]" of the proud mind has nothing to do with "quantity of knowl-edge" but rather with the arrogation of power from God (265). The story of Adam and Eve is not about knowing too much; it's about wanting the wrong kind of knowledge: "an intent in man to give law unto himself and to depend no more upon God's commandments" (265). "Salomon," Bacon writes, "speaking of the two principal senses of inquisition, the eye and the ear, affirmeth that the eye is never satisfied with seeing, nor the ear with hearing; and if there be no fullness, then is the continent [the con-tainer] greater than the content" (265). The emphasis here isn't on endless sources of gratification but rather on the absent threat of overindulgence and the moral problem it poses: swollenness, or pride. The eye and the ear seem themselves to succeed at taking things in by dependably falling short of euphoric fulfillment. In his *Redargutio philosophiarum* (*Refutation of Phi-losophies*, 1608), Bacon associates fullness with false but proud erudition: "Everywhere [in scholastic natural philosophy] you will find endless repeti-tion of the same thing. You have managed to get a sensation of fullness on Lenten fare."[66] He also presents the image of bestial hermeneutic labor we know from the *Advancement* as a parable of overeating: "You would not, I think, seek to deny that the whole of our great learning is merely a surviv-ing fragment of Greek philosophy. Nor was this creature bred and nurtured in the glades and thickets, but in school and cells, like a domestic animal being fattened."[67] Wanting to inhabit a state in which one is "never satis-fied" completely—that desire is met by the experience of affective steadi-ness Bacon locates in "knowledge and learning."

The *Advancement* returns to this idea with some regularity. Speaking more loosely of emotional self-management in book 2, for instance, Bacon rejects the desirability of both unfeeling indifference and overexcitement, opting instead for the gentle rhythm of ongoing gratification. He opposes the posi-tion of Socrates, who "plac[es] felicity in an equal and constant peace of mind," to that of the Sophist, who prefers instead the life of "much desiring and much enjoying" (426). The first enjoys "the felicity of a block or stone"; the second, "the felicity of one that had the itch, who did nothing but itch and scratch" (426). Soon, after a critique of overly "fearful and cautionary" forms of self-discipline, he offers us an image that appears to be an ingenious combination of the Socratic "stone" and the sophistic "itch": the jeweler who rubs away the imperfections in a gem. "Men are to imitate the wisdom of jewelers," he writes, "who, if there be a grain or a cloud or an ice which may be ground forth without taking too much of the stone, they help it; but if it should lessen and abate the stone too much, they will not meddle with it: so ought men so to procure serenity as they destroy not magnanimity"

(427–28). The metaphor indicates that overintense self-discipline (polishing the stone) resembles the "scratching" of the sensualist, making rigor a close cousin of the self-indulgence it aims to obliterate. Thus Bacon withdraws from both the eager pursuit of pleasure and the intensity of desire's suppression. Though the passage seems at first to picture the judiciousness of moderate self-cultivation, it can also be read as an image of delectation. The wish to make something beautiful (a jewel, in this case) is a desire for gratification, but it's weaker and less persistent than the need to satisfy impulse, including the impulse to eradicate impulse. The likelihood of satisfaction is high and the ultimate stakes are low—and so the experience knows nothing of urgency. The passage also clarifies the moral dimension of emotional life, as Bacon understands it. Bacon locates "Magnanimity," the preeminent virtue in his moral universe, in the givenness of disposition, and identifies the exertions of self-cultivation as threats to its survival. In charity, then, we observe the moral importance of dispositional naturalness, but without Montaigne's interest in foregrounding the hazards of social and political life.[68]

As my own metaphor of rhythm has already suggested, Bacon equates pleasure's softness with an escape from the temporality of yearning. When he describes "satisfaction" and "appetite" as "perpetually interchangeable," he superimposes the *before* of desire and the *after* of gratification. Bacon's well-known discussion of knowledge as a "dry light" (*Lumen siccum*) repeats the pattern, conveying his interest in a form of contentment no less dependable for its slightness, and one that reduces the temporal gap between seeking something and catching hold of it—nearly to the point of vanishing (267). "There is no vexation or anxiety of mind," he writes, "which resulteth from knowledge otherwise than merely by accident; for all knowledge and wonder (which is the seed of knowledge) is an impression of pleasure in itself; but when men fall to framing conclusions out of their knowledge, applying it to their particular, and ministering to themselves thereby weak fears or vast desires, there groweth that carefulness and trouble of mind which is spoken of" (266–27). Like the less than satisfying but never underwhelming bid for knowledge Bacon pronounces "good in itself simply," learning is "an impression of pleasure in itself." In both cases, steady gratification persists irrespective of the objects under consideration; such is the force of the repeated locution "in itself." These are different accounts of the same scenario, together performing an ingenious involution of emotion's temporality. Though the Aristotelian tradition defines "knowledge and wonder" by the lag between them, a fact of which Bacon reminds us when he describes wonder as knowledge's "seed," he treats them as transposable, much in the way of "satisfaction and appetite." Once "satisfaction" is understood as arriving reliably in answer to every "appetite," it no longer makes sense to narrate "learning" except as minimal (and, from a narratological standpoint, monotonous) alternation. Indeed, the predictability of that continual oscillation between barely differentiated states of feeling argues for their collapse: "appetite" that need not wait for "satisfaction" barely counts as "appetite" after all. For the schoolmen, "wonder" names a wish for exactly the "knowledge" that would extinguish it, but here desire and gratification are folded together into a subtle but palpable "pleasure" that might only be an "impression" but reliably counters the "anxiety" of "carefulness."[69] If the topos of "wonder" encourages the narration of epistemological hunger (a journey from dumbfounded ignorance to contented understanding), Bacon's interest in the "interchangeable" suspends the enlistment of emotion in narrative development beyond the ongoing exchange of an emergent wish for its near-preemptive satisfaction.

When Bacon speaks of "wonder" in isolation rather than entertaining the thought that it perennially changes places with "knowledge," he makes a swamp of Aristotle's pathway from ignorance to understanding. In the *Sylva Sylvarum: Or, A Natural History in Ten Centuries* (published posthumously in 1627), he characterizes the paradigmatic (Platonic and Aristotelian) philosophical passion as an experience of paralysis: "Wonder causeth Astonishment," he explains, "or an Immovable Posture of the Body, Casting up of the eyes to Heaven, and Lifting up of the Hands. For Astonishment, it is caused by the Fixing of the Minde upon one object of Cogitation, whereby it doth not spatiate and transcur as it useth: For in Wonder the Spirits flie not, as in fear; but only settle, and are made less apt to move. As for the casting up of the Eyes, and Lifting up of the Hands, it is a kinde of Appeal to the Deity, which is the Author, by Power and Providence of strange wonders."[70] Rather than initiate a quest for comprehension, wonder seems here to induce gaping awe.[71] We are made "less apt to move," and we make an "Appeal to the Deity" rather than take action to find something out ourselves. Since "perambulation" is one of Bacon's chief occupations, we are right to detect suspicion—indeed, disapproval—in his description of a passion that "fix[es] . . . the Minde upon one object," ensuring that it doesn't "spatiate and transcur [wander and rove] as it useth."[72] Wonder, in this case, is cognitive obstruction, and Bacon keeps the mind on the move. As chapter 2 shows, the less-than-astonished mind's mobility comes to matter a great deal to later generations of Baconian inquirers.

Falling out of sync with the lockstep teleology he sometimes celebrates, Bacon sometimes offers us an experience of suspended vibrancy—a flicker of gratification that sustains interest without the promise of conclusive fulfillment. Consider his distinction between the "dry light" of the soul at its epistemological best ("Lumen siccum optima anima," A dry light is the best soul) and the blurry or macerated light ("Lumen madidum" or "maceratum") of an overly "careful" mind that cannot achieve clarity (267).[73] "Maceration" is an apt metaphor because it describes what happens when a thing is steeped in water; we "macerate" the light of our understanding when we stew in worry or remain otherwise mired in "affection." The ever-changing but persistent lightness of "learning" offers an escape. The enlistment of knowledge in the fulfillment of ambition ("framing conclusions") generates distortion, unlike the time warp of untainted understanding. Bacon treats the resulting uncertainty about where the inquiring self stands in narrative time as an asset: "Only learned men," he writes, "love business [busyness] as an action according to nature, as agreeable to health of mind as exercise is to health of body, taking pleasure in the action itself, and not in the purchase" (272). Here, we need to recall Montaigne's similar remark: "I aspire to no other fruit in acting than to act, and do not attach to it long consequences and purposes" (601). For Bacon, "maceration" might also call to mind the opposite of what it seems here to suggest; the word can signify fasting and the mortification of the flesh. In that case, the passage would repeat the lesson of the polished jewel: unchecked passion might not be so different from *askesis*. Like Montaigne, for whom principled intensity generates the very thing we expect it to defeat, redoubling violent passion by creating resistance, Bacon seems here to favor the free-floating pleasure of "learning" over purposeful investigation.

With Montaigne's help, we've discovered a new dimension of Bacon's thought: an experience of understanding that lingers with the immediacy of unearned "satisfaction" rather than looking to future gain. Already, then, we may wish to reevaluate Bacon's emphasis on the provisional; perhaps it can be understood not only as the meticulousness of a cautiously progressive thinker but also as a habitual orientation toward the value of unsorted experience. To be sure, that value is nothing other than the possibility of eventual epistemological profit—but might we entertain the possibility that the role of the blossoming future might be to extend a promise that licenses us to enjoy "the meantime" before the flower can be plucked? If we adopt this perspective, we might decide to take Bacon seriously when he casts doubt on the reality of future achievement. "We know not," he concedes, "whether our labours may extend to other ages," though exactly this eventuality would seem to be the premise of the *Advancement* (274).[74] Bacon hopes people can accomplish what time, on its own, never does—but hope is not the same as expectation.

Unwinding the Thread

In "Of Fortune," one of Bacon's additions to his *Essayes* of 1612, he offers words of caution that sound much like the Montaigne of "De l'utile et de l'honneste." "Extreme lovers of their country or masters," he writes, "were never fortunate, neither can they be. For when a man placeth his thoughts without himself, he goeth not his own way."[75] "Extreme" devotion is a liability. Most of the bloody scenes in Montaigne's essay record the unfortunate outcomes of exactly this mistake; the contest between the honorable and the useful pits different feelings of rightness against each other: what feels good without any exertion of pressure against what feels good—*too* good— as an effect of self-persuasion and self-incitement. Like Montaigne, Bacon presents the observational mood as an alternative to "extreme love," though he adopts the language of *disinvoltura*, a conceptual cousin of sprezzatura in the *Cortegiano*, though less clearly freighted with the premise of artificiality.[76] He also reminds us of Montaigne by refraining from naming causes: "There be secret and hidden virtues that bring forth fortune; certain deliveries of a man's self, which have no name. The Spanish name, *desemboltura*, partly expresseth them; when there be not stonds nor restiveness in a man's nature; but that the wheels of his mind keep way with the wheels of his fortune."[77] After seeking a word in a foreign tongue for character traits that otherwise have "no name," Bacon concedes that even his terminological ingenuity only "partly expresseth them." The disposition that best receives Fortune's gifts thus seems itself to be an extraordinary gift of Fortune: a capacity we can barely name, let alone explain. Bacon's Castilian term, *desemboltura*, appeared in Juan Boscán's 1534 translation of the *Cortegiano* (it was already a Spanish word), conveying Castiglione's sense of effortless negligence, as well as the unaccountable looseness or openness signaled by etymology: from the verb *disinvolgere*, to unwind or unwrap.[78] Bacon explains that the lucky are distinguished from the unlucky by freedom from mental blockage and obstinacy ("stonds or restiveness"), keeping steady pace with Fortune's wheel. "Restiveness" echoes Montaigne's "opiniastreté," both of which are warnings about the misfortune of inflexibility, and Bacon quotes Livy's phrase for the salutary alternative: "versatile ingenium," an adaptable nature.[79]

Though, in the *Essayes*, Bacon is not straightforwardly Montaignian— Niccolò Machiavelli is often the more visible presence—this particular essay shows the great tactician of collective and personal ambition setting strategy aside in the interest of carelessness. "It is a greatness in man," he writes, "to be the care of the higher powers"—rather than having to monitor his own behavior too assiduously.[80] Although the essay begins by placing "man's fortune in his own hands" and dismissing "outward accidents," Bacon soon indicates that the capacity to seize opportunity is itself a benefit of enigmatic "desemboltura." Thus he repeats Montaigne's habit of attributing special virtues to "natural" weaknesses, borrowing the expression "Poco di matto" from the Italian to describe his nonspecific protagonist—someone who displays "a little of the fool."[81] Perhaps, in the following lines, Bacon remembers Castiglione's similar suggestion that *grazia* might arrive unbidden "from the stars": "If a man look sharply and attentively, he shall see Fortune: for though she be blind, yet she is not invisible. The way of fortune is like the milken way in the sky; which is a meeting or knot of a number of small stars; not seen asunder, but giving light together. So are there a number of little and scarce discerned virtues, or rather faculties and customs, that make men fortunate."[82] As readers, we lose track of our purpose, looking now for Fortune herself and now for signs of her favor. At first, we assume Bacon explains how to catch sight of *her* (how to be fortunate), but soon he seems instead to advise us on how to identify the beneficiary of her kindness. Perhaps Bacon's point applies to both modes of attention, which would then credit benevolent Fortune for the observational prowess everywhere on display in the analysis of social mores and strategies to which his essays are devoted. When Bacon describes the assembly of stars as a "knot," he calls attention to the metaphor encrypted in the etymology of *desemboltura*, drawing a contrast between the tight coherence of the visual field and the relaxed slackness of the observer. Though Bacon recommends "look[ing] sharply and attentively," he foregoes fineness of detail in favor of the overall impression. When the stargazer is suitably *desenvuelto*, the stars gather together into a light-giving shape. It's by encouraging the softening of the gaze that the observational mood brings the constellation into focus.

Careless perspicuity thus takes shape as the dilation of attention. This is no less characteristic of the *Advancement* than it is of the essay on fortune—a point that shouldn't surprise us, not only because Bacon's whole body of work is cumulative in a more than ordinary sense (he continually recycles and develops his claims), but also because, as R. S. Crane has shown, Bacon writes the *Essayes* in order to fill out gaps in the portion of the *Advancement* that assesses human behavior.[83] In this final section, I explore the set of related themes suggested by Bacon's invocation of disinvoltura. I have argued that Bacon shares Montaigne's premise that unsuppressed but persistently gentle emotion is a pathway for understanding. We have also seen how, for Montaigne, cool carries the connotation of pliancy, and that passion's heat gives it calcifying force, strengthening and solidifying attachment to belief. Though we are nowhere near Montaigne's scene of moral catastrophe, my aim here is to show that the claim holds for Bacon as well, which we might already have gathered from the fungible status of objects of contemplation in his account of the pleasures of "learning." Whatever the matter at hand, Bacon's inquirer reliably enjoys the experience.

This second pass through the *Advancement* begins with Bacon's explicit allusion to the courtier's artlessness in order to establish a connection between what he calls "careless[ness]" and discursive mobility: the practice of smoothly changing subjects.[84] (Over the course of this book, we shall see that such facility of "interdisciplinary" movement remains a distinctive feature of seventeenth-century science.) I also discuss one of the most striking differences between Montaigne and Bacon with respect to our theme: the seriousness with which the latter takes the idea of self-improvement. By addressing this question, we discover our best opportunity to reframe the dominant interpretation of Bacon's philosophy. Bacon minimizes the difference between self-acceptance and self-betterment by describing acts of self-discipline as if they were effortless. I develop my account of Bacon's disinvoltura by exploring two corollaries of ease: dilation and digression, where the latter is understood as the stylistic corollary of the former. It's wonderfully suggestive that the metaphor of being "unwound" rather than tightly knotted implies a distinction between expansive looseness and cramped inelasticity, which corresponds to the difference between Bacon's relaxed open-endedness and a more familiar scholarly emphasis on the vise grip of method. Yet I wish to state plainly that my interest in the language of disinvoltura is heuristic: a useful term for the interpretation of the *Advancement*. It's not this specific vocabulary but the idea of easy flexibility that dependably draws Bacon's attention.

Bacon comes closest to a direct discussion of disinvoltura when he explores the role of "careless[ness]" in self-presentation. In the section of the *Advancement* where he proposes to go some way in "teach[ing] men how to raise and make their fortune," he recommends the "easy and careless" underlining of personal virtue as a useful strategy, but one for which we must be grateful to inborn "nature" (463, 456). For our purposes, the passage is a good point of departure; Brian Vickers notes that Bacon seems here to be thinking of Castiglione and the paradigm of sprezzatura.[85] "Careless[ness]" names a quality of graceful "passage to and from" any subject of discussion or object of attention—which is one good thesis about its role in the *Advancement*.

> If it [self-congratulation] be carried with decency and government, as with a natural, pleasant, and ingenious fashion; or at times when it is mixed with some peril and unsafety (as in military persons); or at times

when others are most envied; or with easy and careless passage to it and from it, without dwelling too long or being too serious; or with an equal freedom of taxing a man's self as well as gracing himself; or by occasion of repelling or putting down others' injury or insolency; it doth greatly add to reputation: and surely not a few solid natures, that want this ventosity and cannot sail in the height of the winds, are not without some prejudice and disadvantage by their moderation. (463)

Bacon here plays the role of the realist, and thus he admits the advantages of tactical self-trumpeting. For Bacon, of course, "ventosity" is rarely a term of praise. Indeed, by this late moment in book 2, it can only call to mind the swollen pride that distorts understanding. In his sudden transvaluation of it, we observe the suppleness of Bacon's mind: his willingness, like Montaigne, to drift from an apparently settled view and inhabit an alternative. His cascading list anatomizes disinvoltura, the several features of which underwrite the unselfconscious quality of his prose: the "easy and careless passage" with which, say, the braggart turns the conversation to the enumeration of his virtues; "natural[ness]" of self-presentation; flexibility; insusceptibility to over-"serious[ness]"; and a dashing air of courage and "unsafety" that makes Fortune's favor palpable. He also draws a telling contrast between the ostentation of "careless" speech and the "moderation" that rules it out. What Bacon's list does best of all is cast "careless[ness]" simultaneously as "nature" and technique, crystallizing a conceptual problem at the heart of this book. Montaigne's "nonchalance" is inseparable from the effortlessness with which he achieves it, but Bacon wants to ponder methods for self-cultivation.

What sort of behavioral training would permit the persistence of affective ease? For Montaigne, the flexibility inherent to "nonchalance" can be read in two ways: as a propensity to yield to passion or a capacity to accommodate all manner of behavior. According to this second reading, even the most strenuous engagements ("the roughest and most violent of human actions") might unfold in an atmosphere of easygoingness. Bacon's description of Adam's work in Paradise depicts knowledge production as just such an instance of accommodation: "There being then no reluctation [struggle, resistance] of the creature, nor sweat of the brow, man's employment must of consequence have been matter of delight in the experiment, and not matter of labour for use" (296).[86] Bacon's paradigmatic scene of understanding affirms the experience it denies, but evades paradox by assuming the coherence of a mismatch between mood and action. Adam works hard, but labor's presiding emotion is easy "delight." As I discuss in chapter 4, this description of innocent experience finds extraordinary expression in Milton's *Paradise*

Lost. As far as the *Advancement* is concerned, the "use[lessness]" of Adam's "employment" lends credence to my argument for the suspended tempo-rality of "learning," which we might in this case recognize as free-floating availability to whatever catches the eye.[87] For Bacon, indeed, that's what the experience of "advancement" is like.

In this light, effortless labor is no paradox.[88] Bacon retains Montaigne's interest in unearned aptitude (hence his enchantment in "Of Fortune" with the mystery of "desemboltura"), and he favors courses of action that seem simply or automatically to bring about the change. His emphasis on ease might be taken as merely rhetorical—it is indeed a form of encouragement—but the conversion of difficulty into ease plays an important role in Bacon's philosophy. From his perspective, to be sure, there's no error in labor itself, but nor is there inherent virtue in it. It depends on the case. "So some measure things," Bacon writes, "according to the labour and difficulty or assiduity which are spent about them; and think if they be ever moving that they must needs advance and proceed" (380). When uncoupled from self-congratulation, however, and when distinguished from achievement itself, "assiduity" receives a positive evaluation. Bacon goes as far as to fault Aristotle for missing opportunities to advise self-correction: "But allowing his conclusion, that virtues and vices consist in habit, he ought so much more to have taught the manner of superinducing the habit: for there be many precepts of the wise ordering the exercises of the mind, as there is of ordering the exercises of the body" (439). Perhaps it's here that we begin to discern the connection between, on the one hand, my unfamiliar portrait of Bacon at rest (and at play) and, on the other hand, the taskmaster we know from intellectual history. Unlike Montaigne, Bacon is very much a theorist of "superinduc[tion]" (he wants to establish new habits of mind), but he likes to remind us how easy it is to bring about the change.[89]

One good example follows immediately from the critique of Aristotle. Here, Bacon recommends arduous self-discipline by explaining arduousness away. He offers this piece of advice as one of several "precepts" for improv-ing the mind: "Another precept is, to practice all things chiefly at two several times, the one when the mind is best disposed, the other when it is worst dis-posed; that by the one you may gain a great step, by the other you may work out the knots and stonds of the mind, and make the middle times the more easy and pleasant" (3:439). What Bacon advises here is to take up a chal-lenge at exactly the moment when you are most likely to fail. His formula-tion of the benefit of the practice resembles the definition of disinvoltura we discovered in "Of Fortune": freedom from "stonds" and "restiveness." Here, the language of "knots and stonds" makes disinvoltura a precise but perhaps befuddling solution. Bacon advises untying mental knots, but "Of Fortune" has prepared us to expect salutary looseness to come "from the stars." That point of contrast is less disorienting when we observe that he in fact collapses these alternatives. By framing the most difficult part of the recommended program as a way to knead or soften the mind, setting aside the ultimate achievement of the desired skill in favor of the present experience of practice, difficulty is recast as an opportunity for grateful pleasure, meanwhile ensuring that every subsequent attempt is "easy and pleasant" by comparison.

Bacon does not always propose a sophisticated solution that reframes difficulty as ease. Sometimes, as in "Of Fortune," he invokes mysterious gifts of character. He attributes disinvoltura's opposite to inborn "nature" as well. Here, we can observe Bacon's habit of drawing a contrast between narrow self-enclosure and ruminative dilation. "These grave solemn wits," he writes, "which must be like themselves and cannot make departures, have more dignity than felicity. But in some it is nature to be somewhat viscous and inwrapped, and not easy to turn. In some it is a conceit that is almost a nature" (465). Montaigne is a good example of a person who escapes the fate of having to be "like himself"—though he would put it differently, understanding lapses and deviations as evidence of being fully and unabashedly exactly who he is. The essayist's "nonchalance" would make a nice remedy for "grave solemn wits," assuming adaptation is possible; they're lamentably "inwrapped," and thus what they lack, thinking now of the word's etymology, is disinvoltura. Bacon ascribes their unhappy state to "nature," but he also mentions cases in which stubbornness is only "a conceit that is almost a nature," suggesting different degrees of inflexibility, and thus different prospects for change.

The "viscous and inwrapped," the sticky and self-enclosed, calls to mind Bacon's image of the schoolmen as spiders (and, for us, Montaigne's self-suffocating silkworm). Bacon dilates on the drawbacks of being tightly wound:

Having sharp and strong wits, and abundance of leisure, and small variety of reading; but their wits being shut up in the cells of a few authors (chiefly Aristotle their dictator) as their persons were shut up in the cells of monasteries and colleges; and knowing little history, either of nature or time; did out of no great quantity of matter, and infinite agitation of wit, spin out unto us those laborious webs of learning which are extant in their books. For the wit and mind of man, if it work upon matter, which is the contemplation of the creatures of God, worketh according to the stuff, and is limited thereby; but if it work upon itself, as the spider worketh his web, then it is endless, and brings forth indeed

> cobwebs of learning, admirable for the fineness of thread and work, but of no substance or profit. (285–26)

If only the schoolmen had as much "matter" as they do "leisure" (available time), they might shed light on creation and give humankind usable knowledge. Instead, they inhabit an atmosphere of "agitation," a kind of unacknowledged claustrophobia, cramped as they are in literal and metaphorical (monastic and perceptual) "cells." The difference between the outward- and inward-looking "mind" is also the one between serenity and "laborious[ness]," "substance" and "cobwebs." This last metaphor recalls several of Bacon's most memorable images. First, it activates the language of wet and dry, "Lumen siccum" and "maceratum"—for what is a spider's web but an apparently dry lattice that turns out to be a sticky trap?[90] Second, a "cobweb" is a tangled "thread," and one impossible to unwind, which indicates our distance from the mental space of disinvoltura.

When Bacon offers a characterological description of "the learned," who display a habit for abstraction that others deem improper, he again suggests a spectrum of experiences of (attempted or successful) "learning" running from the capacious to the cramped. Bacon seems at first to concede the point of the impropriety of their behavior, admitting that they tend to have wandering minds, but soon he reframes that apparent defect as a moral advantage. "Another fault incident commonly to learned men," he writes, "which may be more probably defended than truly denied, is that they fail sometimes in applying themselves to particular persons. . . . the largeness of their mind can hardly confine itself to dwell in exquisite observation or examination of the nature and customs of one person" (279). From this failure, Bacon derives a precept about flexibility: "he that cannot contract the sight of his mind as well as disperse and dilate it, wanteth a great faculty" (279). Yet Bacon is squarely on the side of learned abstractedness; his balancing act is only apparent. A perspective that rightly ranges over the world is less easily drawn to the specific interests of any single interlocutor. Bacon's phrasing suggests his bias: "the largeness of their mind" is unwilling to "confine itself," he explains, casting close attention as constraint. Indeed, we soon discover that we have good moral reasons to dislike "exquisite observation." One advantage of dilated mental "sight," for instance, is that it discourages us from sinister watchfulness: "To be speculative into another man, to the end to work him or wind him or govern him, proceedeth from a heart that is double and cloven" (280). We're reminded here of the opening lines of the *Advancement*, where Bacon assures King James that he looks upon him "not with the inquisitive eye of presumption, but with the observant eye of duty and admiration" (261). Given the itinerary of my interpretation, the language of "wind[ing]" might remind us here of the language of disinvoltura, the opposite of such insidious concentration. Bacon reminds us what it means to be less than at liberty, in this case due to a deliberate effort at manipulation, just as he praises observational openness. We're left with the sense that intellectuals are bumblers: Bacon goes on to acknowledge the predictable failures of the learned "to observe decency and discretion in their behavior and carriage." Perhaps we should reach once more for the Italian phrase "Poco di matto," which suggests both clumsy psychological looseness and a winning attunement to the turns of fortune (280).

I have attended throughout this discussion to Bacon's language (his metaphors, turns of phrase, images, and habits of citation), but I want to close with Bacon's own remarks on form. Famously, he describes knowledge as a "thread to be spun on," inviting us to elaborate it at our pleasure; less often have we noticed that he extends our guiding metaphor by asking us to "unwind" it (404, 413). Here, we should also recall his famous "initiative method"—in which learning something reproduces the experience of first discovering it.[91] Looseness and waywardness, ease and inattention—the range of meanings conveyed by disinvoltura serve well as descriptions of the *Advancement*'s method of conveying knowledge, especially as it meanders in book 2, which is what Bacon suggests in the following passage: "*Antitheta* are Theses argued *pro et contra*; wherein men may be more large and laborious: but (in such as are able to do it) to avoid prolixity of entry, I wish the seeds of the several arguments to be cast up into some brief and acute sentences; not to be cited, but to be as skeins or bottoms of thread, to be unwinded at large when they come to be used; supplying authorities and examples by reference" (413). The parenthetical phrase reminds us of the undecided question of whether such a practice is available to all of us, and Bacon suggests that perhaps he himself is among the unfortunate who find themselves excluded. After all, he expresses only a "wish" for such "brief and acute sentences," and one that perhaps remains outstanding, depending on how we read book 2. The *Advancement* is brief on some matters and wordy on others (more on this below). Perhaps what we have before us is an array of "skeins," variously unwound. What's unambiguous here is that Bacon adopts the metaphor of "unwind[ing]" to depict both the production and dissemination of knowledge, and that it serves as an alternative to "laborious" overexplanation.

Bacon's "wish" for "acute sentences" calls to mind the aphorisms that compose the *Novum Organum*, which raise the same questions we encounter here. "Aphorisms," Bacon writes, "representing a knowledge broken, do invite men to enquire farther; whereas Methods, carrying the shew of a total, do secure men, as if they were at furthest" (405). Bacon favors brevity, but as a point of departure for expansion. Referring to the Aristotelian tradition that understands wonder as "knowledge broken," he casts the aphorism as an "invit[ation]" to inquiry—a brief but suggestive fragment that encourages investigation (405). Yet the aphorism, in its characteristically short and declarative form, also suggests conclusiveness—the delivery of an authoritative pronouncement. The doubleness of the aphorism—what we might call its "initiatory finality"—recalls the affective logic of "learning," where beginnings are endings, and endings, beginnings; we lose track of our position in time and attend to the matter at hand. As Vickers points out, the fragments in the *Novum Organum* are progressively less recognizable as aphorisms in the familiar sense: they are increasingly long and require increasing study.[92] Perhaps we can say that Bacon lets his thinking "unwind," allowing the maxim to meander, ramify, and expand.

The *Advancement* follows its own advice, developing a style that enables the inclusiveness it recommends. Bacon wanders far and wide. When he quotes Heraclitus's complaint that "men sought in their own little worlds, and not in the great and common world," he sets a course not only for the collective project of knowledge production but also for the writing of the book itself (292). Bacon's taste for "variety and universality of reading and contemplation" pertains as much to the *Advancement* as it does to "advancement" (287). When he describes book 2 as a "perambulation of learning," he offers an instructive metaphor for a writing practice that follows the injunction to abandon the intricacy of codified knowledge and look out instead into the multiform world (210). The *Oxford English Dictionary* takes the *Advancement* as an early example when it defines "perambulation" as a "comprehensive relation or description," and notes that "later" the word comes to suggest "circumlocution" and "verbal or literary wandering or digression."[93] Yet Bacon clearly intends both meanings, drawing continual attention to his own digressive dalliances with this matter and that. He ranges over the world of knowledge—tarrying here out of unexpected interest, making haste there with peremptory judgment. "My purpose," he writes, "is at this time to note only omissions and deficiencies, and not to make any redargution of errors or incomplete prosecutions," advertising a desire to do no more than diagnose (210). Such clarity of purpose doesn't rule out the drag of mental tides. "I purpose to speak actively without digressing or dilating," he says, yet throughout he makes a display of his lapses and argues for their inevitability (322).

When, for instance, Bacon turns to a survey of extant knowledge of public life in book 2, he finds himself listing the aphorisms of Solomon—and the list, with enthusiastic commentary, goes on for pages. "Thus I have staid somewhat longer upon these sentences politic of Salomon," he writes, "than is agreeable to the proportion of an example; led with a desire to give authority to this part of knowledge, which I noted as deficient, by so excellent a precedent" (452). Bacon's avowal of "[dis]proportion," which he chalks up to untamed "desire," is less exception than rule. Elsewhere, he makes a point of brevity, suggesting the necessity of leaving things out, as in the following concession to the tightness of time: "Now come we to those points which are within our own command, and have force and operation upon the mind to affect the will and appetite and to alter manners . . . of which number we will visit upon some one or two as an example of the rest, because it were too long to prosecute all" (438). Vickers locates a feeling of headlong momentum in book 2, which I would attribute in part to undisguised waywardness: the embrace of unjustified choices.[94]

Indeed, Bacon makes sure we notice his delinquency. At the beginning of book 2, perhaps the *Advancement*'s most emphatically teleological moment, he acknowledges susceptibility to unseemly passion. We have seen that errant "desire" leads him to tarry with Solomon's wisdom. Here, flights of passion encourage his vehement appeal to King James to make provision for the growth of knowledge: "To your Majesty, whom God hath already blessed with so much royal issue, worthy to continue and represent you for ever, and whose youthful and fruitful bed doth yet promise many the like renovations, it is proper and agreeable to be conversant not only in the transitory parts of good government, but in those acts also which are in their nature permanent and perpetual. Amongst the which (if affection do not transport me) there is not any more worthy than the further endowment of the world with sound and fruitful knowledge" (321). Bacon celebrates the future benef icence ensured by the Stuart monarchy, and expresses a wish that "sound and fruitful knowledge" will count among their gifts to the world. Yet Bacon's parenthetical suggests he is at the mercy of an "affection" that muddies mental clarity.

If we are skeptical of this interpretation, taking Bacon's self-professed "affection" for a simple modesty topos, we should notice that he repeats the gesture at the end of the introduction to book 2: "My hope is that if my extreme love to learning carry me too far, I may obtain the excuse of affection; for that 'it is not granted to love and to be wise.' But I for my part shall be indifferently glad either to perform myself or accept from another that duty of humanity, 'Nam qui erranti comiter monstrat viam, &c.'" (328). Bacon acknowledges his idiosyncrasy by pointing to his aggressive enthusiasm for "advancement," exactly the impulse for which he is now best remembered. Such exorbitant desire, he explains, might be a symptom of "extreme love," which is frankly incompatible with wisdom. Yet he solves the problem of personal passion by aligning himself with the rest of humankind, which can be counted on to moderate his passion if he is in fact out of line. It's this reversal—from intensity to serenity—that interests me here. The Latin phrase, quoted by Cicero in *De officiis*, comes from Ennius: "Who kindly sets a wand'rer on his way / Does e'en as if he lit another's lamp by his: / No less shines his, when he his friend's hath lit."[95] Cicero glosses these lines as a statement about common property: "he effectively teaches us all to bestow even upon a stranger what it costs us nothing to give."[96]

Bacon locates himself in a world of boundless intellectual generosity, characterized by the free exchange of gifts from an unlimited supply. If he cannot point the way or light the torch, others might. He again imagines a scenario in which desire is never likely to last long enough to start feeling like need—and, when it does somehow start to feel that way, importunity won't have much of a lifespan. Though his "indifference" about his role in setting humankind on the right path doesn't evoke unfeeling in the modern sense, his "indifferent" "glad[ness]" either to lend guidance or to be guided does imply equanimity. Both his calm willingness to await an outcome and his helplessness to do anything about the unpredictable welling up of passion in his breast are the conditions under which he finds his way forward.

Conclusion: Philosophy's Echo

At the beginning of this chapter, I alluded to Bacon's version of the story of Proteus, which captures the most familiar aspect of his thought. In his *De sapientia veterum*, a book of ancient myths refashioned as illustrations of his philosophical convictions, he takes Proteus as a metaphor for "Matter" and describes the unrelenting process of interrogation by which the natural world can be coerced into revealing its secrets. We meet this same Bacon, the inquisitor, in the *Novum Organum*, where he offers the following rebuke to the easygoingness I've defended in this chapter: "If in my natural history, which has been collected and tested with so much diligence, severity, and I may say religious care, there still lurk at intervals certain falsities or errors in the particulars,—what is to be said of common natural history, which in comparison with mine is so negligent and inexact? and what of the philosophy and sciences built on such sand (or rather quicksand)?"[97] In these lines, Bacon is emphatic about his "severity": he wants us to know what great pains he has taken—and to observe how laughable, by comparison, other attempts at natural history turn out to be. I don't want to lose track of the confident defiance of tradition conveyed by passages like this one—to forget the enterprising audacity and contemptuousness implicit in Bacon's rhetorical questions. If this chapter has succeeded, I've revealed the strangeness, as well as the disruptive effects, of Bacon's less familiar habits of mind: effortless flexibility and wide digression. Yet I do not wish to deny that he is also an apostle of rigor. The disunity of his body of work forecasts the complexity of the cases I take up in subsequent chapters.

Elsewhere in *De sapientia veterum*, Bacon counters the myth of Proteus with the myth of Pan, who brings together the several themes of this chapter. Taking the jaunty satyr for an emblem of the natural world, Bacon imbues his story with an atmosphere of effortless pleasure. Indeed, the whole collection can be seen as an exercise in carelessness. The sheer exuberance of his amassment of observations (sometimes conventional, sometimes borrowed from other mythographers, and sometimes simply conjured up by his imagination) indicates playful unconcern with the traditional meanings of the ancient tales.[98] He advertises his authorial insouciance when, in the preface, he informs us that "the Wisedome of the Ancients . . . was either much or happy; Much if these figures and tropes were inuented by studie and premeditation. Happy if they (intending nothing lesse) gaue matter and occasion to so many worthy Meditations."[99] Whether he is right about what they mean or incidentally insightful about some unrelated question, he doesn't much care. Thus we should not be surprised to find, as he contemplates the case of "Pan, or Nature," that he makes a display of negligence. I do not refer to his extravagant comparisons, many of which can be found in his sources: that the goat-horned demigod presents an instructive image of nature because the long strands of his beard look like sunbeams, or that his leopard-spotted cloak resembles the flower-strewn earth.[100] What interests me is that he confuses the very terms of the analogical equation in the chapter's title. As if in imitation of his wayward protagonist, his prose meanders, drifting from a description of nature to a portrait of the scientist. The myth recounts how Ceres responds to the abduction of her daughter Proserpina by blighting the earth and hiding herself away. Bacon narrates the story's resolution as follows: "It was *Pan*'s good fortune to find out *Ceres* as he was Hunting, and thought little of it, which none of the other Gods could do, though they did nothing else but seek her, and that very seriously [sedulo]."[101] From this episode, Bacon draws a lesson on the superiority of wide-ranging "experience," represented by Pan, over "philosophical abstractions," represented by the "greater gods," which is also a lesson on the advantages of blithe receptivity over focused intensity. He reiterates the point at the end of the passage: "Oftentimes even by chance [casu] such Inventions are lighted upon [incidere]."[102] Allusions to "falling" (variations on *casus*) in both the verb and the adverbial phrase convey the sheer fortuitousness of Pan's discovery. He falls on Ceres in a spirit of leisurely fun, out on the hunt for something entirely different, and it's the meandering breadth of his search that grants him success.[103] Pan no longer presents us with an image of the natural world; he now plays the role of the investigator of nature—but not the one we know from the Proteus story.

Years later, in the *De augmentis*, Bacon proposes "the Hunt of Pan" as a metaphor for "Learned Experience." This is a broad category that includes all manner of exploratory scientific practices—from the sober attempt to reproduce a natural phenomenon in an artificial setting to the delirious pleasure of making a wild guess: "when you have a mind to try something," Bacon explains, "not because reason or some other experiment leads you to it, but simply because such a thing has never been attempted before."[104] Yet he is careful to draw a distinction between such unreliable procedures and what he calls the "Interpretation of Nature," which is synonymous with method.[105] It's only by way of the latter that the philosopher succeeds at making a "transition from experiments to axioms," thereby arriving on the solid ground of truth.[106] Bacon devotes much of his philosophical career to method's promise of foundational firmness—but we've also seen ample evidence of his interest in casual enlargements of knowledge.[107] The *Advancement* is only the beginning. Consider, for instance, the *Sylva Sylvarum*, his unfinished book of natural-historical observations. Reading through what Bacon's amanuensis and the book's first editor, William Rawley, worried "may seem an indigested heap of particulars," one comes across anecdotes like the following: "I left once by chance a citron cut, in a close room, for three summer months that I was absent; and at my return there were grown forth, out of the pith cut, tufts of hair an inch long, with little black heads, as if they would have been some herb."[108] In moments like this, the *Sylva Sylvarum* savors the at-hand affordances of the senses. It even goes as far as to relay hearsay and old, untested theories. When he isn't insisting on rigor, Bacon reminds us that the world makes a habit of sharing its secrets even when we haven't taken precautions against misunderstanding.

At the end of the story of Pan and Ceres, he raises the question of natural desire. He introduces the "World" as a third gloss of Pan (after Nature and the natural philosopher), absorbing his protagonist's freedom from discipline into the nature of things. (The Latin gives us the word *mundus*, which encompasses a range of meanings from "world" to "universe," but the English version's "World or Nature" indicates that *mundus* here takes the place filled earlier by *natura*.[109]) "For the World or Nature doth enjoy it self," Bacon writes, "and in it self all things else. Now he that loves would enjoy something, but where there is enough, there is no place left to desire. Therefore there can be no wanting love in *Pan*, or the World, nor desire to obtain any thing (seeing he

is contented with himself) but only Speeches, which (if plain) may be inti-
mated by the Nymph *Echo*; or if more quaint [si accuratiores sint] by
Syrinx."[110] Even though Pan inhabits a state of steady "content[ment]," Bacon
stops short of ruling out "desire." On the one hand, the notion of the world's
self-sufficiency, with its Platonic resonance, precludes "wanting love" or longings (*amores*); Bacon describes an experience of satisfaction that admits of no
lack.[111] On the other hand, he avows an exception in "Speeches," picturing
unearned and unsought pleasure as Nature's gift—and "enjoy[ment]" is not
the state of fullness it seems. Pan wants something, and yet his privilege is to
experience desire not as privation but instead as gratification without fullness.
Even without Bacon's eventual admission that something like desire persists
alongside perennial satisfaction, Pan would dodge the charge of solipsism,
since contentment includes the "enjoy[ment]" of "all things else"—one of
the enviable benefits of being "the World." Perhaps, after all, we don't have
to envy it, since dropping the reins of self-management means slipping into
the very state of gentle gratification "the World" knows best.

From the story of Pan and Ceres, Bacon derives a textured concept of
outward-looking satisfaction, not unlike the vibrant indolence of Montaigne's exploratory gambol. The scene of contentment is one of neither
complacency nor lost interest, except in the punning sense of an interest
that wanders abroad. Bacon offers Echo, who generates the "Speeches"
contented Pan desires, as an image of "true Philosophy."[112] Pan speaks, and
the resounding of his voice is an experience of desireless gratification for
which Bacon offers the name "Philosophy." Even the application of care to
the transmission of truth (the *cura* in the "quaint[ness]" [*accuratus*] of poetic
language) leaves Pan's contentment—his heatless desire undisturbed.[113]

One effect of a habit of thinking and feeling that welcomes inconstancy
is that it is especially susceptible to multiple readings. Bacon's interpreters
have not been guilty of inattention, but his most piercingly provocative pronouncements, along with his most violent metaphors, have governed the
interpretation of his philosophy. I've stepped away from a scholarly emphasis on aggression in order to observe Bacon's observational mood and the
variegated textual corpus it animates. The *Advancement* is not alone among
his works in its many-sidedness; nor is it the only one that inhabits an observational mood but also (perhaps, indeed, by virtue of that mood) gives
resistless way to desire's touch. I've defended him from the charge of aggressive teleology and the attendant violence of instrumental reason, but not
because he's immune from prosecution. Like Montaigne, who advertises his
weakness, Bacon is more than aware of his own susceptibility, for good and
for ill, to transformation—both within the space of his writings and over the
careening course of their subsequent reception.

CHAPTER 2

The Angle of Thought

Robert Boyle, Izaak Walton, and the Scientific Imagination

On October 9, 1667, Robert Boyle begins a series of experiments on bioluminescence. With the aid of the air pump, to which he refers affectionately as "one of my little *Engines*," he observes that decaying shards of wood and pieces of "stinking fish" lose their mysterious glow when he deprives them of air.[1] The title of the account he publishes in Henry Oldenburg's *Philosophical Transactions* conveys the simple satisfaction of flipping a switch: "New Experiments, to the number of 16, concerning the Relation between Light and Air (in Shining Wood and Fish); shewing, That the withdrawing of the Air from those and the like Bodies, extinguishes their Light, and the readmission of the Air restores it."[2] Boyle derives pleasure from mastery (with a wave of the hand, moldering matter springs to luminous life), but his introductory remarks are less than masterful.

> Though the main Experiment be but one, I intended to set down what occurr'd to me about it but as several *Phaenomena* of it; yet finding it requisite to acquaint you with some Tryals that are not so properly Parts of it, I shall for distinction sake propose them as several Experiments; the *Narratives* whereof are taken, for the most part, *verbatim* out of the Notes I set down for my own use, when the things to be registred were freshly done. Which Advertisement I give you, both to excuse the carelesness of the Style, and to induce you not to distrust a Narrative that was made only to serve my Memory, not an *Hypothesis*.[3]

Boyle means to describe a single experiment, but he can only present his findings as a sequence of distinct trials. The narrative rattles along in fifteen numbered sections ("Experiment I," "Experiment II," and so on), yet the title speaks of sixteen.[4] Should the reader object to disorder, Boyle asks pardon for laxity. He attributes the treatise's "carelessness" both to his dependence on laboratory notes ("for the most part," he concedes) and to his purely descriptive intention. Were he to make an argument for anything in particular, he explains, he might then feel an obligation to collect his thoughts.

One influential interpretation asks us to translate Boyle's carelessness into subterfuge: a strategy for gaining credence. The artlessness of the natural philosopher serves as evidence that he has no philosophical axe to grind. The kinship of the published description and the private notebook are taken as signs of accuracy: he writes things down as they actually happen.[5] Yet Boyle's treatise makes an ostentatious, even outlandish, display of failure, inattention, and technical malfunction. Crossing the threshold between confessions of imprecision and avowals of chaos, it might just as well erode the confidence of his readers. Reflecting on the thesis that "Shining Wood *will be easily quench'd by Water and many other Liquors,*" for instance, Boyle recounts a strange and revealing episode with disarming unselfconsciousness:

> The same Experiment I tried more than once with high rectified *Spirit of Wine*, which did immediately destroy all the light of the Wood that was immersed in it; and having put a little of that Liquor with my finger upon a part of the whole piece of Wood that shone very vigorously, it quickly did, as it were, quench the Coal as far as the Liquor reach'd; nor did it in a pretty while regain its luminousness: (Which whether it recover'd at all, I know not; for this Trial being made upon my Bed, I fell asleep, before I had waited long enough to finish the Observation.)[6]

Boyle takes his work to bed, and he falls asleep mid-experiment. Unlike other disclosures of misadventure that give the account its air of improvisation (he mentions leaky equipment, lusterless fragments of wood, and much comic fumbling in the dark), the sleepy inattention of the scientist is a personal failing.[7] Perhaps the appeal of the bedroom as a location for the experiment is the darkness it provides, but Boyle takes the opportunity to advertise the psychological limitations of the experimenter.[8] Does Boyle embrace delinquency?

Evoking both the limits of meticulousness and the limits of his desire for it, our narrator sets sober understanding against amusement as competing motivations for inquiry. He describes the reintroduction of air and thus of light to the chamber of his *"Pneumatic Engine"* as an experience of wondrous enjoyment, to which he partly attributes his interest in further trials: "We let in the outward Air by degrees, and had the pleasure to see the seemingly extinguisht Light revive so fast and perfectly, that it looked to us all almost like a little flash of Lightning, and the splendor of the Wood seemed rather greater, than at all less, than before it was put into the *Receiver*. But partly for greater certainty, and partly to enjoy so delightful a spectacle, we repeated the Experiment with the like success as at first."[9] The air pump offers an experience of entertainment like that of a diorama. Boyle and his assistants compose a rapt audience for the brief "spectacle" of "a little flash of Lightning." One can't help but recall the account in Boyle's unpublished autobiography (1647–48) of a terrifying storm that occasions a religious "Conversion."[10] The "considerablest" event "of his whole life" is the clap of thunder that rouses a suddenly vigilant Christ "who had long layne asleepe in his Conscience" but only now commands the scene.[11] It's as if the apocalyptic storm were here restaged as an amusement or novelty: the mother's disappearance and return as a game of *Fort-Da*.[12] Yet the substitution of omnipotence for trepidation is no simple affirmation of power, since pleasure actually interferes with the gathering of information and thus with the establishment of fact. Boyle remarks that he eagerly repeats the experiment "without taking notice how long it lasted," as if the pursuit of "pleasure" distracted him from taking proper notes.[13]

Boyle's mind is no more reliable than his instruments, and he wants us to notice the muddle. Even the proposition that a purely observational purpose legitimates carelessness isn't the bid for objectivity it seems. *"Yet I shall content my self at present,"* he writes, rephrasing his claim to unprocessed description, *"to have faithfully delivered the Historical part of these Apparences, without making, at least at this time, any Reflections on them."*[14] Boyle's abstention from contemplation really is a matter of *"content*[ment]": he doesn't restrain himself so much as set a (gratifyingly) low bar for achievement. He remarks that "Speculation" would be bad for his poor health, on which late-night trials have already taken a heavy toll—as if wayward thought presented a more serious risk to physical wellness than it did to proper understanding.[15] When Boyle's mind wanders, he indulges an unhealthful predilection and thus only incidentally departs from a practice of unembellished observation. Remarking on the loss and return of light to rotting fish, he happily muses, and at some length, on *"short-liv'd Apparitions* of Light" that appear in the "Vaults" of cave systems, returning too late to the theme of bare reportage for it to count as discipline. "As these thoughts were but transient conjectures," he writes, "so I shall not *entertain* you any longer about them"—and yet he has already *"entertain*[ed]" us plenty.[16]

If Bacon's *Advancement of Learning* is the primordial myth of English science, Boyle's laboratory is its primal scene. After the founding of the Royal

Society for the Improving of Natural Knowledge in 1660, Boyle comes to serve as its public face. He embodies the emergent social type of the gentleman-scientist—whose lodestar, perhaps, is conscientiousness, while the air pump Robert Hooke builds for him materializes a second contender: ingenuity. Thanks largely to the work of Schaffer and Shapin, as well as the response their coauthored *Leviathan and the Air-Pump* received from Bruno Latour in *We Have Never Been Modern*, Boyle is again exemplary—though not, this time, of rigor or skill.[17] Instead, he is made to perform the bid for credibility implicit both in meticulous procedure and in displays of aristocratic honor and Protestant virtue. Boyle remains an allegory of rectitude. Yet the episode with which I began conveys the strange plausibility of the contrary formulation. Boyle is also the very picture of discomposure—of the undisciplined mind.[18] Throughout his career, carelessness defines a "way of thinking" in which cognition follows the winding path of errant association.[19] (As far as the term itself is concerned, Boyle uses it both for this distinctive purpose and to issue an ordinary, denigrating judgment.) Often, carelessness suffuses an experience of appreciation prior to methodical study—an admiring awareness of Nature's plenitude that precedes the selection of an object of inquiry. In this respect, it typifies a perspective available to sympathetic readers who happen not to be natural philosophers. Yet the affective distinction between scientific *interest* and investigative *procedure* (or, thinking back to the introduction, between the context of discovery and the context of justification) is far from tidy, which is why I present Boyle's inattentive bedside experiment as a point of departure.

Boyle persistently inhabits a state of affective disorientation, where feeling has yet to achieve definition and imperative force.[20] An experience of effortless indistinction, but one that capitulates readily and unpredictably to passion, Boyle's observational mood presents a challenge to familiar interpretations of early modern science. My basic question is less *what* than *how* Boyle thinks, which is why an available vocabulary of diffidence, nescience, and impartiality, although successfully descriptive of some aspects of his work, is insufficient.[21] Indeed, the neutrality he sometimes performs in the face of philosophical problems conveys something like the opposite of carelessness. Boyle proposes a "corpuscular hypothesis" because he isn't willing to subscribe to the controversial argument that atoms are indivisible; a corpuscle is a cautious atom.[22] Indeed, Boyle's habit of refraining from making unwarranted assertions follows the same logic as the claim to mere transcription in his experiments on glowing matter. It also recalls René Descartes's argument in the fourth of the *Meditationes de prima philosophia* (*Meditations on First Philosophy*, 1641) that all we have to do to avoid error is refrain from making judgments about things we don't understand.[23] This familiar gesture, which we perhaps knew best as the Kantian distinction between *noumenon* and *phenomenon*, creates a halo of self-confidence from which doubt is carefully excluded. By contrast, Boyle's carelessness sends him reeling.

Interestingly, Michael Hunter's thoughtful interpretation of Boyle's philosophical career emphasizes his "legendary" personal piety, the chief feature of which is yet another antonym of carelessness: "scrupulosity."[24] Boyle seems to have been plagued throughout his life by the burden of moral self-examination. As Hunter puts it: "It was here [in this experience of soul-searching] that Boyle's piety was at its most active and stressful, reflecting his concern about the moral and spiritual obligation of every believer to do right and to serve God to the utmost of his ability."[25] Though my argument in this chapter is literary rather than biographical, this view is important because it presents an alternative to Schaffer and Shapin's, establishing a motive for honest reporting other than building credit: an obsession with virtue. Yet several of the details I have already mentioned—a distracting "delight" in experiment, a proclivity for digressive thought—seem less like expressions of Boyle's commitment to integrity than like incidental features of his narrative; they are not straightforward admissions of fault. Whether we take Boyle's scrupulous self as real or fictional (as the actual expression of interiority or the performance of an assumed persona), we might remark that a life beset by relentless self-evaluation and an insistent fear of falling short might also be one in particular need of relief.[26] If Boyle cannot escape the burden of worrying that he's less than adequately good, he can at least find opportunities to think about something else. In an observational mood, attention lingers on features of the exterior world—and it remains unguided by ambitions that would direct it back to the self from which it happily wanders free. Meandering, looking, thinking: these experiences offer freedom from introspection.

This chapter's focus is Boyle's *Occasional Reflections* (1647–65), which models a style of susceptibility that remains a pervasive theme across his body of work, linking his early moral philosophy to his later scientific career. Evoking without adopting the genre of *occasional* meditation as well as the history of *methodical* meditation, the *Reflections* takes its leave from the narrative of self-discipline that defines the tradition, despite an elasticity that accommodates figures as different as Ignacio de Loyola and Joseph Hall.[27] I explore the affinity between Boyle's meandering (and virtually unstudied) fishing narrative (one section of the *Reflections*) and Izaak Walton's enormously popular practical guide, *The Compleat Angler* (1653). My primary interest here is how Boyle's idiosyncratic practice of countermeditation distinguishes him from Walton, but I also offer a sketch of an emergent, nontechnical literature of experiment in which they both participate. The final section of the chapter carries the argument through to Boyle's late career, from *Of the Study of the Book of Nature* (1650–54), which he conceives as an addition to the *Reflections*, to his ruminations on natural theology in *Of the High Veneration Man's Intellect Owes to God* (1685). Thus my account shows how much is at stake in the previous chapter: not just our understanding of Bacon's philosophy but also our sense of his legacy. In the second half of the seventeenth century, England's most prominent natural philosopher is no less interested than Bacon or Montaigne in states of emotional weakness, as well as their cognitive corollaries: clutter, commotion, and confusion. The hero of Baconian science is also its distracted Lord of Misrule.

The Rigors of Meditation

Soon after the publication of Boyle's *Reflections* in 1665, the Puritan divine Richard Baxter writes the author a letter of appreciation, but one that displays his refusal of the book's extravagant departure from precedent.[28] Boyle's collection of meditations, for which he sometimes uses the word "reflections," "discourses," or even "essays," disregards the genre's emphasis on mental guidance, embracing instead an unlikely ethos of deviation—as, indeed, his terminological flexibility suggests, but which Baxter doesn't countenance.[29] "The *Matter* of your Booke having occasioned all these words," Baxter writes, referring to his own commentary on the virtues of meditation, which makes up the bulk of the letter,

> I must thank you for it also as to the *Manner* . . . that you call men to the manly worke of *Meditation*; to waken the sleepy Reason of the world, & bring it into exercise: Most of the world would become much *wiser* (& consequently *better*, & consequently *happyer*) if they could be brought to be more *Considerate*: if they were but shutt up in the Jesuites dungeon one houre in a day, to *thinke seriously* of God, & of their Happynes & Duty, & were forced to give an account of their thoughts, to some sober person![30]

Were it clearly apprehended, Boyle's own "sleepy Reason" could only elicit disavowal from this most punctilious of readers, among whose works one finds an influential adaptation of the spiritual exercise to a Puritan context, *The Saints Everlasting Rest* (1650). Baxter's zeal for discipline goes as far as to animate a wish that "most of the World" might be "shutt up in the Jesuites dungeon," exaggerating the genre's function as a means of constraint. What we can't tell is whether Baxter secretly objects to Boyle's deviant style or actually finds himself persuaded, as subsequent readers often have been, by the philosopher's professions of commitment to self-discipline. When Baxter observes that the *Reflections* "occasioned all these words," he employs the language of slanted thought, but without acknowledging (and perhaps without noticing) that Boyle has radicalized the idea. For Boyle, the connection between a thing and whatever it "occasions" is more than oblique; sometimes, it's close to nonexistent. Baxter champions stringency—but in praise of what he might or might not recognize as oh-so-loosely "occasional."

Indeed, from the very beginning of the letter, Baxter harps on the theme of laborious devotion, speaking of arduousness as if it were characteristic of Boyle's style. Baxter offers a fragment of autobiography as if in affirmation of a book he could only condemn, were he to grant it the same careful attention he gives to the art of meditation: "When God removed my dwelling into a Churchyard, & set me to study bones & dust, & by a prospect into another world, awakend my soule from the Learning of a child, & shewed me that my studyes must not be *play*, but *affective practicall serious worke*, I then began to be conducted by *Necessity*, & to search after *Truth* but as a meanes to *Goodnes*, & to perceive the difference betwixt a pleasant easie *dreame*, & a waking working knowledge."[31] Baxter offers a precise but inverted account of the distinctiveness of Boyle's strange experiment in form, which regularly gets lost in "pleasant easie *dreame[s]*." Even without a sensitive appreciation of Boyle's book, one might nonetheless wonder at the incongruity of Baxter's emphasis on mortality. "Occasional" or "extemporal" meditation, even in the sober manner of Hall's *Occasionall Meditations* (1630), is distinguished by its quotidian subject matter. Baxter arrives quickly at "bones & dust" while supposedly leafing through a book that recounts the unexceptional events of ordinary life. Though the vanity of worldly things is among the genre's themes, Baxter's morbid intensity is less than apposite. He is guiltiest of missing the point, however, when he extols the benefits of "practical serious worke"—as if the author of the *Reflections* were not, as we shall see, a connoisseur of leisure.

Baxter gives us a perfect image of Boyle's spiritual practice, as long as we turn it inside out. Boyle's meditations, I suggest, are barely meditations at all. The *Reflections* wanders across genres (georgic, pastoral, dialogue, essay, and romance), exulting in the freedom of never having to accept the encumbrance of a single one. Though Boyle adopts the language of "assiduous[ness]" from the genre he claims to inhabit, he understands intensity as a matter of frequency rather than dogged directionality. Like Bacon's recasting of self-training as the effortless pleasure of exercise, Boyle finds an easy path through difficulty. In moments, the mere assertion of pious devotion authorizes digressive habits of mind. My purpose in this first section is to describe Boyle's theory of "reflection," especially as he develops the theme of reflective "deviation" in his prefatory *Discourse Touching Occasional Meditations*, and to highlight points of contrast with influential entries in the genre.[32] Occasional and methodical meditations together compose a single negative paradigm that illuminates his purpose. Baxter's image of "the Jesuites dungeon" into which any occasion might be transformed sharpens our sense of the theological and thematic continuity between them. However wide, the world of "occasions" remains, for Baxter, a (valuably) cramped cell. With the conspicuous exception of Boyle's, meditative practice can be counted on to lead the mind in the right direction.

When Boyle denies knowledge of Joseph Hall's book of meditations, a popular and seemingly unavoidable precedent, we should take him at his word—if only in the soft sense of accepting the insignificance of the influence. "Not to Prepossess or Byass my Fancy," Boyle writes, "I purposely (till of late) forbad my self, the perusing of that Eloquent Praelates devout Reflections" (*Discourse*, 10). His comment sounds like a parody of scientific rigor; we know how averse the virtuosi are to "prepossession" and prejudicial attachment, but here Boyle is concerned with protecting "Fancy" rather than reason from contamination. His enthusiastic description of the sheer wildness of thought tells us why: "When the Mind is once set on work, though the Occasion administred the first Thoughts, yet those thoughts themselves, may, as well as the Object that excited them, become the Themes of further Meditation: and the Connection of Thoughts within the Mind, may be, and frequently is, so latent, and so strange, that the Meditator will oftentimes admire to see how far the Notions he is at length lead to, are removed from those which the first Rise of his Meditation suggested" (*Discourse*, 51–52). When Boyle savors "latency" and "strangeness," he is, as far as I can tell, utterly unlike any other author of meditations in the period. Narrating the activity of "the Mind" in the third person, Boyle bears passive witness to a train of thought. Hall's meditations can be supple and surprising, but they move swiftly from the observational to the sententious, reading daily events as readymade moral lessons. For Boyle, the passage of the mind from event to thought is often the first leg of a longer journey or a prospective course that invites interruption.

Hall's paradigmatic meditations make plain the degree of Boyle's extravagance. "Upon the Shutting of One Eye" is a good example of his efficiency. The meditation terminates after a single sentence, albeit one elongated by semicolons. Hall passes seamlessly from observation to precept by way of an adverbial "thus": "When wee would take ayme, or see most exquisitely, wee shut one eye: Thus must wee doe with the eyes of our Soule; When wee would looke most accurately with the eye of Faith, wee must shut the eye of Reason; else the visuall beames of these two apprehensions, will bee crossing each other, and hinder our cleare discerning."[33] Boyle keeps both eyes open; he is more likely to lament the poverty of a sensory experience restricted to only two eyeballs than he is to object to the confusion of multiple lines of sight. (In chapter 3, we find that Henry Power articulates exactly this desire for additional organs of perception.) In "Upon the Variety of Thoughts by Way of Conclusion," Hall again rejects mental disunity in favor of self-possession, giving the very last word in his book to that theme: "My thoughts would not bee so many, if they were all right; there are ten thousand by-wayes for one direct? As there is but one Heaven, so there is but one way to it; that living way, wherein I walke by Faith, by Obedience."[34] If keeping to the road is a hallmark of the genre, as epitomized here by Hall's affirmation of a single virtuous path, Boyle is supremely deviant. In the *Reflections*, "by-wayes" are the way.

Boyle's sister, Mary Rich, Countess of Warwick, who composes 182 occasional meditations between 1663 and 1667, ignores her brother's example, deepening our sense of a continuous tradition from which he conspicuously departs.[35] Like Hall, she displays the genre's facility with the swift translation of occasion into judgment. Indeed, she outmatches Hall's diligence, keeping her eyes resolutely on the prize. In "Upon leaves that fell from a tree in athome [autumn]," for instance, she observes exactly what the title describes and concludes that time is tight: "Dust I am and unto dust I must returne."[36] Here is another of Rich's titles that makes patent, and therefore perhaps superfluous, the forthcoming content of the meditation: "Upon seeing a hog lye under an acorn tree and eate the acorns but never looke up from the ground to the tree from which they fell."[37] That the image recommends gratitude to God comes as no surprise. My point is not that occasional meditations are uninteresting but that the sources of their interest, in addition to the affirmation of devotion that motivates them, are observational detail and analogical skill. For Boyle, however, the practice owes its allure to the irregular path leading from thing to thought—and then to subsequent thought. What matters in Boylean analogy is the contrasting features it generates by way of comparison: X is like Y, but Y looks like X + A, B, C, and D. As the mind proceeds to subsequent analogies, moreover, it sometimes loses track of the original point of comparison.

I have observed that the genre of occasional meditation is single-mindedly directional, but the point can be generalized further. The longer tradition of methodical meditation relentlessly counters distraction. Think, for instance, of Lorenzo Scupoli's *Il combattimento spirituale* (*Spiritual Combat*, 1589), in which the self wages war against the most intimate of adversaries: itself (or its own proclivities). One can quote from almost anywhere in Loyola's paradigmatic *Ejercicios spirituales* (*Spiritual Exercises*, 1548) to illustrate the genre's imperative force: "First, in the morning, immediately on rising, one should resolve to guard carefully [guardarse con diligencia] against the particular sin or defect with regard to which he seeks to correct or improve himself [corregir y enmendar]."[38] Closer to home (in both time and space), Hall's methodical meditations are no less bent on "correction." In his *Arte of Divine Meditation* (1605), he proposes that his words might be "like unto *Goades* in the sides of every Reader, to quicken him up out of this dull and lazy security."[39] "Security," the etymology of which suggests absent *cura*, can carry the connotation of carelessness, as it does in this "lazy" instance.[40]

In his *Meditations and Vowes, Divine and Morall* (1605), Hall imagines a community of reciprocal regulation, exhorting "every Reader" to discipline and asking for them to return the favor: "(*Having after a sort vowed this austere course of judgment and practice to my self*) *I thought it best to acquaint the world with it, that it might either witnesse my answerable proceeding, or check mee in my straying there-from: by which means, so many men as I live amongst, so many monitors I shall have which shall point me to my own rules, and upbraid me with my aberrations.*"[41] Hall intends mutual watchfulness as a supplement to self-discipline, creating concentric circles of "monitors." Combining the language of self-"rule" and unswerving progress, he draws attention to the straight "course" meditation keeps. Directness takes work; any lapse in exertion could mean losing the path. When Hall warns us that "caution is to be had that our Meditations be not . . . too farre-fetcht," and goes on to explain, "farre-fetcht I cal those, which have not a faire and easie resemblance unto the matter from whence they are raised," he anticipates Boyle's transgression precisely.[42] Measured against Hall's criteria for meditative devotion, Boyle is guilty of predictably twinned failures: leisurely cool and blithe disorientation.

As we look more closely at Boyle's account, then, we might observe how powerfully the genre of meditation raises a false expectation. At times, Boyle comes close to adopting a familiar ethos of self-cultivation—but the more we read, the more difficult it is to avoid the conclusion that he swerves from precedent: "The custom of making Occasional Reflections may insensibly, and by unperceiv'd degrees, work the Soul to a certain frame, or temper, which may not improperly be called Heavenly Mindedness, whereby she acquires an aptitude and disposition to make pious Reflections upon almost every Occurrence, and oftentimes without particularly designing it" (*Discourse*, 52).

Boyle recommends practice, but it doesn't much feel like discipline. Though it "works the Soul," the change it effects takes place "insensibly, and by unperceiv'd degrees." Recalling Bacon's recommendation that we "work out the knots and stonds of the mind" without breaking a sweat, Boyle's language suggests an experience of sheer spontaneity we don't experience as what, from a third person perspective, it is: purposeful self-habituation.[43] This is not self-mystification. "Heavenly Mindedness" is an "aptitude and disposition" to let the mind wander "without . . . design," which has been a feature of the experience all along—from the very earliest efforts at learning to meditate. Though here our "custom[ary]" engagement in meditation makes it come naturally to us, "oftentimes" the agency behind reflection is an ordinary "Occurrence." In those cases, meditation is an experience that befalls us.

Boyle again sounds much like other advocates of meditation when he describes it as an antidote to idleness, but "Heavenly Mindedness" grants an air of piety to languor. Consider the following lines, where Boyle asks us to make good use of our time, but can't quite be bothered to make an argument to that effect.

> If I had not elsewhere display'd the Evil and Danger of Idleness, and represented it as a thing, which, though we should admit not to be in it self a sin, yet may easily prove a greater mischief than a very great one, by at once tempting the Tempter to tempt us, and exposing the empty Soul, like an uninhabited place, to the next Passion or Temptation that takes the opportunity to seize upon it: If (I say) I had not elsewhere discours'd at large against Idleness, I might here represent it as so formidable an enemy, that it would appear alone a sufficient Motive to welcome our way of Meditation. (*Discourse*, 22)

Perhaps the playfulness of polyptoton ("tempting the Tempter to tempt us") signals Boyle's heedlessness of danger, and his explicit omission ("If I had not elsewhere" spoken of "Idleness," he says, preferring in this case to skirt the issue) most certainly has that effect. Boyle even "admit[s] [idleness] not to be in it self a sin." We can only be struck by how well his metaphor of the "uninhabited place," though presented to us as the problem meditation solves, suits the very style of reflection he proposes. It's the persistent absence of preoccupation that distinguishes his spiritual practice.

Consider, for instance, the following lines, where he explains how interesting his meditations are to a wide and diverse audience by describing occasional reflection as sheer susceptibility to whatever crosses his path. A mere inclination to goodness, it seems, is defense enough against the Tempter.

If I had written in a more usual or a more solemn way, I should per-
chance have had no Readers but Divines, or Humanists, or Devout
Persons, or Despisers of the World, or (in a word) the Masters, or Lovers of that one kind of Learning, to which my Subject did belong:
But treating as I do, of Whatever chanc'd to come in my way, and
consequently of many very Differing, and Unusual things, Curiosity
will probably invite both the Learned and the Devout; both Gentlemen, and Ladyes; and, in a word, Inquisitive Persons of several Kinds &
Conditions, to cast their Eyes upon these Reflections.[44]

Boyle portrays himself as an utterly indiscriminate observer. When he distances
himself from the "more solemn way" of composing occasional meditations
and those "Despisers of the World" who would most enjoy it, a distinction
between the genre's "usual" form and his own shades into outright distaste
for straitened pathways of devotion. Within the space of a single sentence,
he describes his practice by collapsing two meanings of the word "way." He
refers first to method, which amounts to a lack thereof ("If I had written in a
more usual or more solemn way"), and then to his winding path, both literal
and mental ("Whatever chanc'd to come in my way"), which is the closest he
comes to regularity. We might think here of Montaigne's description of his
writing practice: "I speak to my paper as I speak to the first man I meet."[45]

One conclusion we can draw from Boyle's soft but decisive departure from
the genre is that it foregrounds the question of mood. If the meditator's blithe
inattention were unimportant, he might have adopted a more supple, and
thus less glaringly inappropriate, generic frame: the essay, for instance. (He
does glance in that direction—but he labels the *Reflections* a book of meditations.) Boyle's remarks on meditation suggest a propensity for carelessness
the meditations themselves bear out. Indeed, his conspicuously unmeditative
meditations establish a pattern of unblushing failure. Consider, for instance,
his disheveled introductory self-portrait. I'm wary of making too much of
prefatory apologies (too many seventeenth-century authors signal modesty
by claiming to have rushed their works into print), but Boyle's exaggerated
disarray is an early affirmation of what the book everywhere advertises.

I shall not much wonder to find it said, That the Book is, in general, far
short of being an Exact and Finish'd Piece. For perhaps few Readers
will be more of that mind, than the Author is. But by way of Apol-
ogy, it may be represented, That most of the following Papers, being
written for my own private Amusement, a good deal of Negligence
in them may appear as pardonable, as a Careless Dress, when a man
intends not, nor expects, to go out of his study, or let himself be seen.[46]

The domestic scene excuses the author's "Careless Dress," much as it does in Montaigne's introductory note to the *Essais* ("Au lecteur"). "If I had written to seek the world's favor," Montaigne writes, "I should have bedecked myself better, and should present myself in a studied posture."[47] He goes on to justify his discomposure by explaining that he has "no goal but a domestic and private one," and Boyle follows suit by claiming that he intends the book only for his "own private Amusement."[48] When he eventually tells us that meditation makes the benefits of "Experience" available to us "without grey-hairs" or celebrates a practice in which "we receive the advantage of learning good Lessons, without the trouble of going to school for them," we understand that freedom from laborious care is exactly what the *Reflections* has to offer (*Discourse*, 33, 27). Boyle's self-negligence is a model for the experience of reading the book.

Another early clue that Boyle's rumpled self-presentation is less the pro forma profession of modesty it might seem than his signature style comes at the end of the first chapter of the *Discourse*, which resonates with his knowingly undeveloped remarks against idleness.

> I should judge it a very natural Distribution to divide the following Discourse into two parts, the first of which should contain some Invitations to the Cultivating this sort of Meditations, and the latter should offer something by way of Method, towards the better framing of them. But lest I should at this time be hinder'd from treating of each of them distinctly, I will at present omit that Division, and indeavour in recompence so to deliver the Motives I am to propose, that the first part of the Discourse may not appear maim'd, though it be unattended by the second, and yet the Particulars that might compose the second, may (if it prove convenient to mention them at this time) be commodiously enough inserted in opportune places of the first. (*Discourse*, 21)

Boyle alerts us that a proper guide to meditation would offer both encouragement and instruction, but also that he doesn't bother with the latter. He suggests instead that he gives us the guidance we need when it "prove[s] convenient," but we find, if we read to the end, that the moment never arrives. Indeed, the phrasing suggests that he waves his hand at the issue: he does not promise "Method" but only "Something" like it. Though "Method" should not be understood here as a technical term, it's difficult not to hear in this remark, coming as it does at the beginning of a book that, as we shall see, integrates the genre of meditation and the emergent procedures of experimental science, the repudiation of two distinct but relatedly supervisory practices we expect Boyle to affirm. We might think here of the Greek *hodos*, or "way," contained etymologically in the term. Boyle's departure is unmistakable. The casualness with which he refuses his model redoubles our sense of surprise.

The Science of Digression

Throughout the *Reflections*, the rambling path of the undisciplined mind is the guiding figure of Boyle's scientific imagination. Were we less wedded to the thesis that method defines the Baconian moment, we would have little reason to wonder at the claim. If you wander a landscape, you get to know it better. Though scholars have often relegated the book to a prescientific phase in Boyle's career, its ambitions are Baconian.[49] Boyle's performance of wayward mobility equates easy gratification with receptivity to the surrounding world. In the very different context of his autobiography, he laments his early predilection for "raving" or going astray. The theory of occasional reflection creates a context in which his unfortunate "habitude" need not be regrettable after all. Indeed, it has the salutary effect of maximizing his exposure to phenomena. At Eton College, he explains,

> to divert his Melancholy they made him read the stale Adventures [of] Amadis de Gaule; & other Fabulous & wandring Storys; which much more prejudic'd him by unsettling his Thoughts, then they could have advantag'd him; had they effected his Recovery; for meeting in him with a restlesse Fancy, then made more susceptible of any Impressions by an unemploy'd Pensiveness; they accustom'd his Thoughts to such a Habitude of Raving, that he has scarce ever been their quiet Master since, but they would take all occasions to flinch away, & go a gadding to Objects then unseasonable & impertinent.[50]

Later, he tells us, in Stalbridge, "he would very often steale away from Company, & spend 4 or 5 howres alone in the fields, to walk about, & thinke at Random; making his delighted Imagination the busy Scene, where some Romance or other was dayly acted: which tho imputed to his Melancholy, was in effect but an usuall Excursion of his yet untam'd Habitude of Raving; a Custome (as his owne Experience often & sadly taught him) much more easily contracted, then Depos'd."[51] Boyle's reminiscences include virtually all the motifs of the *Reflections* (which he seems to have begun writing at about the same time): not only "wandring," "gadding," and "think[ing] at Random," but also a state of "unemploy'd Pensiveness" in which he finds himself "susceptible of any Impressions" that happen to strike his "Fancy." Though Boyle implies that he eventually "tam[es]" his proclivity for "gadding," the

Discourse suggests instead that he eventually finds a vocabulary with which to embrace it. It's as if he ultimately rejects the opposition between "restlesse" romance and "quiet" meditation, discovering that emotional "quiet" invites a course of thought no less errant than the pathways of questing knights. By accepting the claims of "unseasonable" "Objects" on his attention, Boyle discovers uses for distraction.[52]

Few scholars have acknowledged the scientific orientation of the *Reflections*; for those who have, it is mostly assumed rather than shown.[53] One initial clue to Boyle's purpose is the central place he accords the image of optics in his book. In particular, he draws an explicit connection between the lens and the drifting mind. Consider his comparison of "scatter'd" moments of time to "grains of Sand," which, by way of reflection, "the skillful Artificer" melts into "Glass" (*Discourse*, 24). Boyle describes several useful applications for the "noble substance," including the construction of an instrument with which to "discern Celestial objects, (as with Telescopes)," by which he refers both to divine themes and to the actual visible constituents of the heavens (*Discourse*, 24). He imagines the collection and transformation of "Intervals of Time" or "Uncertain Parentheses" between "important Occurrences" (fugitive moments at the interstices of events) into a powerful lens with which to unlock the secrets of the universe (*Discourse*, 24). On the one hand, then, Boyle asks us to make good use of fallow time. On the other hand, the "parenthe[tical]" nature of the moment (the free-floating independence of an aside from the purposefulness of a phrase or sentence) evokes the languor of reflection. The temporality of the image likewise unsettles our expectations. We hold out a vague hope for a climactic moment of compilation, transmutation, and understanding (the fashioning of a lens from time's "scatter'd" "grains"), but we experience occasional reflection as perennial drift (one "grain" after another). Elsewhere, Boyle puts the image of the lens to different but complementary use: "Attention, like a magnifying glass, shews us, even in common Objects, divers particularities, undiscerned by those who want that advantage" (*Discourse*, 32). The "attention" required by occasional reflection isn't "attention" after all, but a susceptibility to distraction. Unlike a "magnifying glass" (as we ordinarily employ it), occasional reflection relieves the mind from the burden of sustained focus.[54]

It's no accident that Boyle's lenses are metaphors for mental indirection rather than for the literal enhancement of vision; throughout the *Reflections*, drift is the governing figure of epistemological ambition. Consider the following observation, which reaches for the language of science in order to convey the sheer breadth of meditation's liberty: "In some cases, the Occasion is not so much the Theme of the Meditation, as the Rise. For my part,

I am so little scrupulous in this matter, that I would not confine Occasional Meditations to Divinity it self, though that be a very comprehensive Subject, but am ready to allow mens thoughts to expatiate much further, and to make of the Objects they contemplate not onely a Theological and a Moral, but also a Political, an Oeconomical, or even a Physical use" (*Discourse*, 30). Boyle here takes pride in how "little scrupulous" he is about the subject of his meditative journey. He doesn't "confine" the practice to any one theme—and certainly not to "Divinity," which would serve as a more-than-adequate rubric for Hall or Warwick. Boyle delights once more in the distance the reflective mind can travel, but this time he notices the traversal of distinct intellectual fields. It's fitting that "Physical use" is the last item on his list of possible discoveries. Boyle's crescendo (the word "even" lends emphasis to the final phrase) links scientific thought to the widest curve of the wayward mind.

The dream of perceptual amplitude, in this extravagant sense, isn't unique to Boyle's book. Indeed, one aim of the last chapter was to explore Bacon's underappreciated interest in "perambulat[ing]" an ever-widening world. As the self-evident melts into air, even apparently well-known phenomena await the return of newly sharpened gazes. Robert Burton and Thomas Browne's discussions of an affinity between digression and discernment are instructive points of comparison. Though both authors have a wide influence in the second half of the seventeenth century (and Browne, in particular, is an important point of transfer from Bacon's proposals to Restoration science), I make no claims for direct influence.[55] What interests me is the wide availability of the premise that waywardness aids perception by expanding its reach. Consider, for instance, the fortunes of "expatiation," a now-obscure synonym for "wandering," which we just saw Boyle use to describe the wide ambit of his thought. It features prominently in the works of Bacon's literary progeny.[56] As I discussed in the previous chapter, Bacon sometimes complains (against the Aristotelian tradition) that "wonder" paralyzes the mind so that it "doth not spatiate [expatiate] and transcur as it useth."[57] Burton makes beautiful use of the term when he marries rudderless inertia to definitive assertions of fact. Though it's tempting to ascribe backward-looking bookishness to *The Anatomy of Melancholy* (1621), Burton's interest in the sheer amassment of learning actually accommodates sharp assertions; he takes special pleasure, for instance, in refuting Aristotle.[58] In the *Anatomy*, he initiates his "Digression of Air" by comparing himself to a "long-winged hawk," framing lilting pleasure as a means of discovery: "So will I, having now come at last into these ample fields of air, wherein I may freely expatiate and exercise myself for my recreation, awhile rove, wander round about the world, mount aloft to those ethereal orbs and celestial spheres, and so descend to my former elements again."[59] Far from spinning his wheels, Burton's pleasant "recreation" of "rov[ing]" and "wander[ing]" circumnavigates the globe and voyages beyond the earth's sphere in pursuit of understanding. In *Religio Medici* (1643), Browne offers a similarly positive evaluation of mental digression when he celebrates the space of free inquiry delimited by *adiaphora*—those "things indifferent" about which religion has not laid down clear laws: "There are yet after all the decrees of counsells and the niceties of the Schooles, many things untouch'd, unimagin'd, wherein the libertie of an honest reason may play and expatiate with security and farre without the circle of an heresie."[60] Browne's defense of the propriety of reason's "play[fulness]" anticipates Boyle by identifying mental vagrancy with "libertie" itself.

By cutting loose, Boyle's "expatiations" open new epistemological frontiers. Echoing his precursors, he describes reason as "a *Faculty*, whereby an Inquisitive Soul may expatiate it self through the whole Immensity of the Universe, and be her own Teacher in a thousand cases, where the Book is no less delightful than the Lessons are Instructive" (*Discourse*, 34–35). He also credits reason for "enabl[ing] an Ingenious Man to pry into the innermost Recesses of mysterious Nature" (*Discourse*, 34). Whatever difficulty learning implies is promptly transformed to pleasure. Mysteries are excavated—and hidden corners illuminated. Sounding much like Burton's avian soul, he completes his encomium to reason by describing it as "a *Faculty* . . . by whose help the restless mind having div'd to the lowermost parts of the Earth, can thence in a trice take such a flight, that having travers'd all the corporeal Heavens, and scorn'd to suffer her self to be confin'd with the very Limits of the World, she roves about in the ultra-mundane spaces, and considers how farr they reach" (*Discourse*, 35). The passage "scorns" the narrowness of the circumscribed earth (and the claustrophobia of Baxter's "Jesuites dungeon") in favor of a boundless breadth that reaches as far as the "ultra-mundane." With the assistance of reason, Boyle suggests, we can "travers[e]" the entire universe "in a trice." Though he describes the "mind" as "restless," the feeling of mental omnipotence ensures the endless dependability of perceptual satisfaction. Always questing but reliably pleased by a world that generously yields up its secrets, the "mind" achieves the flickering equilibrium Bacon theorized for us in the *Advancement*.

Boyle counts on continual enjoyment because he assumes a constant inundation of objects for reflection. Recalling the endless alternation of "appetite" and "satisfaction" Bacon locates in the experience of "learning," Boyle both avoids dissatisfaction by dependably finding new sources of pleasure and forestalls overexcitement by dependably losing hold of them. We have already seen evidence of Boyle's emphasis on receptivity, but here we begin to observe that his denial of personal responsibility for even his own thoughts implies an argument for realism. "Since we know not," he writes, "before we have considered the particular Objects that occurr to us, which of them will, and which of them will not, afford us the subject of an Occasional Reflection, the mind will, after a while, be ingag'd to a general and habitual attention, relating to the Objects that present themselves to it" (*Discourse*, 32). We don't choose when to meditate, or about what. From a Baconian perspective, we're exactly where we want to be—though with an avowal of disorder that challenges our ordinary sense of scientific rigor. Hospitable to whatever we don't understand, we listen to nature rather than raising our voices to speak over her.

Boyle goes on to suggest that the *Reflections* owes its very form to whatever happens to transpire, as if recalling the prescriptive force but not the stated goal of Bacon's aphorism: "Nature to be commanded must be obeyed."[61] Boyle offers *"Suppleness of Style"* as evidence of obedience, framing rhetorical invention (*inventio*) as a measure of submission (*Discourse*, 36).

> The Subjects that invite Occasional Reflections, are so various, and uncommon, and oftentimes so odd, that, to accommodate ones Discourse to them, the vulgar and receiv'd forms of Speech will afford him but little assistance, and to come off any thing well, he must exercise his Invention, and put it upon coining various and new Expressions, to sute that variety of unfamiliar Subjects, and of Occasions, that the Objects of his Meditation will engage him to write of: And by this difficult exercise of his Inventive faculty, he may by degrees so improve it, and, after a while, attain to so pliant a Style, that scarce any Thought will puzzle him to fit words to it, and he will be able to cut out Expressions, and make them fit close to such Subjects, as a Person unaccustom'd to such a kind of Composures, would find it very difficult to write of, with any thing of propriety. (*Discourse*, 36)

On the one hand, Boyle's observation recalls the humanist account of verbal dexterity epitomized by the *De copia* (1512) of Desiderius Erasmus, or discussions of decorum and propriety in Cicero and Quintilian.[62] If you've mastered a variety of different ways to say something, the tradition tells us, you are well prepared to choose the one that suits the present occasion. On the other hand, Boyle's focus isn't skill. He savors the sheer strangeness of what happens to language when it bends to meet the occasion. Indeed, he suggests not only that linguistic quick-wittedness should extend to include the most unusual case but also that the unexpected occurrence is exactly the case that best catches hold of the mind. Thus his practice of reflection is predisposed to formal eccentricity, and whatever shape it takes is itself subject to further distortion as the process unfolds. Like a forgotten ancestor of modernist aesthetics, Boyle's description of literary practice inverts Viktor Shklovsky's theory of *ostranenie* (defamiliarization).[63] Where Shklovsky charges literature with the task of casting a strange light on an overfamiliar world, Boyle imagines a world of such persistent strangeness that it defamiliarizes whatever phrases we use as we aim to describe it. It's the revelatory distortions of an ordinary mood rather than the ingenious techniques of artistic labor that bring about the change. For readers, to be sure, these theories resemble each other. Whatever the cause of disproportion or misshapenness, we encounter the world in wildly unusual turns of phrase.

Boyle's account of literary language also conveys a Baconian purpose. He shakes free of the illusions of "vulgar and receiv'd forms of Speech" in pursuit of descriptions that "sit close to" the objects of his gaze. He marries such fidelity to objects with effortless pleasure, commending a "pliant . . . Style" in which "scarce any Thought will puzzle him" as he finds the right "Expressions" and "Composures" (both terms suggesting *phrases* as well as *states of feeling*) without any of the "difficult[y]" that would hinder the efforts of the unpracticed. To be sure, he acknowledges the neophyte's initial experience of "difficult[y]," but this is more a technical than affective problem: he confronts the limits of his rhetorical sophistication rather than the boundaries of his capacity to enjoy himself. After all, the "Objects of Meditation" "invite" him to reflection, seducing the inquisitive mind; he accepts the pleasurable charge of meditation and then faces the rhetorician's problem of finding the right words to accomplish the task. We are again guided by Bacon's account of "superinduc[tion]." We hear evidence of real satisfaction when Boyle informs us that the objects of reflection are "so various, and uncommon, and oftentimes so odd" that they put his wits to the test. Each adjective ups the ante: from the unpredictable to the hard to come by to the frankly peculiar.

What perhaps seems least amenable in the *Reflections* to a scientific project is Boyle's unconcern with results. Such license lends support to my argument that Baconian progress is at times a suppler proposition than scholars have realized: the vagueness of the wish for a better world authorizes the widest scope for exploration. Boyle's fidelity to pragmatism takes an interestingly impractical form. He seems to grant value to failures of efficiency, but only because he disclaims knowledge of where usefulness is to be found. In one of his reflections, he presents medicinal "Syrup of Violets" as proof of the worth of seeming frivolity. When Philip Sidney, in his *Defense of Poesy* (1595), describes the art of verse as a "medicine of cherries," he means quite simply that linguistic pleasure sweetens moral philosophy.[64] In *De rerum natura*,

Lucretius gives us an early version of the figure by smearing the rim of a vessel with sweet liquid, encouraging a curative draught of otherwise bitter medicine.[65] Perhaps with this particular case in mind, Boyle transforms the logic of the image: "One that did not know the Medicinal Vertues of Violets, and were not acquainted with the Charitable Intentions of the skilful person, that is making a Syrup of them, would think him a very great Friend to *Epicurism*: For his Imployment seems wholly design'd to gratify the senses" (*OR*, 148). Boyle might intend a joke: no one is a "great[er] Friend" of "Epicurism" than Lucretius, who makes good use of the metaphor to describe his poetry.[66] What is surprising about Boyle's line of thinking is that he implies that almost anyone who appears to be wasting time on pleasure seeking deserves the benefit of the doubt.

> If I see a person that is Learned and Eloquent, as well as Pious, busied about giving his Sermons, or other devout Composures, the Ornaments and Advantages which Learning or Wit do naturally confer upon those productions of the Tongue, or Pen, wherein they are plentifully and judiciously emploi'd; I will not be forward to condemn him of a mis-expence of his Time or Talents; *whether* they be laid out upon Speculative Notions in Theology, *or* upon Critical Inquiries into Obsolete Rites, or Disputable Etymologies; *or* upon Philosophical Disquisitions or Experiments; *or* upon the florid Embellishments of Language; *or* (in short) upon some such other thing as seems extrinsecal to the Doctrine that is according to Godliness, and seems not to have any direct tendency to the promoting of Piety and the kindling of Devotion. For I consider, that as God hath made man subject to several wants, and hath both given him several allowable appetites, and endowed him with various faculties and abilities to gratifie them; so a man's Pen may be very warrantably and usefully emploi'd, though it be not directly so, to teach a Theological Truth, or incite the Reader's Zeal. (*OR*, 148–49)

Boyle defends the dead end. Occasional meditation falls quite clearly under the rubric of "other Devout Composures," and so Boyle's lenient judgment licenses his own present activity—including his literary style. The cascading list of potentially useful varieties of uselessness shows none of Bacon's impatience with barren argument and oversubtlety (though Boyle isn't immune from the exasperation of the practical-minded), but instead offers blanket approval to "[in]direct tendenc[ies]." After affirming the "Speculative," the "Obsolete," the "Disputable," the "florid," and the aridly "Philosophical," he concludes with a catchall: Nothing is deserving of reproof simply because it "seems extrinsecal." To be sure, he only extends the benefit of the doubt to the "Learned," "Eloquent," and "Pious," but his refusal to draw negative conclusions from actions that seem to cast doubt on those qualities suggests near-limitless liberality of judgment.

If we are to understand how Boyle's book generates understanding by suspending the necessity of doing so (by, more precisely, suggesting the potential for *anything at all* to advance the cause of "instauration"), we must examine specific instances of occasional reflection (in addition to the one on "Syrup of Violets"). Beforehand, however, I propose a detour. We have understood the distinctiveness of the *Reflections* with respect to the genre of meditation, but I also want to show the uniqueness of its success as a participant in the literature of experiment. Thus I hold it up for comparison against the work of a contemporary with similar interests.

"I Heed It Not"

In a letter to Samuel Taylor Coleridge (June 1796), Charles Lamb offers instructions for enjoying *The Compleat Angler*, which seems, in his estimation, to benefit from abridgment: "All the scientific part you may omit in reading."[67] Walton's book is a dialogue, borrowing elements of georgic and pastoral, but it is also, notwithstanding plenty of critical judgments to the contrary, a handbook for catching fish.[68] It's the instructional material Lamb calls "scientific," but scholars have long obeyed this prescription in a second, unintended sense, ignoring the significant extent to which experimental natural philosophy animates the book. Indeed, these two meanings of "science" both refer us to Walton's desire to provide us with usable facts. If you want to catch fish, you need to know what they're like. Much less familiar but perhaps more revealing about the reception of the fishing guide than Lamb's recommendation for selectiveness is the plangent letter he writes to William Wordsworth twenty-six years later (March 20, 1822): "I had thought in a green old age (O green thought!) to have retired to Ponder's End— emblematic name how beautiful! in the Ware Road, there to have made up my accounts with Heaven and the Company, toddling about between it and Cheshunt, anon stretching on some fine Izaac Walton morning to Hoddesdon or Amwell, careless as a Beggar, but walking, walking ever, till I fairly walkd myself off my legs, dying walking! / The hope is gone."[69] Lamb uses Walton's name as an adjective, conjuring forth the *Angler's* opening scene: a "fine pleasant fresh *May day* in the Morning."[70] Yet the name is more than a byword for the book's "pastoral beauties," which he praises in the earlier letter; it also names the atmosphere of "careless" tramping in which Walton's dialogue unfolds. Lamb entertains a fantasy of ambulatory liberty in which walking ("toddling about") is pleasurably endless because it has no destination, and "Izaak Walton" is shorthand for the scene's requisite backdrop: freedom from cold rain and gnawing "care." When Lamb casually idealizes the "beggar's" vagrancy, he inhabits the escapist fantasy for which Walton's book has often been mistaken. Yet the wish for effortless freedom from affective intensity is no stranger to realism—for Walton no less than for Andrew Marvell, whose observational mood I explore in the next chapter (and with whom Lamb associates Waltonian pastoral: "O green thought!").[71] What Lamb doesn't acknowledge is how many seventeenth-century writers understand the self-satisfaction of freewheeling inattention as a dramatic (if dedramatized) advantage for perception.

As I move from the question of theory (how "reflection" is imagined) to that of practice (what it feels like), I enlist Walton as a friendly foil. By juxtaposing Boyle's fishing excursion, *Angling Improv'd to Spiritual Uses*, which he interpolates as Section IV of the *Reflections*, and Walton's much more famous account of the sportsman's leisurely pleasures, I hope to gain interpretative ground through comparison and then, bearing a vivid sense of kinship in mind, finely grained contrast. First, though Walton has never been read this way, we discover his participation in a nontechnical scientific literature that associates leisurely tranquility with the production of knowledge—especially the findings of experimental natural philosophy. Together, these books inhabit a state of calm awareness a scientist might share with a layperson—a common ground Boyle continues to explore as he turns increasingly to science. Second, as we observe that Boyle's narrative is by far the more disorienting, we come to understand that it displays the mechanism by which a careless disposition enables insight. As his theory of occasional reflection predicts, he sets the mind free from orderly management and embraces the unpredictability of chance occurrences. By allowing his narrative to splinter into fragments, and by suspending the moment of synthesis that would reestablish the narrated world's unity, the scientist-in-the-making shows himself much less partial to clarity than the pastoralist—and much more willing to inhabit a state of confusion. If Boyle's convergences with Walton reveal that his observational mood cannot be written off as idiosyncrasy, it remains the case that Boyle deserves credit for inventing a new way of writing. As promised in the *Discourse*, he "exercise[s] his Invention" quite spectacularly. As we have by now come to expect, part of what astonishes us about such "exercise" is how little effort it seems to require.

The genetic relationship between these two documents from the 1650s remains unclear. Almost nobody writes about *Angling Improv'd*, and even the canonical *Angler* seems to have fallen out of favor. Yet the reader who knows both books can hardly miss the resemblance. In 1927, Claude Lloyd describes Boyle's work only as "an obscure analogue" to Walton's, declining to argue for indebtedness. In 1947, H. J. Oliver identifies direct borrowings in Lloyd's enumeration of commonalities ("a dialogue between people going fishing; a clear indication of the passing of time up till noon, when a shower of rain causes an adjournment; and even a singing milkmaid"), using Boyle's book as evidence that "Walton is, if we are to insist on judging him by modern standards, completely unscrupulous."[72] I agree with the spirit of John R. Cooper's 1968 defense of Walton from that charge, which points out significant innovations even in cases of apparent theft, but I cannot agree with his conclusion on Boyle, about whom he is uninformed: "Boyle's narrative and descriptive passages have only an allegorical or illustrative function for his main themes. For all his moralizing, Walton never obeys so narrowly didactic and unimaginative an impulse. Any influence that Boyle had on *The Compleat Angler* (and it is my belief that he had none at all) is at most trivial."[73] This is nothing more than a tendentious effort to defend a properly "literary" author's reputation from association with the "nonliterary" Boyle.[74] My own interest is kinship rather than debt; indeed, the imprecise claim of mere affinity between these books is more in the spirit of Boyle's delectation of "chance" than an argument for influence would be. Yet Cooper argues against affinity as well, on the basis of the false claim that Boyle's book is "narrowly didactic and unimaginative." Even if we translate the evaluation's failure of sympathy into the neutrality of a factual claim, exchanging the imputation of "unimaginative[ness]" for the point that Boyle's fishing expedition is only an excuse for theological reflection, we still find that Cooper is incorrect. Boyle makes a special case for the importance of the narrative itself, irrespective of the pious lessons it draws from individual episodes. Without the frame story, Boyle would lose his claim to a new aesthetic form: what we might call "cacophonous," "inharmonious," or "disaggregated realism."

The best point of departure for comparison is Boyle's remark, which echoes Bacon's desire to marry action and contemplation in experimental practice, that woodland sport offers an image of leisure in which obstructions are transformed into sources of pleasure.

> I Know, it may be objected against the pleasantness of the Mental Exercise I have been speaking of, That to make Occasional Meditations is a work too difficult to be delightful. / In Answer to this, I might represent, That there are employments wherein their being attended with somewhat of difficulty, is so far from deterring us, that it recommends

them: as we see that in Hunting and Hawking, the toil that must be undergone is so much an indearment of the Recreation, of which it makes a great part, that when it happens that we do not meet with difficulties enough, we create new ones, as when Hunts-men give the Hare Law, (as they speak) for fear of killing her before they have almost kill'd their Horses, and perhaps themselves, in following her: Yet I shall rather chuse to make a more direct Answer, by observing, That the difficulties imagin'd in the practice I am treating of, seem to arise, not so much from the nature of the thing it self, as from some prejudices, and misapprehensions that are entertain'd about it. (*Discourse*, 29)

Boyle describes the difficulty of occasional reflection as an opportunity for enjoyment comparable to the challenge of sport or play. Unlike Montaigne, who worries that too challenging a pastime (recall the example of chess) is no pastime at all, Boyle observes that some recreational pleasures depend on exertion. As occasions for the display of skill, "Hunting and Hawking" demand either preexisting virtuosity or the enjoyable exercise of the faculties with which one eventually achieves it. The form of Boyle's description itself attenuates whatever difficulty the latter suggests. He presents the example of the sportsman's arduous effort as an argument he "might" make, but "rather chuse[s] to make a more direct Answer" without the help of illustrations, explaining that actual "difficulties" arise not from the practice of meditation but rather from "prejudices" against it. He goes on to reject the unfounded belief that people should "confine their thoughts to the subject that set them on work," which is "to many men a thing uneasie and tedious enough," and the needless fear that "one is oblig'd to the trouble of writing down every Occasional Reflection that employs his thoughts" rather than letting any of them fade from memory (*Discourse*, 29–30). Boyle explains that the less captivating reflections, the ones we don't record, "[have] perform'd all the service that need be expected from them within the mind already," suggesting the usefulness even of those thoughts we are only minimally aware of entertaining (*Discourse*, 31). By framing the claim about "Hunting and Hawking" as one he doesn't ever quite make (adopting a rhetoric of paralepsis or praeteritio), he evades even the minimal intensity of framing an argument.

These are exactly the sports Walton juxtaposes with fishing in the opening pages of the *Angler*. (I write just barely against probable chronology, Walton before Boyle, because my central purpose is to derive from the comparison an account of Boyle's literary distinctiveness that bears on his later writings.) In the opening pages of the first edition, Piscator crosses paths with Viator, who is on his way to join a hunting party. (A subsequent revision gives his companion

a more informative name, Venator, and adds a falconer, Auceps, to the group.)
Walton's mouthpiece extols the virtues of fishing over hunting, after which Viator asks permission to call himself "Scholer" and Piscator "Master," a relationship that defines a book devoted almost entirely to pedagogical dialogue.[75] As
readers, our role is to listen like students as Piscator explains the art of angling.
I recount the basic features of the narrative because they fill out our understanding of the recreational scenario Walton and Boyle share. In Walton's case,
the encounter with the hunter (and the falconer) defines a buffered world of
leisurely peace (fishing) the outside of which is leisurely too; indeed, fishing is
defined by its peacefulness even in comparison with other forms of recreation
(hunting and falconry). In one of the revisions, Piscator distinguishes himself
from the hunters, however ineptly from a fish's point of view, with the following observation: "I am not of a cruel nature, I love to kill nothing but Fish."[76]
Walton's piscatory fantasy, like Boyle's collection of wartime reflections, can
be understood as a commentary on the violence of civil war: a wish for a less
bellicose world. My interest, however, is how Walton coordinates emotional
peace and cognitive violence: a tranquil but forceful encounter with the intransigence of the natural world.

Emotional quiet is central to the book's ambitions. In his preface, Walton
writes, "I wish the reader also to take notice, that in writing of it, I have made
a recreation, of a recreation; and that it might prove so to thee in the reading, and not to read *dull*, and *tediously*, I have in severall places mixt some
innocent Mirth; of which, if thou be a severe, sowr complexioned man, then
I here disallow thee to be a competent Judg."[77] As if in negative imitation of
Hall's request that his readers mirror his stringency, Walton asks for an audience as easygoing as he is. Discarding the challenge of converting the "sowr
complexioned" to his own frame of mind, he writes for those who share his
"Mirth." By "[making] a recreation, of a recreation"—by understanding literary composition as an opportunity for recreational pleasure—he ensures that
reading about fishing is much like fishing itself. If Walton cares not to cajole
the "sowr complexioned," he nonetheless suggests that disposition is as much
a question of context as it is of physiology. The "severe" reader or fisherman
may find himself softened by circumstance. "The whole discourse is a kind of
picture of my owne disposition," Walton explains, "at least of my disposition
in such daies and times as I allow my self, when honest *Nat.* and *R.R.* and I go
a fishing together."[78] Thus Walton identifies the book's central theme as an
easy "disposition" that finds expression in serene conviviality (here, he refers
with abbreviations to his kinsmen) and quiet exploration, owing as much to
the circumstances of "recreation" (or the redoubled "recreation" of writing
about "recreation") as it does to inborn or ingrained mental habit.

"A picture of my own disposition": Walton's affective self-portraiture recalls no one so much as Montaigne—a resemblance of which he is himself aware. He elaborates an account of his persona through an encounter with the essayist. When, before he has had time to be impressed by Piscator and become his eager student, Viator remarks with approval that he has heard "many grave, serious men pity [anglers]," Piscator parries the blow with a famous passage from Montaigne, somewhat expanded, in which he momentarily inhabits the perspective of his cat while they "entertain each other with mutual apish tricks": "Who knows but that I make my Cat more sport than she makes me?"[79] He refashions the episode as an image of fishing, deviating from Montaigne's words by specifying the kind of play he has in mind. Dangling a "garter" above his cat, he plays the teasing role of fisherman.[80] "There are many men," Piscator explains, performing the 180-degree turn with which Montaigne relativizes judgment, "that are by others taken to be serious grave men, which we [anglers] contemn and pitie."[81] We mock them, then, just as surely as they mock us—and the difference between them and us is readily grasped as one of disposition.

Walton's objects of ridicule are "mony-getting men, that spend all their time first in getting, and next in anxious care to keep it: men that are condemn'd to be rich, and always discontented, or busie."[82] For Walton, as for Boyle, the rich man serves as a negative paradigm, revealing for us what the carefree life of the angler evades. The easy denigration of wealth might seem too familiar to count for much, but Walton and Boyle's shared refrain that Mammon is the god of care is doubly instructive. In addition to clarifying, by way of contrast, the nature of their carelessness, it offers an alternative to the influential argument that greed is the model for scientific desire.[83] In Boyle, as here, "anxious care" or "discontent" is the mood of the moneyed as well as the would-be. Like Bacon's image of the "ambitious prince" who "turn[s] melancholy" because his worldly objectives turn out to be "deceits of pleasure, and not pleasures" after all, the rich man is also the "busie" man, striving to obtain wealth in the first place and then having to keep vigilant watch over his possessions.[84] The theme recurs, reminding us by contrast of Walton's unguardedness—his freedom not to worry about self-protection. Much later, for instance, he finds himself "joying in [his] own happy condition, and pittying that rich mans that ought [owned] this, and many other pleasant Groves and Meadows about [him]."[85] He owes the pleasure he takes in the pastoral scene to the fact that the land does not belong to him. "I did thankfully remember," he continues, "what my Saviour said, that *the meek possess the earth*; for indeed they are free from those high, those restless thoughts and contentions which corrode the sweets of life."[86] In contrast to the "restless[ness]" of Boyle's adventurous mind, which finds sources of satisfaction everywhere it goes, the "restless thoughts" of the "rich man" rule out the enjoyment of Nature's "pleasant" provision. Like the "carefulness" that makes a *lumen maceratum* of Bacon's *lumen siccum* (water-damaged light from the pleasantly dry illumination of uncorrupted knowledge), anxiety "corrode[s] the sweets of life," which are only ours to enjoy insofar as they *aren't* actually ours: *"the meek possess the earth."*[87] In one of Walton's revisions, he offers an instructive emblem of acquisitiveness, depicting the "[busyness]" of making and saving money as an agony without respite: *"I have a rich Neighbour that is always so busie, that he has no leasure to laugh; the whole business of his life is to get money, and more money, that he may still get more and more money."*[88] Like a nightmare version of Bacon and Boyle's shared fantasy of everlasting but painless appetite, the rich man's quest for wealth is torturous. Walton goes on to express pity for the *"weary days and restless nights"* of the well-to-do, and he repeats his lament for the rich man's "corroding cares," comparing him to *"the* Silk-Worm, *that, when she seems to play, is at the very same time spinning her own bowels, and consuming herself."*[89] Another place we have seen this image is Montaigne's *Essais*, where it serves as an image of the interpreter who claims to produce knowledge when in fact (like Bacon's scholastic spider) he's only drawing conclusions from groundless premises, mistaking his insides for the outside world.[90]

The spaciousness of Walton's book is a matter of slowness and digression: the time of going nowhere. Walton advertises his freedom from haste ("But whither am I strayed in this discourse?" Piscator asks with backward-looking self-satisfaction), linking style and disposition. By ensuring that he retains the *"leasure to laugh,"* he inhabits a world the rich man doesn't know.[91] Piscator describes his discursive "free[dom]" as an effect of disposition: "I find my Scholer to be so sutable to my own humour, which is to be free and pleasant, and civilly merry, that my resolution is to hide nothing from him."[92] At the end of the first chapter, he names the interval of his conversation with Viator as the "time of leisure." Though it's an unexceptional turn of phrase, it draws our attention to the atmosphere in which the dialogue takes place: "I must be your Debtor (if you think it worth your attention) for the rest of my promised discourse, till some other opportunity and a like time of leisure."[93] Piscator's attitude toward his lesson suggests negligence, since he seems not to care when, or indeed whether, it continues. At an unspecified future time ("some other opportunity"), he intends to carry on, but only if his student thinks it "worth [his] attention."

Of all my claims about Walton, perhaps the most surprising is that his carelessness animates a program of experimental research. My particular interest is the continuity between pastoral tranquility and the trials of knowledge production. Yet given the current state of scholarship, even a minimal acknowledgment of a Baconian intention is news. Signs of Walton's deep sympathy with the ambitions of experimental natural philosophy, which sit right on the surface of the dialogue, have garnered little commentary.[94] The book is littered with references to Bacon. Furthermore, Bacon's kinsman, Henry Wotton, is the genius loci of Walton's landscape. A natural philosopher in his own right, Wotton is also Walton's fishing companion—as well as, incidentally, the provost of Eton College during Boyle's time there as a student. (One scholar speculates that Boyle might have been among the boys at Eton who accompanied Walton and Wotton on fishing excursions![95]) Piscator's fishing lessons brim over with the language and evidence of natural philosophy. In one of the book's revisions, he passes from rivers to fish as objects of praise, referring to them as "Monsters, or Fish, call them what you will."[96] "The waters are natures store-house," he explains, "in which she locks up her wonders."[97] Everywhere, we're asked to examine fish as specimens. "Nor are the number, nor the various shapes of fishes, more strange or more fit for *contemplation*," he writes, "then their different natures, inclinations and actions."[98] He includes scarcely credible accounts of fantastical specimens like the "Sargus," who "courts *She-Goats* on the grassie shore" and makes cuckolds of "their husbands."[99] Since here he quotes from Joshua Sylvester's translation (1604) of the *Sepmaine* (1578) of Guillaume de Salluste Du Bartas, we may wonder at Walton's credulity, but my point is not that Walton wields the cutting edge of Baconian science. What Walton continually shows us is interest in the newest methods and experimental findings.

Although Walton cites traditional natural-historical authorities (Aristotle and Pliny among them), he shows a persistent engagement with the contemporary scene, portraying himself as an enthusiast of the latest findings who actively seeks them out and puts questions to scientists for experimental resolution. He consults "chimical men" (Wotton and George Hastings) about the efficacy of his bait, and alludes to the discoveries of dissection to buttress his account of animal anatomy.[100] He also depicts fishing as an experimental trial in which claims (about, say, the proper strategy for snagging particular kinds of fish) are affirmed or falsified—not just a metaphor for experiment, then, but an instance of it. "Give me your hand," Piscator says, "from this time forward I wil be your Master, and teach you as much of this Art as I am able; and will, as you desire me, tel you somewhat of the nature of some of the fish which we are to Angle for; and I am sure I shal tel you more then every Angler yet knows."[101] Piscator's lessons are field trips, and they're less concerned with generalities than "the fish which we are to angle for."[102] He

doesn't teach from books but proffers knowledge in view of particular specimens. Indeed, his citations are less bids for authority than points of departure
for further reflection; as Austin Dobson points out, and as we already saw
in Walton's allusion to Montaigne's cat, his quotations are often loose adap-
tations, and on occasion they're pure inventions.[103] One particularly good
illustration of his attitude toward established knowledge is his discussion
of underwater sound. Quoting from Bacon's *Sylva* with unacknowledged
alterations, he proposes *"that if you knock two stones together very deep under
the water, those that stand on a bank near to that place may hear the noise without
any diminution of it by the water."*[104] From this, he surmises that fish can hear
us as we move about on the river's bank: "And this reason of Sir *Francis Bacon*
(*Exper.* 792) has made me crave pardon of one that I laught at for affirming
that he knew *Carps* come to a certain place in a Pond, to be fed at the ring-
ing of a Bell or the beating of a Drum: And however, it shall be a rule for
me to make as little noise as I can when I am fishing, until Sir *Francis Bacon*
be confuted, which I shall give any man leave to do."[105] Walton revises his
own incredulity about the auditory experience of fish on the basis of Bacon's
observation, but then, like a good Baconian, promises another self-revision
if anyone manages to "confute" Bacon. By giving "any man leave to do"
so, he wrests this capacity to make knowledge from established repositories
of wisdom and places authority in experimental findings. For the moment,
but perhaps only for the moment, his fishing practice obeys the latest word:
he walks with quiet steps along the water's edge. As he elsewhere puts it:
"Angling may be said to be so much like the Mathematicks, *that it can ne'r be fully
learnt; at least not so fully, but that there will still be more new experiments left for
the trial of other men that succeed us."*[106]

Throughout, Walton takes up this theme of experiment by describing the
kind of person an angler is or ought to be. We should not be surprised to
find that this characterological type has an easy disposition; Walton's casual
reference to "the undisturbed mind of Sir Henry Wotton" is a synecdoche of
his general estimation of anglers.[107] As he prepares to present the practice of
fishing as the happy marriage of action and contemplation, he offers experimental science as his sole example of the former: "And contrary to these
[who favor the contemplative life], others of equal Authority and credit, have
preferred *Action* to be chief, as *experiments in Physick, and the application of
it.*"[108] We might expect martial heroics as the prime example of action, but
Walton substitutes the achievement of public good through experimental
trial. The marriage of action and contemplation is no union of opposites: it
combines quiet stillness and quiet endeavor. "Both these meet together," he
writes, "and do most properly belong to the most *honest, ingenious, quiet,* and

harmless art of *Angling*."[109] Interestingly, then, the synthesis seems already to "belong" to the angler's innocent character. Yet character, Walton explains, is only one tributary to the experience of angling: "*Angling* is somewhat like *Poetry*, men are to be born so: I mean, with inclinations to it, though both may be heightned by discourse and practice, but he that hopes to be a good *Angler* must not only bring an inquiring, searching, observing wit; but he must bring a large measure of hope and patience, and a love and propensity to the Art it self."[110] Fishing is an "Art" both like "Physick" and "like *Poetry*." It's neither simply an inborn skill nor an achievement of self-cultivation, but it requires aspects of both—to say nothing of the importance we have already seen Walton attribute to circumstance. Though "discourse and practice" serve it well, "inclination" and "propensity" are fundamental. What all of this amounts to is a portrait of the angler that could also be a portrait of the scientist: "an inquiring, searching, observing wit" alongside a "love and propensity" for experiment itself, irrespective of findings.

In the *Angler*, images of immersive effortlessness are not hard to come by. Walton recommends, for instance, that we obtain "Gentles," or maggots, for bait by burying a dead cat in the earth and waiting for them to appear, after which he offers a less gruesome alternative method "if you be nice to fowl your fingers (which good Anglers seldome are)."[111] The unflustered angler is happy to get his hands dirty. After describing adjustments in fishing strategy in accordance with changes in wind direction, he shrugs at the instructions he's just finished unfolding: "He that busies his head too much" about the wind creates needless trouble; "let the Wind sit in what corner it will," he writes, "and do its worst I heed it not."[112] When, in one of the revisions, he describes the baiting of a river and remarks, "whilst the Fish are gathering together . . . you may take a pipe of Tobacco," we understand that he invites us to adopt his attitude of worldly contemplativeness, with or without the puffing of pipes.[113]

Idylls of the Mind

Whether or not Boyle's *Angling Improv'd* inspires Walton's *Angler*, they can fairly be said to inhabit the same world, marrying the quiet, ambulatory pleasures of the fishing excursion and the aims of scientific experiment. I have described Boyle's idiosyncratic brand of counter-meditation as capitulation to wayward thought; I want now to show that the practice of occasional reflection explains the mechanism by which dispositional ease lends assistance to understanding. Walton describes the mutual reinforcement of casual serenity and edifying receptivity, but without Boyle's emphasis on self-abandonment. The difference is not simply thematic; the *form* Boyle's narrative takes is an illustration (or succession of illustrations) of thought's waywardness, as he describes it in his prefatory *Discourse*. Boyle's "analogue" of Walton's *Angler* is thus a hectic affair; unlike the *Angler*, it careens between unrelated themes. Boyle takes every opportunity to alert us to the unpredictable agency of "chance," and develops a writing practice that makes a display of constant interruption: "I chanc'd to stop," "I chanc'd to look," "chancing to tread," "chancing to express"—and so on (*OR*, 99, 106, 112, 131). His trains of thought are as discontinuous as one would expect in a world presided over by sheer contingency. Though Boyle certainly takes for granted the role of providence in ordering these events, he asks us to share the finite human perspective from which they look random. Thus he displays his willingness to accept the failure (or endless deferral) of synthesis, attending throughout to the dispersive proliferation of thought, both within the individual mind and across the divided consciousness of an intellectual community. Walton and Boyle both link emotional quiet neither to withdrawal nor to detachment but rather to involvement with the world. Looking ahead to the next chapter, I suggest this equation is distinctive of mid-century Baconianism— but only Boyle details the intellectual process that *binds* undermotivation to hyperreceptivity.

Angling Improv'd is alone among the six sections of the *Reflections* in receiving its own "Advertisement," inserted between the *Preface* and the *Discourse*.[114] Before we even get to the *Reflections* proper, we learn that the fishing expedition deserves special attention. What Boyle tells us in his introductory note is that his narrative recounts true events. With the relieved hindsight of a royalist in the immediate aftermath of Restoration, he explains that he disguised his identity behind fictional characters not because he wished to depart from reality but rather because these particular reflections "were written several years ago, under an Usurping Government," and they "contain some things not likely to be Relish'd by those that were then in Power" (*OR*, 93). Yet beneath the self-protective fictionalization, he insists, one should discern a real-life situation. The reflections on fishing "come forth . . . in such a way as will make most Readers look upon them as containing a Story purely Romantick: Yet they may have in them much less of Fiction, than Such will (tis like) Imagine. For being really a great Lover of Angling, and frequently diverting my self at that sport, sometimes alone, and sometimes in Company; the Accidents of that Recreation, were the true Themes, on which the following Discourses were not the Only Meditations I had made" (*OR*, 93). Though he seems here to use the term "romance" as a general term for something like "fiction," the remark calls to mind his autobiographical account of "raving" under the bad influence of Amadis. I suggested above that the *Reflections* creates a context in which Boyle can reclaim the symptoms of his literary distemper. Worth noting as well is Boyle's claim, also in the autobiography, that only mathematics brings his mind under control—and yet the mature Boyle is notably *not* a mathematician.[115] On his own account, then, the specific brand of scientific realism that captures his interest, experimental natural philosophy, would do nothing to settle his mind. He assures us that *Angling Improv'd* grants us access to reality even as it shows him "think[ing] at random" about "Accidents."

Earlier, in the *Preface*, Boyle singles out this particular narrative, along with Section II, the book's only other interpolated frame story—about his experience of illness, which recalls John Donne's *Devotions upon Emergent Occasions* (1624)—as evidence that his reflections derive from actual events. He understands continuity itself as proof of the reality of reported incidents, deflecting the charge that he dreams up the occasion after the fact to match whatever lesson he already wishes to draw from it.

> Having observed Men to be inclinable, either openly to Object, or at least tacitly to Suspect, That in Occasional Meditations, that may hold true, which is (perchance not altogether undeservedly) said of Epigrams, That in most of them the Conceits were not Suggested by the Subjects, but Subjects were Pretended, to which the Conceits might be Accommodated; I thought, that to manifest, that (at least, some) Writers of this kind of Composures need not have recourse to the suspected Artifice; the fittest way I could take was, By putting together what the Accidents of my Ague, and of my Angling Journy, had suggested to me, to shew, that 'tis very Possible for a person, that pretends not to a very pregnant Fancy, to Discourse by way of Reflection upon the several Circumstances that shall happen to occurr to his Consideration. (*OR*, 11)

The passage attributes the kind of (literary) inventiveness we might associate with "Fancy" to "Circumstances" instead, making a case for realism in the unusual sense that actual events play a role in the narrative's composition. What Boyle doesn't do is defend the truth of the content of the reflections generated by those events. He wants us to believe that his "Conceits" derive from unwilled incidents, but he doesn't insist that we accept the claims he makes via reflection. He treasures the results of mental randomness, but he does so without conferring credibility on them. He treats them instead as knowledge-in-the-making: contributions to understanding in an importantly vague sense. Should we wish to make a practice of occasional reflection ourselves, he suggests, we need not think ourselves handicapped if we lack literary skill or rational insight. (*Angling Improv'd* is less interested in inborn disposition than Walton's *Angler*.) Whatever events befall us can be counted on to set the mind in motion. Thus Boyle places trust in "the several Circumstances that shall happen to occur to his Consideration." By creating a continuous world (the day's journey), he curates an experience of ordinary steadiness reliably interrupted by "random" flights of imagination.

Interestingly, *Angling Improv'd* doesn't try to preserve the distinction between an outward event and the thought to which it gives rise. Like Descartes's concluding words in his *Meditations*, which dismiss his original fear that perhaps he has mistaken a dream for reality by pointing out the enduring continuity of his real-life experience (which cannot, he explains, be found in dreams), what matters to Boyle is a certain narrative coherence that grants experience the quality of actuality.[116] For Boyle, scenic continuity is also important for the specific reason that it serves as the stable background against which the wild discontinuity of thought can be observed. We might notice, then, how far away we are from a much more familiar impulse to distinguish a naked event from an impression of it (as in the philosophical distinction between primary and secondary qualities).[117] Boyle seems unconcerned when narrated reality bends under the pressure of thought. How could it not, given the narrator's carelessness? In one emblematic scene, Eugenius, a member of the fishing party, "to whom his Rambling up and down, added to the heat of the Day, had given a vehement Thirst," kneels by the river to have a drink.

> He took up with his Hat, which by Cocking the Brims he turn'd into a kind of Cup, such a proportion of Water that he quench'd his Thirst with it; and carelessly throwing the rest upon the Ground, quickly return'd towards the Company, which he found he had not left so silently, but that our Eyes had been upon him all the while he was absent; and that sight afforded *Eusebius* an occasion to tell us, Our friend *Eugenius*, might, if he had pleased, by stooping lower with his Head, have Drank immediately out of the entire River; but you see he thought it more safe, and more convenient, to Drink out of a rude extemporary Cup; and that this way suffic'd him fully to quench his Thirst, we may easily gather, by his pouring away of some remaining Water as superfluous. (*OR*, 141)

Boyle's narration of the event is as "rude" and "extemporary" as the fisherman's makeshift "Cup." Eusebius observes that Eugenius drinks from his hat rather than lowering his lips to the running river. Slurping on all fours like an animal seems less an actual likelihood than a counterfactual Eusebius dreams up in order to make his point. Notwithstanding Boyle's protestations, he looks guilty of inventing the "subject" to match the "conceit." After satisfying his thirst, Eusebius explains, Eugenius makes a casual display of satisfaction by "pouring away" the "superfluous" remnant—and Eusebius himself behaves no less "carelessly" when he weighs the gesture against a fanciful alternative. Boyle underlines the simultaneity of these events when he remarks that Eugenius "return'd towards the Company, which he found he had not left so silently"; they are *already* engaged in transforming the event into moral philosophy—no need to wait for it to be over. The passage unfolds into a long conversation on the merits of moderation and the dangers of overindulgence. As with Walton, the rich man serves as negative exemplar, the rendering "necessary" of the "superfluous" creating a striking contrast with Eugenius's "careless" indication of surfeit: "The frequent and sad Complaints of the Rich themselves sufficiently manifest, that 'tis but an uneasie Condition, that makes our Cares necessary for things that are meerly superfluous" (*OR*, 142). Notice how the antinomy between sufficiency and glut, apparently the fruit of occasional reflection, actually shaped the initial description of the incident. Boyle begins with the end of meditation; he passes quickly to the "quench[ing]" of thirst, making the ample draught an emblem of easy gratification rather than narrating a scene that happens to conclude on that note. In that specific respect—in its settled focus on carelessness itself—the passage doesn't seem careless after all, and yet the awkwardness with which Eusebius makes his point, molding the event ham-handedly into a lumpy pedagogical shape, more than does the trick.

"Upon ones Drinking water out of the Brims of his Hat," the nineteenth of the reflections that compose *Angling Improv'd*, is only an especially instructive illustration of the larger narrative's characteristic mood. *Angling Improv'd* begins with an experience of dreamy inattention it only seems to leave behind.

> The sun had as yet but approach'd the East, and my Body as yet lay moveless in the Bed, whilst my roving Thoughts were in various Dreams, rambling to distant places, when, me-thought, I heard my name several times pronounc'd by a not unknown Voice; This noise made me, as I was soon after told, half open my Eyes, to see who it was that made it, but so faintly, that I had quickly let me Self fall asleep again, if the same Party had not the second time call'd me louder than before, and added to his Voice the pulling me by the Arm. (*OR*, 94)

The passage goes on to describe Eusebius's persistent efforts to rouse the oversleeping Boyle for their excursion. (In *Angling Improv'd*, Boyle goes by

"Philaretus," the same name he adopts in his autobiography—but I have chosen to continue to refer to him as "Boyle," in keeping with his encouragement in the *Preface* to identify this persona with the meditator who composes the rest of the *Reflections*.) "And thus," Boyle eventually writes, after much resistance to the prospect of rising from bed, "whilst I was making him my Apologies, and he was pleasantly reproaching me for my Laziness, and Laughing at the disorder I had not yet got quite out of, I made a shift hastily to get on my Cloaths, and put my self into a condition of attending him and the Company to the River-side" (*OR*, 94–95). Boyle elongates the experience of semiconsciousness, climbing away from the "roving Thoughts" of "Dreams" only to make it as far as "half open" eyes before he "fall[s] asleep again." Even when he is finally out of bed, he acknowledges, in his semiclothed state, "the disorder I had not yet got quite out of." Nor does he ever quite do so. Boyle's sleepy "Laziness" suffuses the subsequent reflections.

Eusebius responds to Boyle's somnolence with a reflection on the theme of sleep's pleasures, and it seems an atypically conservative specimen: a set of coherent thoughts on a single theme with no dissenting commentary from the others. He reflects on the frequency with which "Man is lull'd asleep by sensual pleasures" rather than directing his thoughts toward God with "the exercise of Reason" (*OR*, 95). Yet Boyle's experience of the day's magnificence has already called our attention to "sensual" if not sinful "pleasure," aligning the waking world with the dubious pleasures of dreams rather than rational escape from them: "I was delighting my self with the deliciousness of that promising Morning, and indeed the freshness of the Air, the verdure of the Fields and Trees, and the various and curious Enammel of the Meadows, the Musick of the numerous Birds, that with as melodious as chearfull Voices welcom'd so fair a morning" (*OR*, 95). Perhaps the most sententious moment in *Angling Improv'd* extols the supremacy of reason over mere sense in the midst of a scene of luxurious sensuality. Boyle invokes the five senses: "deliciousness" (taste, metaphorically speaking), "the freshness of the Air" (either touch or smell), "the verdure of the Fields and Trees" alongside "the various and curious Enammel of the Meadows" (sight, perhaps with intimations of texture), and "the Musick of the numerous Birds" (sound). It's difficult to accept Eusebius's rejection of "sens[uality]" when at least one of his companions immerses himself so fully in the experience of bodily gratification.

When, in the following reflection, Boyle refers to the "costless, and yet excellent Musick" of birdsong, we understand that unearned pleasure, conveyed by the motifs of pastoral, is a central theme. Boyle's characters are everywhere finding gratification, as when, in a reflection entitled "Upon the sight of ones Shadow cast upon the face of a River," they marvel at the easy diversions of shadow puppetry. "I make this Shadow here," Eusebius says, "without the least pains to do so"—"and," he concludes, "with as little toyl God made the World" (OR, 108). Pious reflection on divine omnipotence shades into the fun of playing God, and Eusebius marvels at an exact correlation between power and will—to which his present activity is just about the closest a human being can come: "I was also taking notice . . . that to produce what changes I pleas'd, in all, or any part of this Shadow; I needed not employ either Emissaries, or Instruments, nor so much as rowse up my self to any difficult Exertion of my own strength, since, by only moving this or that part of my own Body, I could change at pleasure in the twinkling of an Eye, the figure and posture of what part of the Shadow I thought fit" (OR, 109). Like anyone cavorting freely under a sunny sky, Eusebius manipulates his shadow with perfect facility. As if to underline the effortlessness of the experience, he imagines and then dispels the support on which he would have to depend in a counterfactual world of recalcitrant shadows: "Emissaries," "Instruments," and "Exertion." The last thought in the sequence blurs the distinction between minimal effort and serious strain: the strange phrase begins with an emphasis on the absence of any effort at all ("nor so much as rowse up my self") but then identifies mere "Exertion" with actual "difficult[y]." For Boyle's protagonist, even the least taxing application of force already feels like a laborious burden that would disrupt his pleasure. The doubled adverbial phrase lends emphasis to ease as it creates the distortion of near-redundancy: "at pleasure in the twinkling of an Eye." These are imperfect sentences, but—if I am permitted sheer evaluative enthusiasm—ravishingly so. Perhaps occasioned by Boyle's indolence-savoring assertion that he need not "so much as rowse" himself to accomplish his purpose, he gestures not only to the creation of the world but also to no less extraordinary a feat than the Resurrection of the Dead: "in the twinkling of an eye . . . the trumpet shall sound, and the dead shall be raised incorruptible" (1 Cor. 15:52). In shadow puppetry, the absence of constraint feels like evidence of perfect power.

For Boyle, mental wandering unravels experience into threads that unspool in different directions. Though his Baconian heroism invites us to integrate his reflections into a coherent vision, often his thoughts are far too scattered for us to take their eventual synthesis seriously. Where Walton associates the green world with natural philosophy, for instance, Boyle stages a conflict between affective green and pastoral green; the observational mood cuts through an idealized world. When Boyle's fishermen (like Walton's) come upon a singing "Milk-Maid," the encounter occasions a reflection on rustic innocence (*OR*, 99). At first, we seem to know exactly where we are. The milkmaid sings with "a native sweetness" in which "Art was absent without being miss'd," and Lindamor, like the Montaigne of "Des cannibales," recalls "what *Ovid* and other antient Poets had in their strain deliver'd concerning the felicity of the Golden Age," after which Boyle, perhaps already with a hint of irony, remarks that Lindamor goes on "to apply as much of it as the Matter would bear, to the recommending of a Rural life" (*OR*, 99, 101). Even if we hear the implication in Boyle's remark that Lindamor is overreading, making the occasion "bear" as much fanciful imagery as possible, we are nonetheless surprised to find that the subsequent dialogue dismantles the fantasy. "Nor is there any Cottage so low, and narrow," retorts Eusebius, bringing his rebuttal to a high rhetorical pitch, "as not to harbor Care, and Malice, and Covetousness, and Envy, if those that dwell in it have a mind to entertain them" (*OR*, 102). He continues in this vein, smashing an idol by casting doubt on the idyll: "Spiders and Cobwebs are wont to abound more in thatch'd Cabbins, than in great Mens houses" (*OR*, 102). Eusebius's contempt for the poor is repellent, but Boyle pushes back, using the occasion to dispel the myth of rural tranquility unmarked by toil without substituting a negative caricature for the positive one. Though Boyle describes Lindamor as a "Learned Youth" and otherwise indicates his status as a student, anticipating Walton's Viator, the reflection doesn't endorse the teacher's perspective (*OR*, 10). Boyle, "with a low Voice," suggests to Eusebius that perhaps he has argued with such vehemence not because his perspective is true but instead "to check *Lindamor* a little, and keep up the Discourse" (*OR*, 104). Eusebius concedes that Boyle is "not altogether mistaken," acknowledging that "some Airs," including that of rural "quiet," are "very much wholesomer than others," but then changes direction, turning back to "the other side" of the matter in order to restate and intensify his critique of pastoral fantasy ("purposely raising his Voice") (*OR*, 104). The dispersiveness of thought devolves into outright contention as multiple voices respond to the occasion—but one likely outcome is suspicion of both sentimental and slanderous truisms about rural life.[118]

Boyle's *Reflections* puts careless pressure on fictions in order to enable freshness of perception—even at the expense of "the freshness of the Air" (rural beauty giving way to the possibility of "Cobweb"-clogged "Cabbins"). Thus *Angling Improv'd* achieves demystifying force by coming apart. This is often the effect of disagreement, as between Eusebius and Lindamor, but it's also the effect of "chance." The sheer unpredictability of occasions divides Boyle's world into distinct, unassimilated movements of thought. Indeed, we discover disagreement downstream from reflection's randomness; how could easy agreement be possible in a world where thought spirals heedlessly away from its point of departure? Boyle stages dividedness as social solitude; his characters are continually lost in thought, abstracted, together in their mental separateness. The following two examples suffice to establish the pattern:

> Not finding him [Lindamor] there, I hastily cast my Eyes all over the Field, till at length they discover'd him a good way off, in a Posture that seem'd extremely serious, and wherein he stood as immoveable as a Statue. (*OR*, 99)

> I went softly towards *Eusebius*, to see what it was that made him so regardless of his Sport, whilst yet, by the posture he continu'd in, he seem'd to be intent upon it: But approaching near enough, I quickly perceiv'd, That instead of minding his Hook, his Eyes were fixt sometimes upon his own Picture, reflected from the smooth Surface of the gliding stream, and sometimes upon the Shadow projected by his Body, a little beside the Picture upon the same river. (*OR*, 107)

Both passages are taken from episodes we have already explored. The first describes Lindamor's absorption in the milkmaid's singing, and the second occasions Eusebius's meditation on shadows. In both cases, Boyle's characters abandon the world they otherwise share, Lindamor in his abstraction turning metaphorically to stone, and Eusebius losing track of the matter at hand when he forgets to "[mind] his Hook." In both cases they are asked to explain their absorption, and thus each reports back from his respective experience of private abstraction. Like an inversion of Hall's imagined community of meditators who police each other's thoughts, Boyle's fishing party shares the unshared experience of wayward attention.

In the eighteenth reflection, "Upon a Giddiness occasion'd by looking attentively on a rapid Stream," *Angling Improv'd* comes close to presenting a theory of perception that explains its pattern of departure and return. Notice how the reflective exercise immediately prior to this episode gives rise to individual acts of "silent" self-"entertain[ment]."

> These thoughts of *Eusebius* suggested so many to *Lindamor*, and me, that to entertain our selves with them, we walk'd silently a good way along the River-side; but at length, not hearing any more the Noise his Feet were wont to make in going, turning my self to see what was become of him, I perceiv'd him to be a pretty way behind me upon the Rivers brink, where he stood in a fixt Posture, as if he were very intent upon what he was doing. And 'twas well for him, that my curiosity

prompted me to see what it was that made him so attentive; for, before I could quite come up to him, me-thought I saw him begin to stagger, and though that sight added wings to my Feet, yet I could scarce come time enough to lay hold on him, and, by pulling him down backwards, rescue him from falling into the River. (*OR*, 139–40)

After Boyle saves Lindamor from falling dizzily into the river, Eusebius issues a warning about the "unwary consideration of some sorts of sinfull Objects, especially those suggested by Atheism and Lust" (*OR*, 140). Like a perfect foil to the Milton of *Areopagitica* (1644), he cautions us against mere exposure to danger: "we may unawares fall into the Mischief, even by too attentively surveying its greatness, and may be swallowed up by the danger, even whilst we were considering how great it is" (*OR*, 140).[119] As we've seen, drifting away into trancelike abandon is exactly what happens in "occasional reflection." Indeed, Lindamor's "giddy" attention is an image of Boyle's style of thought. Why, then, does Boyle seem to caution us against it? The answer, I suggest, is the duration (which is to say, the immobility) of Lindamor's gaze. Danger lies in the fixity of his attention. Boyle's original question about what Lindamor is up to is about "what it was that made him so attentive," and Eusebius attributes our peril to the mistake of "too attentively surveying" the object in question. When the unfixed attention of reflection settles into place, only then do we find ourselves perilously magnetized. Lindamor's account of his experience confirms the point:

As I was thinking, *Eusebius*, on your last Reflection, I was diverted from prosecuting my Walk in *Philaretus*'s Company, by happening to cast my Eyes on a part of the River, where the Stream runs far more swiftly than I have all this Day taken notice of it to do any where else, which induc'd me to stop a while, to observe it the more leisurely: And coming nearer, I found the Rapidness of the Current to be such, notwithstanding the depth of the Water, that I stood thinking with my self, how hard it were for one to escape, that should be so unlucky as to fall into it; But whilst I was thus musing, and attentively looking upon the Water, to try whether I could discover the Bottom, it happened to me, as it often does to those that gaze too stedfastly on swift Streams, that my Head began to grow giddy, and my Leggs to stagger towards the River. (*OR*, 140)

The passage begins with the chance swerve of attention we have come to understand as the engine of Boyle's narrative: "I was diverted . . . by happening to cast my Eyes on a part of the River, where the Stream runs far more swiftly." When the danger of transfixion presents itself, it's quite precisely an effect of care: the sudden thought of the risk of falling into the river and the unavailing labor (redoubled care) of trying to save oneself. Lindamor's "musing" is here a form of focused "attent[ion]" (rather than a sequence of cascading distractions), and when he "grow[s] giddy," we learn that this "happen[s] . . . often . . . to those that gaze too stedfastly on swift Streams." Interestingly, Bacon describes Montaigne's Pyrrhonian skepticism, his refusal to make truth claims, as epistemological "giddiness": the vertigo of cognitive vagrancy rather than overconcentration.[120] Boyle seems not to have been a great reader of Montaigne, though he happily avails himself of the shape-lessness of the essay (or, more precisely, of a differently shapeless form he sometimes *calls* an "essay"), but here, where temperamental cool encourages extravagance, and where "stedfast[ness]" rather than inconstancy poses a risk to both tranquility and understanding, he adopts a habit of shrugging impro-visation that might also have earned Bacon's censure. This would depend, as I argued in the previous chapter, on the philosopher's mood.

The Butcher and the Anatomist

One last convergence between Walton and Boyle's literary experiments is an emphasis on the sheer profusion of natural phenomena. In both cases, one important reason to grant value to cognitive suppleness is that Nature is various enough to demand it. In order to observe this aspect of Boyle's literary writing, we have to look beyond *Angling Improv'd*, in which form, disposition, and meta-phor, rather than explicit attention to nature, confirm Boyle's Baconian orienta-tion. In Walton's remarks on the worms and flies he uses as bait, we begin to discern the features of a shared perspective from which nature endlessly unfolds.

> There are also divers other kindes of worms, which for colour and shape alter even as the ground out of which they are got: as the *marsh-worm*, the *tag-tail*, the *flag-worm*, the *dock-worm*, the *oake-worm*, the *gilt-tail*, and too many to name, even as many sorts, as some think there be of severall kinds of birds in the air.[121]
>
> Now for *Flies*, which is the third bait wherewith *Trouts* are usually taken. You are to know, that there are as many sorts of Flies as there be of Fruits: I will name you but some of them: as the *dun flie*, the *stone flie*, the *red flie*, the *moor flie*, the *tawny flie*, the *shel flie*, the *cloudy* or black-ish *flie*: there be of Flies, *Caterpillars*, and *Canker flies*, and *Bear flies*; and indeed, too many either for mee to name, or for you to remember: and their breeding is so various and wonderful, that I might easily amaze my self, and tire you in a relation of them.[122]

Walton savors the simple fact that we have numerous terms for worms and flies. In each example, he leaves off listing kinds and concedes that there are "too many to name." He passes in both cases from an awareness of the diversity of a single species to some unrelated instance of wild proliferation: from worms to birds, from flies to fruit. In the case of worms, Walton makes their diversity an indication of environmental variation: their "colour and shape alter even as the ground out of which they are got." Like Bacon in his *Sylva Sylvarum*, the list conveys admiration for the simple fact that natural variety overwhelms our capacity to describe it. An awareness of diversity is also an acknowledgment of nature's mystery: enumeration fails.

When Boyle (in, for instance, other sections of the *Reflections*) turns his attention to the natural world, he likewise takes pleasure in the world's many-sidedness. In "Upon being presented with a rare Nosegay by a Gardener," for instance, he suggests that rhetorical sophistication is no match for natural variation: "Yes, here are Flowers above the flattery of those of Rhetorick; and besides, two or three unmingled Liveries, whose single Colours are bright, and taking enough to exclude the wish of a diversity; here is a variety of Flowers, whose Dyes are so dexterously blended, and fitly checquer'd, that every single Flower is a variety" (*OR*, 158). The nosegay mitigates our desire for variation and then gives it to us in spades, producing an experience of overabundance. Its monochromatic flowers are too wonderful for us to wish for any other ones, but these are juxtaposed with individual flowers so pleasingly mottled that each achieves "variety" on its own. Boyle's comment that natural beauty is "above the flattery of [the flowers of] Rhetorick" suggests that Nature's copiousness is one reason for the linguistic ingenuity demanded by "occasional reflection," which doesn't depend on rhetorical topoi but finds itself called upon to invent ever-new but inevitably inadequate "Expressions." "We may now proceed a little further," Boyle writes in the *Discourse*, turning from the consideration of the night sky to that of the multifarious earth, "and add, that if we suppose our Contemplator's thoughts to descend from Heaven to Earth, the far greater multitude and variety of Objects, they will meet with here below, will suggest to them much more numerous Reflections" (*OR*, 45).

As I now extend my interpretation of Boyle's career, exploring his interest in "occasional reflection," I take his delight in "inexhaustible fecundity" as an entry point. There is nothing especially novel about the motif, which calls to mind earlier traditions, including pre-Baconian natural history and neo-Platonic descriptions of nature's plenitude.[123] My interest is less the fact of Boyle's interest than his persistent indication that carelessness is the characteristic mood of the distributed awareness variety demands. Two of

Boyle's treatises on natural theology—*Of the Study of the Book of Nature* and *Of the High Veneration Man's Intellect Owes to God*—are separated by a gap of more than thirty years (from the early 1650s to the mid-1680s) but nonetheless delineate similar perspectives. By juxtaposing them, I show that the theory of "occasional reflection" and its attendant mood continue to animate Boyle's thinking as he turns from moral philosophy to science—or, more precisely, as he makes a less decisive transition than the scholarship often suggests. Walton the angler and Boyle the philosopher converge on the matter of Nature's plenitude because, as he acknowledges, immersion in the splendor of natural variety is exactly the experience the layperson shares with the virtuoso. Though we might expect the mature scientist to discard his interest in negligence in favor of the rigor for which he earns his reputation (to turn from ordinary appreciation to intensive study), the truth is different. Carelessness belongs to the moment of awareness that precedes methodical inquiry, yet the transition from serene delectation to disciplined indifference is less than fully assured. What distinguishes easy abandon from careful disinterest is effort, and one gives way to the other just as easily as tightened concentration collapses into distraction.

The truth of my basic claim for continuity between the *Reflections* and both treatises can be easily demonstrated; *Book of Nature* makes it especially plain, beginning with the headnote, "*For the first Section of my Treatise of Occasionall Reflections.*"[124] Elsewhere in *Book of Nature*, Boyle employs an idiom we recognize from the *Reflections,* observing that "the World instructeth us in many particulars . . . by suggesting unthought Rises & Reflections to us" (*Book of Nature*, 164). *High Veneration* adopts the same vocabulary. "The Creatures are but Umbratile (if I may so speak)," Boyle writes, "and arbitrary Pictures of the great Creatour: of divers of whose Perfections though they have some signatures; yet they are but such, as rather give the Intellect rises and occasions to take notice of and contemplate the Divine Originals, than they afford it true Images of them."[125] The "umbratile" creatures cast shadows in the manner of Eusebius, who isn't interested in his shadow's human shape but rather in the thought it obliquely occasions of the effortlessness of God's activity. Similarly, the creatures do not resemble divine "Perfections" but instead present opportunities to reflect on them. Thus the great champion of the corpuscular philosophy recommends a turn away from material bodies to divine truths even as he saves creation from being explained away as mere representation. Though he wouldn't put it this way, the emphasis of his natural theology is nature rather than theology. Given our knowledge of "occasional reflection," we understand that the movement of thought from the creatures to their Maker is no sure thing. It's just as likely that the meditator tarries with the creatures.

Both books savor cornucopian abundance. *High Veneration* celebrates the *"Immense Quantity* of Corporeal Substance" and adopts a similar tone of admiration when it describes the "part of the Universe which has been already discovered" as "almost unconceivably Vast" (*High Veneration*, 164). Later, Boyle assures us that "we need not . . . fear our admiration of God should expire, for want of Objects fit to keep it up" (*High Veneration*, 193–94). Speaking in a similarly reassuring vein of the heavens, for which he adopts the metaphor of a "boundless ocean," Boyle explains that it "contains a variety of excellent Objects, that is as little to be exhausted as the Creatures that live in our sublunary Ocean or lie on the shores that limit it, can be numbred" (194). *Book of Nature* is no less sanguine about the endless treasures that await our delectation. In the following passage, Boyle imagines understanding as an experience of eagerness: "Whereas other Creatures are content with those few obvious & easily attainable Necessarys that Nature has almost every where provided for them; In Man alone every sense has numerous greedy Appetites, for the most part for Superfluitys & Daintys; that for the Satisfaction of all these various Desires, he might be oblig'd with an inquisitive Industry to range, anatomize & ransake Nature, & by that concern'd survey come to a more exquisite knowledge of the Workes of it" (*Book of Nature*, 156). "Greedy appetite," "industry," and "concern" are obstacles to my argument for emotional cool; they show Boyle's willingness to adopt a vocabulary of excitement. One reasonable response to this passage is simply to observe the disunity of his thought, and I want to acknowledge the plausibility of that sensible judgment. Sometimes, indeed, Boyle performs the role of the natural philosopher we recognize from extant intellectual histories: the rapaciously curious investigator who wants to devour the world. Yet the account I have developed in these pages suggests the possibility of an alternative. We can find a precedent for a different interpretation by thinking back to Bacon's reflections on the dependable alternation between desire and gratification that distinguishes "learning" from other experiences. Perhaps Boyle's "greed" can here be read as a metaphor for the insatiability of constantly answered desire. When Boyle explains that what "Man alone" seeks out are "Superfluitys & Daintys," one possible effect of this phrase is the attenuation of propulsive "appetite." Despite the violence suggested by the effort to "ransake Nature," the "appetites" of sense perception pursue the passing pleasures of surface enjoyment. Indeed, that such an appetite belongs to the senses rather than the inquirer suggests something like a function rather than a wish. Boyle says humankind is "greedy," but greed isn't ordinarily taken for an experience of seamlessly getting what you want; his version seems to be missing the quality of importunity that defines the vice.

Were Boyle describing concupiscence, we might read these lines as a critique of human desire, lamenting how ardently we want what we don't need. Yet Boyle wishes to affirm humankind's unflagging interest in understanding God's creation, suggesting that we "range" over Nature's wonders with an enthusiasm we might think considerably softened by both the immediacy of satisfaction and the expectation that whatever we find is enjoyable rather than rapturous. There's something almost comic about the humming along of "numerous greedy Appetites," their seemingly outsized wishes reduced upon reflection to ordinary desires to see, smell, taste, and so on.

The best indication of Boyle's midlevel desire is his affirmation of the endlessness of gratification. However impatient his attitude, he pictures a world replete with occasions for perceptual pleasure, making patience unnecessary. When, in *High Veneration*, Boyle speaks of "a perpetual vicissitude of our happy acquests [acquisitions] of farther degrees of knowledge, and our eager desires of new ones," we understand that such "eager[ness]" is of the recreational kind; as in the "Hare Law" of the *Reflections*, though without taking credit ourselves for the creation of obstacles, our enthusiasm is rendered peaceful by an atmosphere of dependable satisfaction (*High Veneration*, 193). The book teems with the language of sheer abundance ("innumerable multitude," "unspeakable variety," "stupendious number," and so on) and uncountable proliferation ("so many distinct Engins," "complicated ones," "sundry subordinate ones," and so on) (*High Veneration*, 175). Boyle offers the following gloss of Aristotle's definition of infinity in the *Physics*: "'Tis that of which how much soever one takes, there still remains more to be taken" (*High Veneration*, 194). "If the Intellect," he continues, "should for ever make a farther and farther Progress in the knowledge of the Wonders of the Divine Nature, Attributes and Dispensations; yet it may still make discoveries of fresh things worthy to be admired" (*High Veneration*, 194). The persistence of desire is met at every moment by "fresh" gratifications. "Admir[ation]," we might also notice, usually a synonym for wonder (*admiratio*), is not the motive for interpretation; as in his *New Experiments*, Boyle's pleasure affirms an experience of inquiry that does not proceed simply from question to answer but instead makes every answer into a question: a new "discovery" dependably calls for further investigation.[126]

Both treatises seek access to nature by way of receptive digression. We cannot help but recall Boyle's *Discourse* when, in *Book of Nature*, he remarks, "And certainly to an ingenious Man . . . it cannot but delight him to expatiate his Thoughts thorogh the whole Immensity of the Universe" (*Book of Nature*, 170). Once again, the errant agency of the occasion impels reflection's "expatiat[ion]." "There is scarce any Instructions," he writes, "more likely to prove effectuall than those whose unexpected novelty engages our Attention, & whose pleasing surprises preventing our Prejudices are receiv'd as the Silent voices of Nature hir selfe" (*Book of Nature*, 171). "Nature" outsmarts us, dodging our defensive "Prejudices," when it strikes our "Attention" with a "pleasing surprise." In *High Veneration*, we discover similarly unembarrassed defenses of lost focus: "I know you may look upon a good part of this excursion as a digression; but if it be, 'twill quickly be forgiven, if you will pardon me for it, as easily as I can pardon my self" (*High Veneration*, 195). Boyle's longstanding interest in digression here takes novel shape as permission to entertain thoughts far outside the bounds of current scientific understanding. Retrieving the figure of the parenthesis from the *Reflections*, he writes, "Notice is to be given, that those other long Passages that are included in *Paratheses*, may with the Author's consent (or rather by his desire) be skip'd over; being but *Conjectural thoughts*, written and inserted for the sake of a *Virtuoso*, that is a great Friend to such kind of adventurous speculations" (*High Veneration*, 159). Boyle thus imagines the scientist as susceptible to patently unwarranted flights of fancy. By inviting us to skip over his digressions (the most risky of them), he marks them as spaces of free exploration rather than authorial guidance—not unlike those "many things untouch'd, unimagin'd" in which Browne invites the "libertie of an honest reason" to "play and expatiate." Between parentheses, Boyle considers the possibility of extraterrestrial life, as well as strangely specific fantasies like that of "an inestimable multitude of Spiritual Beings," some of whom inhabit "a probational state, wherein they have *free-will* allow'd them; as *Adam* and *Eve* were in *Eden*" (*High Veneration*, 186). He even considers the possibility that different natural laws apply in different regions of the universe.[127]

At times, Boyle opposes careless awareness to cautious investigation—but he cannot always sustain that difference. Unlike, say, the distinction between "wit" and "judgment," Boyle creates a conceptual opposition in which one term describes frank contempt for the other. Unconcern with regulatory proposals is often definitive of his perspective. He is given to declarations of boundless freedom. In *Book of Nature*, he explains that he "scorne[s] to suffer [his thoughts] to be confin'd within the very Limits of the World, beyond which they wander in those Imaginary Spaces; where even they may loose themselves: And unlocke the closest Cabinets of lurking Nature, & picke out thence retir'd & undiscovered Truths" (*Book of Nature*, 170). Rigor folds into recreation—and experimental procedure into untrained appreciation. On this matter, the key passage comes from *High Veneration*, where Boyle adopts the same voice that, in the *Reflections*, promises "something by way of Method" before advertising its failure to deliver.

The Wisedom of God which Saint *Paul* somewhere justly styles πολυποίκιλος, manifold or *multifarious*, is express'd in two differing manners or degrees. For sometimes it is so manifestly display'd in familiar Objects, that even superficial and almost careless Spectators may take notice of it. But there are many other things wherein the *Treasures of Wisedom and Knowledge* may be said to be hid; lying so deep that they require an Intelligent and attentive Considerer to discover them. But though I think I may be allowed, to make this distinction, yet I shall not solicitously confine my self to it; because in several things both these Expressions of the Divine Wisedom, may be clearly observ'd. (*High Veneration*, 168)

God makes the wondrous diversity of creation available to the eyes of the "almost careless Spectator" just as he does to the "Intelligent and attentive Considerer," but he does so in different ways. We might expect Boyle, speaking as the celebrated scientist he has by now (in the 1680s) become, to affirm his identification with the latter category, focusing his attention on the advantages of scrupulous study, but instead he takes his distance from "solicitous" "self-confine[ment]" and proceeds to explore "both . . . Expressions." In other words: Boyle absolves himself at present of the responsibility of preserving the distinction between surface and depth. As we might expect from someone who has just cited scripture by waving vaguely in the direction of it (quoting from "somewhere" in the Pauline epistles), Boyle again confirms his comfort with imprecision.

Book of Nature is more concerned to denigrate the undisciplined, but, as an addendum to the *Reflections*, it can only apologize for its fidelity to a "careless" perspective.

I dare not at present insist on those particulars wherein Nature is more Mysterious, because she is so; & because consequently, nothing can give us a true Account of That Theame, but an Exact & scrupulous Survey; & such a Notion of the Principles of Nature as must be prerequir'd, & in few Readers can be presuppos'd. I must therefore content my self to instance in such Applications, as are less suited to the merit of the Subject then accommodated to uninitiated Capacitys: & to make my Reflections Intelligible, must make them oftentimes derogatory; For 'tis not a Cursory View, or carelesse Prospect of the Creatures, that will enable us to discover in them, those curiouser Lineaments that an Omniscient hand has traced there; No; it must be a piercing & concern'd Inspection of them that must enable to say truly with the Psalmist; Marvellous are thy Works, & that my soul knoweth right well: & in another Place, O Lord how manifold are thy Workes, in Wisdom hast thou made them all! (*Book of* Nature, 168–69)

Though here Boyle rejects the "carelesse Prospect of the Creatures" as an appropriate method for "discover[ing]" their "curiouser Lineaments," his purpose is to defend his present failure to accomplish that goal. He speaks to "the uninitiated" in whom "a Notion of the Principles of Nature" is lacking, and thus he leaves an "Exact & scrupulous survey" for some other time. On the same page, he explains that "The Common Gazer, nay the vulgar Philosopher, & the true curious Naturalist, do as differingly improve their Admiration in considering the Creatures; which the one cuts out (not so mangles) grosly & the other dissects skilfully; as dos a Butcher from an Anatomist." Here we detect both reverence for the finer "skil[l]" of the "Anatomist" and affirmation of his current "Butcher's" art, which gives little in the way of detail but avoids the charge of "mangl[ing]" God's creation. To be a "true curious Naturalist" is a worthy aspiration, but for the moment we are all "Common Gazer[s]"—author and readers both.

Boyle's distinction between a "carelesse Prospect" and "a piercing & concern'd Inspection" suggests that the former presents us with endless candidates for the latter. Under Boyle's gaze, everything starts to look like a doorway to discovery. Recalling the metaphor that transmutes grains of sand (individual meditations) into a perspective-bending lens, Boyle exclaims, "What Creature is there more despicable then Sand & (what are but the natural Loaves of it) Flints: & yet have I made of them, lasting & orient Gems; & yet they are the true Metallicke Wombes & Paps, & not unfrequently praegnant with the pretiousest of them" (*Book of Nature*, 159). Thus Boyle extends the metaphor of "inexhaustible fecundity" to the most "despicable" matter. "And so many fine things," he explains, "are to be done with what we thus slightingly trample under Foot" (*Book of Nature*, 159). Boyle goes on to generalize the point: "We tread upon nothing we should not have occasion to kneele for; were we perfect in all it's Propertys & Uses" (*Book of Nature*, 160). What most interests me about Boyle the "Common Gazer" is that he recognizes the importance of paying careful attention to the apparently trivial, but simultaneously affirms that most of our encounters with things are necessarily "slighting." Everyone's eyes—even his—are untrained with respect to *most* things, which means attention to nature's profusion (rather than its individual parts, about some of which he knows a great deal) requires the embrace of ignorance. To inhabit an endlessly variegated world is to accept as a matter of course that it teems with proliferating opportunities, not all of which we can take. Nothing has to happen now.[128]

Conclusion: Out of Time

One of Baxter's misapprehensions about the *Reflections* is that writing is an antidote to "raving," which contradicts Boyle's suggestion that recording wayward thought widens its curve. That point of contrast is a fitting place to end, since it focuses our attention on a distinctively literary question that remains of central importance to this book. Baxter thinks the permanence of the written word, its enduring availability to careful perusal, exposes the error of insufficiently attentive thought; Boyle suggests, on the contrary, that writing invites the participation of readers in digressive reflection. Here is Baxter on writing as a form of discipline: "Men are such Hypocrites & Atheists, that while their cloathes, & words, & all that is seen of men, are composed as beseemeth men awake; their *Thoughts* which are seen to God alone, are willfully left so discomposed & distracted, as if they were dreames or bedlam ravings: And if one daies thoughts were written downe, & read over before a sober company, they would meet with more such censurers than me."[129] "Discompos[ure]" and "distract[ion]," "dreames" and "ravings"—these are the signatures of Boyle's meditations. He is often the lazy fisherman who doesn't want to get out of bed. Where Baxter imagines readers whose souls are just as handsomely dressed as their bodies, Boyle embraces dishevelment. Unlike Walton, a fellow traveler if ever there was one, Boyle discovers the formal means by which to usher his readers into his "dreame"-world—where interpreters of Nature's book are too busy exploring its voluminous pages to straighten their collars or button their coats. In "De l'oisiveté" ("Of Idleness"), Montaigne tells us that his mind "gives birth to so many chimeras and fantastic monsters, one after another, without order or purpose, that in order to contemplate their ineptitude and strangeness at my pleasure, I have begun to put them in writing, hoping in time to make my mind ashamed of itself."[130] The sentiment anticipates the difference between Baxter and Boyle; to put things chronologically backwards, we might say that Montaigne presents himself as an ironic Baxter who quite obviously enjoys the Boylean pleasures of mental vagrancy. In a book like the *Essais* that makes a sport of wandering off-course, we don't take seriously Montaigne's "hop[e]" that "writing" down the "fantastic monsters" of undisciplined imagination "shames" him into orderly sobriety.

I close by looking forward to the next chapter, which explores the role of a new technology in the scientific imagination: the microscope. About vision, Boyle has a great deal to say—one example of which prepares us for what comes next. In *High Veneration*, he explores the tradition in which God sees every last aspect of creation in a single glance. We recognize this image from, for instance, the *De consolatione philosophiae* (523) of Boethius, who explains that God's knowledge "is permanent in the simplicity of his present, and embracing all the infinite spaces of the future and the past, considers them in his simple act of knowledge as though they were now going on."[131] Boyle describes human attention, here understood as visual focus, as the absence of exactly this capacity to "embrac[e] . . . the infinite": "We Men can perceive and sufficiently attend, but to few things at once; according to the known saying, *Pluribus intentus, Minor est, ad singula sensus* [Attention to many things is less than awareness of single things]. And 'tis Recorded as a Wonder of some great men among the Ancients, that they could dictate to two or three Secretaries at once. But God's knowledge reaches *at once* to all that He can know; His penetrating Eyes pierce quite thorough the whole Creation, at one look" (*High Veneration*, 189).[132] Boyle's "*at once*" conveys his wonder at God's capacity to receive every possible stimulus at the same time and to understand them clearly. (We have seen him imagining what such mastery would be like when he creates figures from shadows "in the twinkling of an Eye.") "The whole Creation" is at God's perfect disposal, and so he never has to bring any particular thing into view. For God everything is equally "clear and distinct"—to borrow Descartes's criteria for certainty. For humankind, however, there's an inverse relationship between the number of things to which we pay attention and the adequacy of our focus. The greater the number of objects we try to keep in mind, the less successful we are at "perceiv[ing]" them. Given Boyle's belief that awareness of creation's manifold is an expression of devotion, he cannot recommend that we concentrate on every individual thing that strikes us as important. The best we can do is capitulate to whatever comes to mind. The observational mood translates divine simultaneity into the realm of temporal succession, and Boyle drifts from one object to another—not *ad infinitum* but *usque ad mortem*. Yet the emphasis falls on modesty rather than morbidity: contentment with whatever comes to light before time runs out.

CHAPTER 3

The Microscope Made Easy

Andrew Marvell with Henry Power

In 1742, Henry Baker, a fellow of the Royal Society, publishes *The Microscope Made Easy*, an early work of "popular science" that takes its bearings from two familiar Baconian theses: first, that the success of scientific inquiry depends on wide and active participation; and second, that the best way to recruit new participants is by making experimental practices accessible to all. As he puts it: "The likeliest Method of discovering Truth, is, by the Experiments of Many upon the same Subject; and the most probable way of engaging people in such Experiments, is, by rendering them easy, intelligible, and pleasant."[1] Yet Baker also counts the effortless pleasure of the microscope's operation among the difficulties that impede its use: "Many, even those who have purchas'd Microscopes, are so little acquainted with their general and extensive Usefulness, and so much at a Loss for Objects to examine by them; that after diverting themselves and their Friends, some few times, with what they find in the Sliders bought with them, or two or three more common Things, the Microscopes are laid aside as of little farther Value."[2] Baker blames the public's ignorance for its impatience with the device: they just don't know what to do with it. By the time they discover the insufficiency of their knowledge, they have already gotten started; the experience can only be one of swift disappointment. So it goes, Baker suggests, when "diversion" comes before true understanding. We might state the problem this way: *the microscope is too easy to use.* Before we have a chance to figure out what we're up to, we've already been up to it for long enough to lose interest. Elsewhere, Baker speaks directly to the concern that a dignified scientific instrument has been misconstrued as a cheap novelty: "Others have considered [the microscope] as a meer Play-thing, a Matter of Amusement and Fancy only, that raises our Wonder for a Moment, but is of no farther Service."[3] After an initial flicker of "Wonder," grown-ups have little use for toys.

Going back to the seventeenth century, the microscope shows a disconcerting liability to embarrass its champions—devolving from a cutting-edge technology into a silly gadget, from a weapon for conquering hidden worlds into an easy way to waste an afternoon. In his 1665 *Micrographia*, Robert Hooke sounds the trumpet of lens-assisted optimism without a care for such eventualities, but, by the early 1690s, we find him anticipating Baker's concern with precision: "I hear of none that make any other Use of that Instrument, but for Diversion and Pastime, and that by reason it is become a portable Instrument, and easy to be carried in one's Pocket."[4] Complaining that only Antonie van Leeuwenhoek still finds bona fide scientific uses for the microscope, Hooke vaguely attributes its neglect to the absence of "inquisitive Genius" in the "present Age."[5] He also proposes, much more instructively, that the device's pocket-sized convenience gives the false impression of frivolity; an invention that once promised the triumph of new discoveries now invites mere "Diversion."[6] Hooke's worry is that user-friendliness is deflationary—for those very users who should both benefit from "accessibility" and, in turn, make themselves beneficial to the collective project of "advancement."

There is a secret knowledge, I suggest, in Hooke and Baker's frustration. Easy pleasure is central to the first accounts of microscopic observation in England; the frivolous "Play-thing" is a tarnished version of a technology of emotional buoyancy. What Hooke and Baker dislike about the microscope's misuse is the dark side of the fantasy originally materialized in it, the flip side of the coin of effortless accessibility. In this chapter, I explore the literary life of this fantasy—not as it ends up curdling into an anxiety about inconsequence and meaninglessness, but as an expectation of visual pleasure. Whatever the *actual* technical difficulties Hooke encounters in making the microscope work, he gratefully *imagines* the device he holds in his hands as child's play. This is not a simple counterfactual; the device's conceptual elegance survives the practical problems inherent to the earliest experiments with it. The idea of the microscope is beguilingly simple. By bringing the lens to the eye, a new world comes into view. In this way, Hooke employs the instrument as the concrete realization of the very thing I have set out in this book to explore: the equation of effortlessness and perceptual amplitude.

This chapter takes up the experience of vision, especially as transformed and enhanced by optical technology. Across the many genres of Baconian writing, however, thought bleeds into sight—and sight into thought; it is not surprising that an effortless practice of visual inspection sometimes forgets to distinguish between the seen and the imagined. Even a privileged mode of scientific perception (meticulously managed vision) shows a propensity for deviation. The literature of microscopy shows an appreciation for ocular wandering—even for drifting away from what actually enters the observer's field of vision. Because the story of the rise of experimental science is often narrated as a series of technical advancements (with Galileo's telescope at the center), my interest here is the capacity of technology to materialize fantasy. Like no other scientific device (not even the telescope), the microscope transforms the premise that experimental labor is easy into an object, an operation, and a fait accompli.

In pursuit of the instrument as both usable technology and metaphor of facility, I take the poet Andrew Marvell as my guide. In his strange (indeed, unprecedented) country-house poem, *Upon Appleton House* (1651), Marvell both inhabits the observational mood and uses the lens as an emblem of it. There, the natural philosopher's exertions are almost literally a walk in the park—and the microscope is an image of the near-automatism with which conventional appearances dissolve in the absence of fixed (because, much of the time, eagerly sustained) premises. Taking cues from Marvell, I proceed to a new interpretation of the first English contribution to the genre of microscopic observation, Henry Power's *Experimental Philosophy* (1664).[7] Thus I reverse the usual practice of literary-critical contextualization in which a poem is explained by an adjacent historical development. The history of optical technology does not tell us very much about *Upon Appleton House*, but the poem, if we give it the close attention its complexity demands, tells us something about the microscope as an idea—both for those who merely ponder its powers and those who actually peer through it. The microscope is not so much this chapter's topic as its destination. Attending to larger questions of visual distortion and scalar transformation as they inflect the experience of reading, I explore Marvell's response to a variety of overlapping intellectual, social, and political contexts (not only the rise of Baconianism but also issues as seemingly disparate as the violence of civil war and the appeal at midcentury of apocalyptic rhetoric) before turning to his representation of lens technology—and, ultimately, to Power's. I do not argue that Marvell's visual imagination is specifically microscopic (that he is preoccupied with the device) but that his representation of both naked and technologically enhanced vision helps us make sense of the microscope's appeal

to the imagination—both within the space of the poem and in scientific treatises like Power's. I show that carelessness is not only a "way of thinking," as
Boyle puts it, but also a way of seeing.

The most influential interpretations of the new technology and the new
literary genre to which it gives rise foreground the microscopist-author's
ambition to dazzle his readers with wondrous appearances, but *Upon Appleton House* exemplifies a different wish: not having to lift a finger.[8] We've
already seen, not only in Baker's late-Baconian musings but in the works of
Bacon himself, that "wonder" can be a name for cheap and fleeting sensation;
Marvell's observational pleasures are much softer and longer lasting than
that. The microscope holds out the promise of sudden, graceful, miraculous
discovery—but it's a standing offer. The poet describes himself as an "easy
philosopher" for whom novel appearances catch the eye without capturing the imagination; the experience of discovery goes on and on. The same
expectation defines Power's observations, which trace a drifting path from
one object to the next, including, in some cases, fantasized objects of speculation. This pattern of sustained anticlimax helps explain my focus on Power,
who is often overshadowed by Hooke in the scholarly literature.[9] Attending
to the very first English entry in the genre needs no special justification—but
it's worth noting that the astonishing engravings in *Micrographia* partially
account for the usual scholarly emphasis on emotional drama. Though I suggest that Hooke shares Power's (and Marvell's) understanding of microscopy
as a gently pleasurable exercise, his images almost inevitably speak at a higher
volume than his prose. To be sure, even the images might lend themselves
to the leisurely interpretation on offer in this chapter, if only because our
second and third encounters with Hooke's detailed renderings can hardly
continue to startle us with the intensity of the first. Often, however, the
forcefulness of that initial encounter seizes hold of the critical imagination.
Micrographia's first readers treat it like a magical *Wunderkammer*; it keeps
Samuel Pepys, who declares it "the most ingenious book that ever I read in
my life," up until two o'clock in the morning.[10] A sudden confrontation with
the human-sized face of a drone fly, staring back at you with hundreds of bul-
bous eyes, will do that. So marvelous are the insect denizens of *Micrographia*
(to say nothing of the other minuscule objects that find visual representation
there) that thinking about the microscope in England after 1665 almost necessarily means thinking about Hooke. Thus an incidental advantage of my
account is that it (mostly) inhabits an earlier moment (Hooke's images have
yet to enter the popular imagination), granting us access to an experience of
scientific vision less dazzling than stimulating, more continuous than punc-
tuated, and no more arduous than a stroll.

Disarmament as Induction

In the penultimate stanza of *Upon Appleton House*, Marvell draws a perplexing contrast between the orderliness of his patron's country seat and the chaos of the wider world. After having followed the poet's errant passage across the ample acreage of Thomas Lord Fairfax's estate (95 of 97 stanzas), most readers share the feeling of disorientation that earns the poem its dubious distinction as the "most complicated topographical poem in English"—unless such confusion collapses into the outright frustration we hear in one critic's pronouncement that it's simply "too long."[11] Either way, we're unlikely to take the speaker at his word when he speaks of a "decent" manmade "order" that puts wild Nature to shame.

> 'Tis not, what once it was, the world;
> But a rude heap together hurled;
> All negligently overthrown,
> Gulfs, deserts, precipices, stone.
> Your lesser world contains the same,
> But in more decent order tame;
> You heaven's centre, Nature's lap.
> And Paradise's only map.[12]

It's hard to think of a less apt description of Marvell's Nun Appleton, from which he has withheld even the semblance of equilibrium—unless he means to display the difference between the reader's perspective and the credulous cartographer's. Like any number of seventeenth-century poems, this one seems at last to affirm neat symmetries between human and superhuman scales. It so closely identifies Fairfax with his property that together they play the role of microcosm, "centr[ing]" the "heaven[s]" and partaking in the harmony of the universe.[13] But Marvell has dizzied us with such success that our only remaining certainty is of puzzlement. Perhaps he rehearses, but does not endorse, the metaphysical assurances of a Christianized Ptolemaic cosmos.

I begin with the poem's intimation of cosmic asymmetry because it links one of the classic topoi of the New Science—which increasingly casts doubt on easy parallels between microcosm and macrocosm—and the neglected question of neglect itself.[14] When Marvell supplies the phrase "negligently overthrown" to describe the ravaged "world," he offers us just the adverb we need to approximate his style; it's the poet's insouciantly curatorial spirit rather than the Fall of Man that lays everything waste. Since defamiliarization turns out to be the poem's raison d'être, Marvell's apparent lamentation might be read instead as a contented backward glance at the work these verses accomplish: "'Tis not, what once it was, the world; / But a rude heap together hurled." The poet's casual torsions of perspective unmake "the world"—but only the artificial one we know from other country-house poems.[15] While Marvell writes this poem, adventurers with the microscope are breaking the world up into shards of insight, promising a future of total comprehension without coming close to delivering it. The poet's practice is similar: by offering a series of incommensurable views, he offers up a "rude heap" with the implication but not the illustration of future rectification. It's with disabused realism rather than gaping admiration that Marvell refers to the estate as "Paradise's only map"; such schemata are the best we can do if we place our hope for utopian transformation in images rather than experiment. Thus we discover an apt description of the observational mood; the poet's inquisitorial intensity is an expression of disregard.

Ordinarily, and especially with respect to its mid-seventeenth-century reception, Bacon's thought is understood as a wellspring of vanguardist aggression against the unearned assertions of scholastic natural philosophy. Without the argument I have unfolded in these pages, we could only narrow our eyes at Marvell's embrace of "negligence" in the service of epistemological progress. Many brilliant readers of this poem have drawn a thick line between the poet's temperament and his interest in the New Science, tipping the scale in one direction or the other. Describing the forest sequence in *Upon Appleton House*, for instance, and thus observing Marvell's speaker at his most conspicuously wayward, Harry Berger Jr. writes, "Marvell's experiments . . . are not marked by the scientific ideal of 'objectivity.' His *indecus* behavior in the forest mocks the very order—sober, constrained, and neat—which protects him from the world."[16] In Berger's reading, an attitude of self-abandonment rules out "science" no less than moral "decency." Picciotto, who demonstrates Marvell's commitment to Baconian values, risks the opposite formulation. Taking up the metaphor of the lens, she places Marvell among "experimentalists" whose "instruments of truth" transform the world by "looking through" it—in contrast to earlier Renaissance poets whose "fictions" celebrate the "gratuitous leisure" that grants them independence from "applied forms of knowledge."[17] This is a beautifully persuasive argument, except that "leisure[liness]" calls no one to mind so much as the poet who "languishes with ease" under the shady trees of Nun Appleton (593). Marjorie Hope Nicolson, in her classic work on science and poetry in the seventeenth century, draws a similar distinction between the "contentment" of "Elizabethans" who prefer the safety of the closed universe ("agoraphobia") and the "restlessness" of their "modern" contemporaries who long to break its boundaries ("claustrophobia").[18] She counts Marvell among the former—one of two available conclusions, both of which leave something out.

I suggest we dispense with the premise that prohibits questing "contentment." Marvell integrates carelessness and vehemence with perfect seamlessness, intensifying both experiences even as he draws them together. In *Upon Appleton House*, we observe extremes of cognitive disarray and investigative momentum, languid self-indulgence and purposeful relentlessness. Picciotto is alone in pursuing this apparent paradox, describing an "almost inhuman combination of immersion and detachment."[19] I avoid the vocabulary of "detachment" altogether, however, since the observational mood is a mode of engagement—and there's nothing "inhuman" about that. Marvell achieves perceptual intensity without burdening the buoyant mind with cares. The fantasy of casual indifference does not flit into his view or passingly capture his interest; nor does it have to contend with diligent sobriety. It's purely and simply his premise.[20] I would go as far as to say that it needs no explanation (that the poem explains itself), were it not the case that almost every critical interpretation is evidence to the contrary. Perhaps the difficulty is the set of interpretive habits I discussed in the introduction: above all else, a literary-critical emphasis on self-fashioning that rules out the possibility of artlessness.[21] Careless receptivity is the poem's through-line, integrating its disjunctive contexts. I consider the most pressing of these, the trauma of civil war, before taking up the conceptual proximity of optics and images of apocalypse. From there, we can look beyond Marvell's poem to the wider metaphorical uses of the magnifying lens.

The poem's title is metonymy. The "house" isn't the half of it. Nor does the poet simply write about other things; he writes about too many of them. Like the eponymous estate, the poem is a site of near-impossible convergence. Marvell speaks of natural history and family history, apocalypse and agriculture. We might answer overabundance with Michel Foucault's vague but evocative language of "heterotopia"—that strange "counter-site" in which "all the other real sites that can be found within the culture are simultaneously represented, contested, and inverted."[22] The figure captures the reader's experience of prismatic discontinuity: myriad things and incommensurable kinds. Yet the speaker's description of Nun Appleton says little, however obliquely, about the totality of England's "other real sites." The estate is less an alternative to the wider world than a point of departure for its investigation. Alongside Foucault's image, then, we might place Bacon's epistemological fantasy, everywhere challenged by his own insistence on the frailty of the senses, that "all depends on keeping the eye steadily fixed upon the facts of nature and receiving their images simply as they are."[23] States of minimal feeling are apertures for endlessly hospitable "reception," but affective waywardness can only flout Bacon's call for "steadiness." Indeed, it's the wavering quality of Marvell's casual indifference that clears him of the charge of naïveté, unless it convicts him of a truly radical simplicity.[24] False appearances are artifacts of desire for which the observational mood is the perfect solvent, and yet it fails to stabilize any single experience as a realist alternative to myth.

The overabundance of the poem's contents lends credence to the speaker's observational mood. It makes good sense for an unknown and unsorted world to flood the sensorium, especially when subjected to strange perceptual experiments. Marvell refrains from deciding ahead of time where to level his gaze; it's in this specific sense that the poem has no object. Yet many critics agree that the opposite is true: the poet's specific interest is the event of civil war. I suggest that nothing separates Marvell's epistemology of carelessness from his specific response to a nation at arms. In this respect, he recalls Montaigne's irenic perambulations. Carelessness disarms the self, exposing it to novel sensations and perceptions—and one of Marvell's disarming achievements is the seriousness with which he treats this apparently metaphorical connection. For him, the war would have been felt not only as England's collective trauma but also as the specific background to his patron's retirement at Nun Appleton. Fairfax gave up his command of the army because he shrank from a preemptive strike against the Scottish.[25] This poem about almost everything is also a poem about this one thing.

An affinity between Marvell's experiments and those of the virtuosi should come as no surprise. The poet and Katherine Jones, Lady Ranelagh (Robert Boyle's sister and a leading intellectual in her own right), both spent time at John Milton's house in Petty France, and it was Marvell who wrote her daughter's epitaph.[26] John Aubrey reports that Marvell counted John Pell, the mathematician and eventual Fellow of the Royal Society, among his friends.[27] But the case for Marvell's interest in the New Science has never rested on his personal acquaintance with its luminaries; poems like this one, along with The Last Instructions to a Painter (1667) and The Garden (1668), pay explicit attention to its premises and practices. If Marvell "does more than any poet before him to reproduce the sensory disorientation and richness of experimentalist insight," I would supplement that judgment with the assertion that his generative confusion is a corollary of the observational mood.[28] Carelessness is freedom from defensive vigilance, one variety of which is the fearful preservation of the world's consolatory guises.[29]

James Grantham Turner offers the right point of departure for this view, suggesting that Upon Appleton House should be read as a squarely pacifist exercise.[30] As he memorably puts it, the poem "taught the Fairfaxes to see life at Appleton as a beginning and not an end, to use these warlike studies [the phrase is Marvell's] to defeat the legacy of war."[31] Much ink has been spilled about this poem since Turner wrote these lines (some of it by Turner himself), but the persuasive simplicity of the observation bears repeating. In keeping with the critical paradigm that can only locate hidden agitation behind a gentle disposition, this poem about peace is often understood as a poem about war. Tranquility becomes a disavowal of actual conflict or a fragile state of affairs over which the poet wrings his hands.[32] If we assume anxious ambivalence and then set about locating it, we are not likely to read the poem as a straightforward bid for anything at all—let alone the laying down of arms, as apparently naïve a demand as any.

But the poem's objective is exactly this one. The garden sequence (stanzas 36 to 40) happily enacts the restoration of peace, notwithstanding Turner's argument that it depicts "the military world in its most pleasant aspect, 'in sport,' bustling, decorative and painless," so that Marvell seems to present only the alluring aspects of martial activity—one deceptively attractive side of the sordid business of waging war.[33] In fact, the most basic achievement of the sequence is to reinvent guns as flowers—a gesture not unlike the iconic image of 1960s antiwar activism in which a flower is placed in a gun barrel. Marvell's "flower power" implies a different vantage than its late modern successor, however, since the miracle of peace is more an achievement than a wish. The machinery of war is simply absent; floral profusion has taken its place. Unlike the peacefulness of a military parade, which makes a spectacle of power, these botanical exercises are annulments of force.

It's common enough to link Marvell's reflections on warfare to the arrangement of the gardens in the shape of a fortress—but the nature of that relation is more precisely an inversion than critics have noticed.[34] Like Shakespeare's Feste, who quips, "A sentence is but a cheveril glove to a good wit; how quickly the wrong side may be turned outward," Marvell delights in easy reversals, but he takes his pleasure in ease rather than wit.[35] The watchman's vigilance is only a few tonal shades away from the natural historian's attention, and yet the difference is the stark one between security and susceptibility.[36] The best entryway to the strange logic of Marvell's poem is the proposition that disarmament and induction are synonyms. I want to show how Marvell's garden conveys this equivalence, after which I turn to the poem's second emblem of inversion, the forest (stanzas 61–78), where he invites us to inhabit the experience (of awareness as danger) itself. There we discover that restless paranoia and leisurely scopophilia are not only different modes of looking but also different moods; it's the bare change of tone that makes all the difference.

Clues to my interpretation sit right on the surface of the poem. In the garden sequence, the speaker explicitly identifies floral "bastions" with the five senses (287). Sir Thomas Fairfax, he explains (not Marvell's patron but a forebear of the same name),

> . . . when retirèd here to peace,
> His warlike studies could not cease;
> But laid these gardens out in sport
> In the just figure of a fort;
> And with five bastions it did fence,
> As aiming one for ev'ry sense. (283–88)

The rhyming of "fence" and "sense," which sonically underscores the replacement of a defensive posture with a receptive one, strengthens the analogy. Turner is right that Marvell's "warlike studies" intend to "defeat the legacy of war," but these lines treat such an event as an achievement we get to enjoy rather than a lesson the Fairfaxes need to be "taught." After all, the original creator of Nun Appleton's garden, who is also the subject of the long sentence that fills out the stanza, receives credit for outfitting it for receptivity.[37] However ambiguous the agency behind the analogy ("as aiming one for ev'ry sense"), the garden's design seems to offer up the comparison rather than awaiting its superimposition by the poet.

Marvell's freedom from responsibility grants him the effortless energy to recount correspondences between the garden and the sensorium. While the stanza partitions the sensorium into "five bastions," encouraging awareness of the modularity of perception—its composition from distinct sensations— the following one underlines the point by spreading different kinds of sense perception out in time. Because nothing in this sequence actually succeeds at isolating individual senses, it continues nonetheless to suggest an integrated experience. Like Boyle taking in the varied "deliciousness" of the "promising morning" at the start of his fishing adventure, the speaker here stimulates each of the senses in turn.[38]

> When in the east the morning ray
> Hangs out the colours of the day,
> The bee through these known allies hums,
> Beating the *dian* with its drums.
> Then flowers their drowsy eyelids raise,
> Their silken ensigns each displays,
> And dries its pan yet dank with dew,
> And fills its flask with odours new. (289–96)

Marvell's dawn is the dawning of perception. The "morning ray" that "hangs out the colours of the day" in the first couplet asks us to see, while the "bee" that "hums" and "beat[s]" the "drums" in the next one asks us to hear. The flowers mimic the reader's newfound awareness, unfurling themselves and "rais[ing]" their "drowsy eyelids." The final couplet presents a simple image that appeals simultaneously to our senses of touch, smell, and taste. We feel the wetness of the petals as the flower "dries its pan yet dank with dew," before it "fills its flask with odours new," granting the nose its particulate pleasures. Answering the drum roll of the bee, this final gesture offers the jaunty insect odoriferous nourishment. We imagine the bee's proboscis sipping nectar with satisfaction. Yet it's vision that folds the experience together: as both background and foreground, "[hung] colours" and "rais[ed]" "eyelids," it centers multisensory perception.

When the speaker bids farewell to the garden, he continues to draw attention to both the experience of sight and the defenselessness of the open eye. He refashions a gun sight as a light beam. The stanza describes the inspection of foreign objects with repurposed technology that would once have threatened to destroy them.

> The sight does from the bastions ply,
> Th'invisible artillery;
> And at proud Cawood Castle seems
> To point the batt'ry of its beams.
> As if it quarrelled in the seat
> Th'ambition of its prelate great.
> But o'er the meads below it plays,
> Or innocently seems to graze. (361–68)

The speaker performs the wary circumspection of the besieged before releasing the gun sight to "play" over the landscape, ocular aggression giving way to visual exploration. One early manuscript has "gaze" instead of "graze," which makes the conversion of aiming (with a gun) into observing (with an attentive eye) as plain as day.[39] But "graze," the more likely reading, accomplishes the same feat. The term refers to the shining of a beam of light.[40] The object of the gaze, "the meads," is exactly the speaker's destination; the light beam anticipates the action of his footsteps when, in the next stanza, they "play" across the meadows, carrying the speaker further along in his project of making them visible.

The garden's images of defenselessness, with their mood of "retire[ment]" and "peace," suggest an atmosphere of tranquility, but the speaker's woodland wanderings surpass mere suggestion, inviting us into the observational mood

(283). The wood first comes into view as a fortified enclosure, but we soon find the opposite to be the case, just as we did with the garden's "bastions."

> When first the eye this forest sees
> It seems indeed as wood not trees:
> As if their neighbourhood so old
> To one great trunk them all did mould
> There the huge bulk takes place, as meant
> To thrust up a fifth element;
> And stretches still so closely wedged
> As if the night within were hedged.
>
> Dark all without it knits; within
> It opens passable and thin;
> And in as loose an order grows,
> As the Corinthean porticoes . . . (497–508)

The perimeter seems unassailable, the trees having grown together into "one great trunk," but all it takes is a quiet approach to discover the mistake of the "first" impression: "within / It opens passable and thin." We're reminded at first of Spenser's Garden of Adonis, where Venus seeks protection with her "wanton boy" in an "arbour" where the trees "[knit] their rancke branches part to part" so that "neither *Phoebus* beam could through them throng / Nor *Aeolus* sharp blast could worke them any wrong."[41] No sooner does the image come to mind, however, than we join the speaker in "thronging" "through" the welcoming wood. The ageless "bulk" buttresses an antiquated cosmos, "thrust[ing] up a fifth element," until the speaker discovers it's all illusory; the wood is open and "loose"—and only architectural in the sense that it resembles those manmade "porticoes" modeled after organic forms (the Corinthian style).[42]

The end of the sequence similarly transforms a defensive brace against the world into a generous prospect. Fearless in his defenselessness, the poet's casual indifference shrugs off caution.

> How safe, methinks, and strong, behind
> These trees have I encamped my mind;
> Where Beauty, aiming at the heart,
> Bends in some trees its useless dart;
> And where the world no certain shot
> Can make, or me it toucheth not.
> But I on it securely play,
> And gall its horseman all the day. (601–8)

The sentiment of "safe" "encamp[ment]" reveals itself to be only that: freedom from fear is different from actual security. We hear the speaker's carelessness in the "or" that splits the phrase: "And where the world no certain shot / Can make, or me it toucheth not." Either the trees shield him from the world's volleys, or the arrows happen not to have hit their target. The preceding lines hint at the reversal, permitting allegorical Beauty to breach the perimeter as soon as the speaker celebrates its "strength"; like the archers of exteriority, Cupid's "darts" might be "useless" for some reason other than any actual barrier. Indeed, the vagueness of the reason for their "uselessness" indicates the speaker's carelessness. When he mimes aggression, claiming to "play" on the "world" like well-mounted cavalry, he again defangs a military metaphor, exchanging strategic advantage for mere perceptual purchase.

Following the example of "passable" trees, the speaker's senses drop their guard. Playing the role of natural historian, he makes a display of formal devices by which we too can taste, and perhaps even inhabit, his characteristic mode of feeling. Of the insights such intimate familiarity with Marvell's observational mood affords, I want to underscore two, both of which we recognize from Walton and Boyle, Bacon and Montaigne: (1) it's a state of immersion rather than detachment, and (2) it's an experience of flexibility. An early example—Marvell's first use of the first person singular—initiates the pattern I have in mind.[43]

> And now to the abyss I pass
> Of that unfathomable grass . . . (369–70)

The speaker introduces himself as an agent of continuity who bridges distinct scenes of observation. In the same breath, a grammatical conjunction draws phrases together into an unfolding experience of multiplicity: "and now." The perspective contracts from a general view of the estate's location and design (in the previous stanza) to the specific moment in which our tour guide saunters both into the poem and into the meads. Indeed, the poem consists of nothing but such "passages"; as the speaker traverses the variegated grounds of the estate, he grants himself license to comment on whatever comes to mind and whatever comes into view.

Such couplets proliferate in the forest sequence, with the difference of explicit attention to emotional tonality. Our privilege here is to observe the unfolding of the experience with all the precision Marvell's formal habits convey. A conjunction (or conjunctive adverb) announces a shift in the speaker's attention from one scene to another. His perspective scales down from descriptive breadth to momentary experience. Neutral absorption in time's present unfolding makes the sensorium available for new discoveries.

Perspective broadens as it narrows. We can observe the pattern in the following selection of extracts from Marvell's sylvan promenade.

Then as I careless on the bed
Of gelid strawberries do tread . . . (529–30)

Thus I, easy philosopher,
Among the birds and trees confer . . . (561–62)

Then, languishing with ease, I toss
On pallets swoll'n of velvet moss . . . (593–94)

We can hardly miss that these couplets narrate transition, since each appears at the beginning of a stanza. "Thus" and both "thens" are conjunctive adverbs, describing the manner in which the speaker passes from one experience to another. Most interesting of all, for our purposes, carelessness is now explicit; the speaker is precisely "careless" in the first couplet and "easy" in the second, while he "languishes with ease" in the last. We have to drop our expectation that these terms carry a negative connotation, unless what we find instead is that the poet's charged vocabulary intensifies the impression of waywardness he everywhere asks us to savor.

Furthermore, Marvell's casual conjunctions mark transitions from observation to participatory discovery. Indeed, each couplet interrupts a more comprehensive view of a scene with the speaker's deliberate passage into a new one. The first couplet carries us out of the speaker's rhetorical dilation on "stock-doves," which marries the language of pastoral to that of natural history, and into the auditory and tactile experience of "treading" on "strawberries." The second narrows the speaker's perspective from his patient account of woodpeckers' woodwork to a "confer[ence]" with the surrounding wood in which he becomes the grammatical subject of the sentence. The third interrupts a sequence in which flora engulf and assimilate him ("And ivy, with familiar trails / Me licks, and clasps, and curls, and hales"), permitting the enactment of a swoon that makes him an agent (if not *the* agent) of his absorption (589–90). In each case, the speaker generates an image of himself as an engaged observer whose awareness now includes something new: the underfoot crunch of cold strawberries, birdsong, and the softness of mossy beds one only discovers by lying in them. The reader might also hear a pun in "gelid," taking it for "jellied" and feeling the squish-squish of berries beneath her feet, in which case she herself has taken on the role of the solitary explorer, undergoing a distinctive experience of her own.

My collation of couplets makes plain how far away we are from the rubric of disinterest.[44] Marvell's observational mood is an experience into which any feeling might swim. The most impressionistic account of the poem argues for such impurity, since casualness accommodates (tonally) diverse experiences, but these lines make a precise display of the blur.[45] To be sure, I do not wish to muddy the concept of the observational mood, which here refers, as it has throughout this book, to an experience of dispassion without the stringency of self-control—and thus to an experience of (relative) emotional quiet. What I do wish to suggest is that it represents a default resting place on an affective spectrum, one that accommodates different shades of feeling, and one from which the speaker sometimes departs but to which he reliably returns. We should also notice that the emotion-words Marvell offers us in these couplets turn out to be less than constant. "Careless[ness]" is not simply a synonym for the observational mood. Consider the "careless[ness]" of strawberry stomping, which seems, when the speaker immediately proceeds to bird-watching, to name the mood of natural history.

> And through the hazels thick espy
> The hatching throstle's shining eye . . . (531–32)

Afterward, we return to the affective vocabulary most often associated with early modern science, and from which I have taken some distance in these pages: "But most the hewel's wonders are," the speaker says, nominalizing "wonder" as a finding rather than the initial impetus to inquiry (537). However distinctly "careless[ness]" seems to name a perceptual freedom that supports the interpretation of nature, it also points elsewhere. Earlier in his journey across the estate, the speaker describes the mowers in the meadows as "careless victors" (over cut grass) whose merrymaking bears only a faint resemblance to the inquisitive interest he brings to the forest (425). Similarly, the "easy" transitions I've discussed are not identical. The "easy philosopher" stanza describes an awareness of kinship with the wood, the speaker's disposition sending his imagined self "floating on the air" and into an alternate vegetal body (566).

> Or turn me but, and you shall see
> I was but an inverted tree. (567–68)

"Languishing with ease," however, extends such identification into the complete emptying out of the interior self. One recalls the famous moment in *The Garden* when the speaker's botanical reverie "annihilat[es] all that's made / To a green thought in a green shade."[46] The self-satisfaction of one who shares the powers of the natural world but simply lacks the equipment to make them manifest ("wings," the ability to stand on his head) gives way to the pleasure of utter self-abandonment, the wind "winnow[ing]" the "chaff" from the speaker's mind. Marvell hasn't piled up examples of a

specific emotion, narrowly defined; he's narrated the unpredictable affordances of an appealing mood, one of which is exactly his readiness to depart
from it. "Or turn me but," Marvell proposes, and these words signal an attitude of sheer willingness—to follow the path, variously tinged with feeling,
from looking aimlessly around to a jolt of surprise.

The sequence confirms that Marvell's observational mood is no "art-
concealing art" by discovering something like it in the natural world. We
saw in chapter 1 how Bacon, in exactly the moment in *De sapientia veterum*
when he describes scientific discovery as an accident, gets confused about
whether Pan is an image of Nature or of the natural philosopher. We also
saw that he understands philosophy itself as Nature's single desire (an underexplained exception to her general desirelessness, and thus an invitation to
contemplate the nonparadoxical complexity of passionless passion). Marvell
resembles Bacon the mythographer insofar as he describes the observational
mood as capitulation to natural process, suggesting a formula like the following: "I am most like Nature when I understand her." Note how the natural
historian's duties come more easily to birds than they do to the speaker:

> And where I language want, my signs
> The bird upon the bough divines;
> And more attentive there doth sit
> Than if she were with lime-twigs knit. (571–74)

When, five stanzas earlier, the speaker's inquisitive gaze caught sight of the
"throstle's shining eye," we could have guessed that the bird was the more
perceptive party. Here, the total stillness of sustained attention is no feat of
self-discipline but simply what birds do. Similarly, however far the speaker's
awareness falls short of the bird's, we might suppose he owes his powers
of attention to Nature's gratuitous provision. When the "cooling" "wind"
comes "through the boughs" and the speaker gives thanks for "cool zeph-
yrs," we again discover the unthinking ease with which Nature exhibits the
speaker's signature qualities (595, 598). If he experiences the novelty of an
unexpected perception, it's not so much an achievement as a capitulation to
the nature of things. We might think back to the flower's "drowsy eyelids" in
the garden sequence, which convey not sleepiness but the state of newfound
alertness for which we owe daily gratitude to the rising sun.

Scholars in the humanities and social sciences are used to feeling suspicious about calling things "natural," since the claim so often smuggles in
assertions of universality and permanence. As we observed in Montaigne, it
might instead achieve the much narrower effect of making receptivity look
easy. Indeed, the speaker's incomplete knowledge of the birds' language

reminds us that he is no authority on natural phenomena. "Natural" means a lot of things we don't yet understand, the poem seems to say, but one thing it does mean is "effortless." Like blooming flowers that prove but don't earn the absence of war's ravages, the speaker wanders as if by accident into an affective opportunity for vision's liberation. Seeing the world afresh is as simple as surrender.

Eschatology as Optics

In 1649, two years before Marvell writes *Upon Appleton House*, Abiezer Coppe issues a fierce warning about the coming apocalypse. "High mountains!" he exclaims. "Lofty cedars! It's high time for you to enter into the rocks and to hide you in the dust for fear of the Lord and for the glory of his majesty. For the lofty looks of man shall be humbled and the haughtiness of men shall be bowed down, and the Lord ALONE shall be exalted in that day."[47] Though Coppe's place in history is among the Ranters, the vehemence of his rhetoric is not atypical of mid-century millenarianism.[48] The prophetic "shall" staves off catastrophe, but only by the slimmest of margins. With a mere instant's reprieve, we tremble in the face of disaster. "It's high time," Coppe proclaims, commending what he predicts, for the Lord to exert His destructive power over all creation. The "humbl[ing]" of "lofty looks" seems not to convey sufficient violence; even the "high mountains" and "lofty cedars" must be laid low, as if to redouble the sense of human smallness they ordinarily impart by standing tall.

For Coppe's contemporaries, more is in store than brimstone; the end of days is good news for some. As Margarita Stocker has pointed out, apocalyptic rhetoric permeates English society with such completeness by the time of the Civil War that both royalists and parliamentarians can speak of their causes in terms of sacred history.[49] Tonal diversity does not follow from ideological breadth, however—for how but with declamatory intensity is one to speak of imminent destruction? Marvell is the rare case of one for whom this question need not have been rhetorical. Coppe's rousing admonitions find their antitype in the near-impossible coolness of Marvell's intimations of apocalypse. Along with the history of the New Science, the recovery of the millenarian imagination has proved indispensable to readers of his poems. My view is that these contexts compose a single one.[50] Just as Marvell's epistemology of carelessness forms the matrix of his pacifism, it explains the mysterious unconcern with which he evokes the world's end.[51]

It's well known that millenarianism suffuses Baconianism's first half-century, but my interest here is the less familiar (if plainly etymological) sense in which apocalypse is above all else a matter of vision.[52] *Apokalupsis*, the first word and de facto title of the book we call "Revelation," describes the lifting of a veil.[53] Bacon himself expresses a wish for a specifically visual apocalypse: "May [God] graciously grant to us to write an apocalypse or true vision of the footsteps of the Creator imprinted on his creatures."[54] Rather than approach Marvell's poem as if it conveyed systematic eschatology, I understand his language of apocalypse as the figurative means by which he imagines discovery.[55] More specifically, Saint John's revelation displays the powers of prosthetic vision. Indeed, we come close to grasping Marvell's poetics if we understand eschatology as optics. As we've seen, effortlessness is the affective medium by which the world comes into view. Thus we understand what's at stake when we apprehend that both the distortions of glass and the disclosures of Providence release the speaker from the labors of disclosure. One reason Marvell is interested in revelation's supernatural agency is that it isn't his own. The lens materializes visual power, the sheer hardness of the cause (convex glass) ensuring the permanence of the effect (magnification). All the speaker has to do is look.

Marvell synthesizes distortion and revelation. The astonishing disclosure of hidden realities is not the confirmation of truth-claims about the world; nor does it restore order to a "world turned upside down."[56] Instead, it throws the poet into a state of confusion—but one he seems to value for both its epistemological promise and sensory richness. Here, I examine the poem's optical images and then go on to explore apocalyptic moments that extend their powers. Yet I also suggest that Marvell runs these (fused) images together with still other ones; he teaches us to notice continuity across various metaphorical vocabularies. We notice this pattern as soon as we focus our attention on optical technology. Picciotto follows Pierre Legouis in proclaiming Marvell, two hundred years ahead of Walt Whitman, "the poet of grass," but she complicates matters by noticing that Marvell's interest in grass is only a single stage in a process of transformation.[57] "The effort to reveal as much of the world in the grass as possible," she writes, "finds satisfaction in the effort to transform grass into a perceptual medium, which can be done with the alteration of a single letter; grass becomes glass."[58] I propose a third term: Marvell figures grass as flowing water, and when the world turns to glass at the poem's end, metamorphosis begins with the "jellying stream" (675). The river runs through the estate as a transforming and transformative medium of observation—a lens *in potentia*.

Let's look first, then, at the water's surface, which comes to resemble a mirror. Even when the technology in question is as old-fashioned as the looking glass, Marvell curates an unlikely combination of confusion and

Archimedean vision. In stanza 80, the river renders itself reflective through self-delectation.

> See in what wanton harmless folds
> It ev'rywhere the meadow holds;
> And its yet muddy back doth lick,
> Till as a crystal mirror slick;
> Where all things gaze themselves, and doubt
> If they be in it or without.
> And for his shade which therein shines,
> Narcissus-like, the sun too pines. (633–40)

We might here recall Bacon's suspicion that "human understanding" is "like a false mirror" that "distorts and discolours the nature of things by mingling its own nature with it," though he is equally capable of wide-eyed admiration for the mind's mirror-like impressionability: "God hath framed the mind of man as a mirror or glass capable of the image of the universal world, and joyful to receive the impression thereof, as the eye joyeth to receive light."[59] In this more sanguine mood, Bacon sounds almost like Leonardo da Vinci in his notebooks dispensing advice that chimes with Marvell's invitation to receptivity: "The mind of the painter should be like a mirror which always takes the colour of the thing that it reflects, and which is filled by as many images as there are things placed before it."[60] In Marvell's image, receptivity depends on the observational mood, a state of feeling that both encourages and shrugs off "distort[ion]." The "wanton harmless" river "holds" the "meadow" it wanders much in the way of a looking glass, but only because it indulges in pleasurable self-"lick[ing]." The self-satisfaction of the self-tasting tongue generates awareness rather than solipsism, just as the sun's "Narcissus-like" gaze is unexpectedly compatible with a broad view of creation. No less multitudinous an audience than "all things," including celestial bodies, confronts a comprehensive image of itself. Better still, they encounter themselves and each other—the plural subject and object of the phrase suggesting inclusive vision. "All things gaze themselves," Marvell writes, rather than picturing each separate thing engaged in self-inspection. The stanza imagines cosmic communion as panoramically mutual regard, yet it blurs the distinction between *looking at* and *looking through*, since the creatures (I use this term, in the early modern sense, to refer to all created things) "doubt / If they be in it or without." One can imagine "Narcissus-like" confusion about whether the objects of the gaze are terrestrial or aquatic.

The confusions of the object-world serve not as signs of failure but as proof that it's worthy of further contemplation. When Marvell turns to the microscope and telescope in stanza 58, he offers a dense set of references to other moments in the poem, encouraging the reader to circulate through the speaker's world. We find ourselves directed recursively to the poem's teeming contents rather than toward a delimited thing we take away with us as knowledge. Every couplet seems to point us to another of the poem's passages, so that these eight lines function much like the technologies they describe, bringing the distant and intangible within reach. Note that the subject of this first sentence is a herd of grazing cattle:

> They seem within the polished grass
> A landskip drawn in looking-glass.
> And shrunk in the huge pasture show
> As spots, so shaped, on faces do.
> Such fleas, ere they approach the eye,
> In multiplying glasses lie.
> They feed so wide, so slowly move
> As constellations do above. (457–64)

As with the reflective river, Marvell begins with the simple image of the mirror but soon describes an enhancement of multidirectional vision. The transformation can only disorient, the cows taking the redundant but defamiliarizing form of cow-shaped blemishes ("spots, so shaped") before turning to magnified "fleas" and then to stars ("constellations"). The rapid transition from "fleas" to "constellations" scrambles microscopic and telescopic scales; we examine the infinite space of the sky just as soon as we descend beneath the threshold of perception.[61] The "polished grass" in the first line is a concise formula for the "vitrifi[cation]" we witness at the poem's climax, while the "constellations" in the last line point back to the "vigilant patrol" of stars in the earlier garden sequence (688, 313). The stanza begins by pointing forward and ends by pointing backward.

At the risk of silliness (does this poem not show an appetite for it?), we might understand these ruminants as figures for the speaker's habitual activity, and not only because they ruminate. As Mary Carruthers has shown, tracing the image of digestion and regurgitation from medieval mnemonics to Milton, *ruminatio* could serve as a figure for both reading and writing.[62] It's not overly imaginative to observe the implied pun at work here as well, especially since the speaker's own "slow" "grazing" earns him bovine bona fides.[63] The serene receptivity of the poet's eye is not unlike the tranquility of the cow, gently working the earth with its mouth. We already found evidence for such "joco-serious" self-presentation twelve stanzas earlier, where "graze" suggested casting one's gaze into the distance by shining a beam of light on a foreign object.[64] If now we glance back to the poem's opening sequence, we notice that the speaker attributes to himself exactly the sort of "slow" "move[ment]" we here discover in cattle.

> While with slow eyes we these survey,
> And on each pleasant footstep stay,
> We opportunely may relate
> That progress of this house's fate. (81–84)

It's exactly that "fate" the two middle couplets of the "magnifying glasses" stanza recall, confirming the point that it redirects our attention to other passages in the poem. In this case, we're sent back with "slow eyes" to the story of Isabel Thwaites's imprisonment in the nunnery that used to occupy one of the buildings of the estate—a prehistory of the Fairfax line, their place of residence, and the Reformation of the church. The integration of that past in the poem's hyperbolically inclusive vision suggests a form of "progress" much closer to Montaigne's sense of "progrez" (simply a temporal trajectory or mutation) than it is to the teleology with which that term is habitually associated, especially in the context of Baconianism.[65] Marvell's "slow" traversal of space and time suggests the value of a return even to those scenes the Reformation wishes to have transcended. But this is not the nostalgia we know from influential strains of New Historicism; it's awareness rather than wistfulness.[66] For a perspective the virtue of which is breadth of vision, the inclusion of even the obsolete language and practice of unreformed religion is a measure of its good faith. The cows' resemblance to "spots . . . on faces" recalls the nuns' attention to Isabel's visage.

> But much it to our work would add
> If here your hand, your face we had. (129–30)

The nuns treat the face as a sign of purity, displaying simultaneously the body's physiological and moral constitution.

> And holy-water of our tears,
> Most strangely our complexion clears. (111–12)

In the seductive words of the Prioress, "complexion" binds together the appearance of the face and the discourse on the balance of humors, both of which Roman devotion "clears." The meadow seems here to require such clarification—or rather its reformed counterpart, magnification—but first the cows must take the shape of "fleas." We think back to the diminution of the nuns' ritual instruments when William Fairfax rescues Isabel by "waving [them] . . . aside like flies" (257). It's as if the poem is taking back its own trivializing vocabulary. One lesson of the lens is that mere smallness fails as evidence for insignificance. The opposite is true, since the inaccessibility of the world of insects makes it exactly the sort of tantalizing object for which we need the microscope.

With these images of glass technology in mind, we can observe eschatology's perfectly continuous participation in the process of making things visible. The abortive apocalypse of the flood, where God seems to renege on His promise to Noah, presents another occasion for Marvell's readers to inhabit an observational mood. Paralepsis or praeteritio, the rhetorical figure whereby one introduces an idea in the very act of omitting it, is the formal means by which we apprehend the speaker's insouciance. He admits hallucinatory images into the stanza by describing them as less than worthy of admission. Marvell's style of carefree enumeration, listing rather than describing, contributes to the effect.

> Let others tell the paradox,
> How eels now bellow in the ox;
> How horses at their tails do kick,
> Turned as they hang to leeches quick;
> How boats can over bridges sail;
> And fishes do the stables scale.
> How salmons trespassing are found;
> And pikes are taken in the pound. (473–80)

The speaker claims he wants to "let others tell" about such fantastical transformations, but "tell[s]" about them himself in the act of making this claim. It's only by way of casual disregard for such wonders that they make their way into the poem. Aquatic creatures take up residence on land (eels in the belly of an ox, all manner of fish in the "stables" and the "pound"), while "horses" experience a similar inversion, finding their "tails" transformed to "leeches."[67] In a poem in which perspective is frequently in bewildering flux ("grasshoppers" appear as "giants" atop the "spires" of the grass), the "sail[ing]" of "boats . . . over bridges" epitomizes the pattern with elemental force (372, 376). As Walton says: "The waters are natures store-house, in which she locks up her wonders."[68]

Rosalie Colie notes that the poem seems to be building toward climax before undermining that expectation with an inconsequential flood, but she assumes disorientation generates discomfort.[69] It's certainly true that the flood parodies violence by claiming only a figurative victim: "the river in itself is drowned" (471). Yet the event presents a good opportunity *not* to experience the unexpected as "disturbing."[70] Several stanzas earlier, the speaker figures the flowing grass as flowing water, making the swamping of the meadow seem less transgression than repetition:

> To see men through this meadow dive,
> We wonder how they rise alive.
> As, under water, none does know
> Whether he fall through it or go. (377–80)

Marvell prepares us for a noncatastrophic flood by "preflooding" the meadows.[71] The apparent end of sacred history seems no different from the worldly present—even before we realize an ark would serve us little. The flood's effects are beneficial—leaving the meadow "fresher dyed," as the speaker later puts it (626). Furthermore, the speaker disorients us with such persistence that we lose the coordinates of anticipation that would justify a vocabulary of climax and anticlimax. These lines grant their constituents freedom from gravity, anticipating the hallucinatory perceptual advantages of the flooded landscape, where things are swept from their hiding places and float into view: "eels," "fishes," "salmons," "pikes." These are not the "fat aged carps that run into thy net" at Ben Jonson's Penshurst; the only feast the speaker here anticipates, and it's a Rabelaisian one, is visual.[72]

The closest the poem comes to climactic fulfillment is the eventual appearance of Fairfax's daughter, Maria, who, invoking the Book of Revelation, turns everything to glass.[73] A composite of scientific virtues, she shatters the idols of the mind by placing them under the lens. Her "judicious eyes" cause "Nature" to "recollect" "itself," calling the natural world to attention by turning it upside down (653, 657–58). Things blur into view, and the first of these is the halcyon in flight.

> The viscous air, wheres'e'er she fly,
> Follows and sucks her azure dye;
> The jellying stream compacts below,
> If it might fix her shadow so;
> The stupid fishes hang, as plain
> As flies in crystal overta'en;
> And men with silent scene assist,
> Charmed with the sapphire-wingèd mist.
>
> Maria such, and so doth hush
> The world, and through the ev'ning rush.
> No new-born comet such a train
> Draws through the sky, nor star new-slain.
> For straight those giddy rockets fail,

Which from the putrid earth exhale,
But by her flames, in heaven tried,
Nature is wholly vitrified. (673–88)

Maria appears in the final image as an agent of divine judgment who turns the world to glass, recalling the words of Saint John: "And I saw as it were a sea of glass mingled with fire: and them that had gotten the victory over the beast, and over his image, and over his mark, and over the number of his name, stand on the sea of glass, having the harps of the Gods" (KJV, Revelation 15:2).[74] Maria's "victory" over satanic "image[s]" depends on the power of emotional peace to generate awareness. The speaker contrasts her ability to "hush / The world" with the spectacular violence of "new-born comet[s]" and dying "star[s]," and he draws a contrast between the "train" of soothing pacification she carries with her and the pyrotechnics of meteors.

The effect of Maria's presence is, above all else, visual. The beginning of the second quoted stanza compares her passage to that of the halcyon ("such, and so"), identifying the blurred world with the vitrified one. The "stupid fishes" for which "flies in crystal" are figures have been rendered "plain," but vitrification muddles as it hardens; the "air" is "viscous" and the "stream" is "jellying" (another convergence of the "gelid" and the "jellied"). The blue of the bird's feathers bleeds into the atmosphere, eroding the line between fore- and background until the air itself displays properties that should belong to the bird: a "sapphire-wingèd mist." Maria effects a torsion of perspective that blurs the edges of things while making them available for inspection. Thanks to her arrival, the fish that might have been hidden by the rushing of Appleton's river can now be closely observed, but they are metaphorically transformed to "flies in crystal" as the surrounding world blurs.

Though climactic in its synthetic relation to the foregoing stanzas, Maria's blurred apocalypse is no more final than the flood. The poem ends at sunset— like an ordinary day. Marvell's reflections on Last Things aren't about last things, just as his epistemological fantasies know nothing of Bacon's dream of "arriving finally at the most general axioms" or even drawing the interim conclusions of "axioms from the senses."[75] "It became impossible," Charles Webster explains, describing mid-century Baconianism as a species of apocalyptic fervor, "to approach the future with equanimity, once it was clear that the world was approaching a cosmic watershed."[76] Yet Marvell makes the unprecedented disclosures of the end times dependable features of ordinary time. Apocalypse is no longer an object of intense desire and fear; it's just a name for the gradual discovery of God's creation. Marvell finds satisfaction in dismantling the given world without breaking a sweat. The effortless labor of distortion, which here goes by the name of Revelation, achieves the effect.

The Philosopher as Polyphemus

On June 28, 1680, Hooke makes the following report to his diary: "Spent most of my time in considering all matters."[77] It's tempting to adduce this observation as loose characterological evidence for the virtuoso's intellectual range; today, after all, he is sometimes known as the "Leonardo of London."[78] Yet rather than take the opportunity to celebrate the talents of the polymath, I suggest we follow his lead into the forking pathways of his prose—and even more deeply into Power's. Have we not seen, in Marvell's careless persona, a style of thought that makes a virtue of sheer expansiveness? Have we not discovered, in Marvell's lens, a technology that renders such perceptual breadth an actual fact rather than an outsized ambition? Range of interest, I suggest, is better understood as a feature of the observational mood than as a special privilege of the gifted. It's less characteristic of Hooke the man than of the larger discourse in which he participates. I refer not only to the wide field of Baconian-inflected writing but also to the genre of microscopic observation in particular. With Marvell's lenses in mind, my purpose here is to describe the wide curve (rather than the cutting edge) of the optical imagination. Because Hooke figures prominently in the scholarship, a glance at *Micrographia* (before turning to Power) also helps us identify those interpretive questions that have shaped interpretations of microscopy coming from both historians and literary critics.

Hooke's complaint that "the Science of Nature" is usually "only a work of the Brain and the Fancy" would seem to make him an unlikely candidate for mental vagrancy.[79] Yet the appeal of effortlessness in *Micrographia* encourages flights of imagination. At the outset of his first "Observation," Hooke expresses a methodological desire. He wants "to follow Nature in the more *plain* and easie ways" before taking up more complicated matters—but soon these "natural way[s]" prove no less winding than Nun Appleton's.[80] Just as in geometry we must first understand the concept of "a Mathematical *point*" before attending to other figures, he says, so in natural history our literal point of departure is a *"Physical"* one.[81] (His discussion resonates with Margaret Cavendish's reflections on the mathematical point in her *Philosophical Letters* [1664].[82]) Once he places the "point of a sharp small Needle" under the microscope, however, he discovers an "irregular and uneven" surface—and in only an instant the train of his thought carries him off into imaginative excess.[83] He remarks offhandedly that this newly discovered "surface" is "big enough to have afforded an hundred armed Mites room enough to be rang'd by each other without endangering the breaking one anothers necks, by being thrust off on either side."[84] As Hooke launches a seemingly accidental critique of mathematical abstraction (to the delight, we might suppose, of Boyle, the self-professed bad mathematician), we should recall the "negligence" with which Marvell "overthrows" his "world." Yet the more surprising point of convergence here is the return of exaggerated "Fancy" in the midst of sober contemplation—the sheer wildness of which flouts the ideal of meticulousness: "an hundred armed Mites"! Finding ourselves back in the unbalanced realm of Marvellian entomology ("The bee through these known allies hums, / Beating the *dian* with its drums"), we rediscover the intimacy of minimalist affect and maximalist poetics. Hooke seems here to embrace sheer make-believe, capitulating without embarrassment to what Milton calls Fancy's "wild work."[85]

As with the specimen on the slide, the lens magnifies the observer's literal and figurative prospect on the world; it opens a wide path that makes room for errant trains of thought as well as literally visible matter. Yet scholarly appraisals of microscopy affirm the contrary. The microscope, we are told, pierces the scrim of deceptive appearance. The effect is perhaps most striking as a political metaphor. In *The Last Instructions to a Painter* (1667), for instance, Marvell the Restoration satirist (as opposed to Marvell the Interregnum pastoralist) exemplifies exactly this familiar rhetoric of incisive penetration:

> With Hooke then, through the microscope, take aim:
> Where, like the new Comptroller, all men laugh
> To see a tall louse brandish a white staff.[86]

This same intensity of purpose has been attributed to the natural philosopher's quest for knowledge. The theme of aggressive intervention has been central to the interpretation of the microscope's cultural history. In her path-breaking study of literature and science in the period, Mary Baine Campbell notes affinities between books like *Micrographia* and contemporary genres of pornography; Hooke, she explains, eroticizes invasive vision, ushering us into Nature's private chambers, the better to catch her unawares.[87] The point is well taken. Sometimes, indeed, Hooke adopts a sharply aggressive observational style: "I took a large grey Drone-Fly," he writes, "that had a large head, but a small and slender body in proportion to it, and cutting off its head, I fix'd it with the forepart or face upwards upon my Object Plate."[88] Campbell has singled out an important aspect of the genre, but it need not orient our interpretation. Hooke's celebration of the roominess of the topographical point of the needle, which he populates with anthropomorphized insects, has already revealed to us that whatever air of mastery he adopts is only one of several attitudes that compose his scientific persona. Taking Marvell's lead (I'm thinking here of *Upon Appleton House* rather than *The Last Instructions to a Painter*), we notice that Hooke and Power also invite us to join them on the winding path that leads into the landscape of the newly visible world—and then, at the expense of the distinction between description and speculation, off into the realm of thought. Although Hooke and Power are never as self-conscious as Boyle about the permission they grant themselves to let their minds wander, the object of microscopic observation ends up playing the role of an "occasion" in his idiosyncratic sense. The louse or spider is "not so much the Theme of the Meditation, as the Rise": the contingent point of departure for a reflection that travels some unpredictable distance in search of who knows what.

As in Campbell, scholarly attention to the penetrating gaze implies an argument about emotion; the eroticized object is one from which the eye cannot easily detach itself. Daston and Park make exactly this point: "Curiosity had to keep the gaze glued to the object of observation, when boredom or distraction might have lured it elsewhere."[89] Here, they describe the circumstance in which microscopic vision seems most likely to succeed: the scientist has the best chance of discerning the hidden properties of the object under inspection if he keeps his eyes trained upon it. Yet microscopists are less interested in success than one might suppose; it's not clear they even know what success would mean.[90] The effect of books like *Micrographia* is not only to ignite epistemological desire (Daston and Park speak of a "flash of interest") but also, to the contrary, to diffuse it across a wide field.[91] Most everything is marvelous under the lens, and overabundance softens the power of individual objects to keep readers focused.[92] Like people wandering a museum for more than a few minutes, readers of microscopic observations do not find themselves in awe as often as they lose themselves in drifting interest. "In order to rivet the attention upon a common fly," write Daston and Park, "Hooke had to transform it into a marvel by means of the microscope. Unmagnified, the fly barely registered in the observer's consciousness. Magnified and marvelous, it became the quarry of avid curiosity, pursued with a '*material* and *sensible Pleasure*.'"[93] Their description applies very well to the magnificent moment of first encounter, and yet one might spend hours with Hooke's descriptions. Wonder can't last forever—not for the microscopist-author and not for his reader.

In addition to focused intensity (conveyed in terms of time and depth by motifs of fixation and penetration, respectively) and the aggressive stance it implies, the literature of microscopy embraces the opposite. Throughout the genre, we find evidence of the observational mood in the passing glance and the wayward thought, the casual turn of phrase and the delectation of gentle pleasure—all of which direct our attention back to the claim with which this chapter began. The microscope itself is a perfect emblem of effortlessness. Unlike other new technologies, or at least more dependably, the lens activates a fantasy of instant gratification. Plenty of work goes into the description and illustration of miniscule objects, to say nothing of the procurement and preparation of specimens. I speak only of the ease with which anyone (a reader of the genre, for instance) might succeed at inhabiting the microscopist's point of view. As we observed at Marvell's Nun Appleton, transforming the world is no more difficult than lifting a lens to the eye. No great feat of imagination is necessary to imagine oneself with augmented vision: the lens is a portal to a new scale of perception. Power's book of observations asks nothing of us beyond the simplicity of looking.[94] Look: a water spider. Look: a drop of quicksilver. As a literary artifact, the microscope bids us welcome by creating a foreshortened geography: nothing separates here from there. The individual observation usually begins with a description of the world on the other side of the lens—after which Power gets his bearings and responds (quite variously) to whatever he sees. Unlike, say, Francis Godwin's journey to the moon (which relies on a kind of makeshift chariot drawn by birds, linking scientific perception and the problem of travel), Power's observations grant us the privilege of having already arrived at our destination; the voyage is over as soon as it begins.[95] (Interestingly, Hooke frames his *Micrographia* with the image of the telescope—suggesting this very comparison between enhanced vision and the crossing of distances.[96]) With nothing left to accomplish except the minimal task of describing newly perceptible objects, Power occupies himself with whatever he wants. Sometimes he works toward wondrous effects, and sometimes we begin with aesthetic fireworks that taper off into descriptive flatness. The microscopist holds one thing still in order to keep everything else in motion: obeying the imperative to describe, everything else he does is freely chosen. To be sure, some observations make straightforward argumentative use of the magnified image—to disprove, for example, Thomas Browne's claim that snails do not have eyes at the ends of their horns (Browne was Power's mentor).[97] But that's only one of the many feats they handily accomplish.

Power both relishes the ease with which the eye surveys literal, hypothetical, and purely imaginary worlds and offers an implicit defense of the confusions that follow from such an approach.[98] Observing, in anticipation of Hooke, the "flaws and deficiencies" of a straight line drawn on paper, his mind soon turns, again in anticipation of Hooke, to similar defects in literally microscopic writing: a "Rarity" produced by a "famous Scrivener" (53).[99] He recalls having seen the Lord's Prayer and the Ten Commandments squeezed into "the compass of a single penny," and yet still "distinctly writ" and legible (53). Under the microscope, "the Letters appeared (as we have observed of the line) crooked and unhandsome; so Inartificial is Art when she is pinched and streitned in her Workmanship" (53). Perhaps the distortion of literal miniaturization stands as a metaphor for the twisting and turning of Power's descriptive prose as he seeks to describe the world beneath the lens. In that case, the image would at first imply a struggle. Under the constraint of an unfamiliar scale of perception, both the scrivener and the microscopist produce misshapen forms. Yet the scrivener doesn't worry about the imperfection of his art; he means only to achieve unmagnified legibility. The bar is low—and the exercise is forgiving. Irregularity implies neither distress nor awareness of underperformance; it suggests only the "crooked[ness]" and "unhandsome[ness]" inherent to artifice.[100]

The microscopic world deprives the philosopher of familiar perceptual coordinates, presenting him with novel appearances for which he needs a new descriptive language. In Power's analogies, we find evidence of the casual attitude with which he faces the challenge. He reaches for comparisons (usually by way of simile) that allow us to visualize hitherto unseen forms, and yet he takes the translator's task as an occasion for playful reflection. Comparisons do much more than explain. Like Boyle, who describes an experience of receptivity in which one "accommodates ones discourse" to natural (and other) phenomena, constantly coining new expressions, the microscopist makes an experiment of language itself. Furthermore, it's the only resource Power has at his disposal, which is not true for Hooke the visual artist. Yet he sees little conflict between the challenge of realism (reproducing the experience of microscopy for readers) and the pleasures of purely associative (Boylean) reflection. When he describes a fly's "Bristles" as "Porcupine quills" (Hooke uses the same simile, though to describe the flea), perhaps we conclude that the image is chosen for its explanatory power (5).[101] When he explains to us that a butterfly's eye looks like "checker'd Marble," while the "Probe" in its mouth resembles "the twining tendrils of the Vine," I am less inclined to think so (8). A locust's mouth, we find, is like "a pair of closed Compasses in her snout"—an image that calls to mind the figures of mathematical abstraction for which the microscopist finds resources for critique in the rugged imperfection of the (magnified) material world (27). Fancy itself offers resources for the correction of our usual sense of things.

Power's digressiveness isn't only a response to the necessity of a pliant descriptive practice. I have suggested that the genre's sole responsibility is the description of magnified surfaces, thus setting the philosopher at liberty to say and do whatever else he likes. Microscopic vision is proof of the elasticity of the visual field: a theme Power wants us to remember. Recall Boyle's description of the mind: "having travers'd all the corporeal Heavens, and scorn'd to suffer her self to be confin'd within the very Limits of the World, she roves about in ultra-mundane spaces, and considers how farr they reach." Power shows a similar interest in giving the eye endless places to go. Once again, Boyle is a good guide to the fantasy: "If the Intellect should for ever make a farther and farther Progress in the knowledge of the Wonders of the Divine Nature, Attributes and Dispensations; yet it may still make discoveries of fresh things worthy to be admired."[102] In much the same way, Power assumes an infinite universe that forestalls the consummation of intellectual hunger; surrounded in every direction by endless expanses, all we can hope to do is wander the region of the world in which we presently find ourselves. Yet Power ensures that such delimited knowledge feels endlessly expandable. At first, the point is literal: a statement about reality. Like Margaret Cavendish, who affirms the "probable" "opinion" that "there may be animal creatures of such rare bodies as are not subject to our exterior senses," Power envisions a universe that outdistances our powers of perception.[103] Catherine Wilson observes that the appeal of atomism for early modern microscopists lies in the "hope that the limits of visibility could be pushed back indefinitely far."[104] For exactly this reason, Power pictures an infinity of ascending and descending scales of perception: "It hath often seem'd to me beyond an ordinary probability and somthing more than fancy (how paradoxical soever the conjecture may seem) to think, that the least Bodies we are able to see with our naked eyes, are but middle proportional (as it were) 'twixt the greatest and smallest Bodies in nature, which two Extremes lye equally beyond the reach of humane Sensation" (sig. b1r–v). Just as Boyle shrugs at the assertion that the atom is the smallest possible unit of matter, leaving open the possibility of bottomless descent into the infinitesimal, Power poses a rhetorical question that conveys his premise that the universe is inexhaustible: "Who can set a *non-ultra?*" (sig. c3r). Power's question interestingly recalls both the optimism of Bacon's phrase, "Plus Ultra," which he proposes as a replacement for the ancient, "non ultra," and Montaigne's shrugging motto: "What do I know?"[105] Throughout the book, Power describes all manner of small things as "atoms" (the eyes of a grasshopper, for instance, and droplets of quicksilver), as if to remind us, technically speaking, that there's no such thing (26, 43). No particle, however small, is fundamental; the conquest of once-invisible spaces is less a triumph than a first step.[106] When we notice Power's refrain that whatever object he observes under the microscope is "a very pleasant Spectacle," we should attribute his pleasure, at least in part, to the endlessness of objects for contemplation (see, e.g., 14, 50). Unlike Blaise Pascal, who trembles with terror at the "double infinity" of a nonanthropocentric sense of scale—the possibility (to borrow a cinematic metaphor) of a limitless capacity to zoom both in and out—Power savors the promise of inexhaustibility implicit in his disorientation.[107]

By analogy with vision, he describes thought as a widening perceptual frontier.[108] He ignores the distinction between the limits of the thinkable and the limits of the visible, treating them as aspects of a single question. Indeed, he takes microscopic observation as an occasion for meditation on the eye—both its anatomy and its (sometimes metaphorical) powers. In the very selection of objects for examination, Power is guided by his predilection to think *about* vision—and to treat such thinking *as* vision. As we make our way through his book, we quickly notice how many pairs of eyes stare back at us from Power's cabinet of miniatures (and, as we shall see, "pair" is not always the right word).[109] Here is the first paragraph of the first observation, which concerns the lowly flea: "It seems as big as a little Prawn or Shrimp, with a small head, but in it two fair eyes globular and prominent of the circumference of a spangle; in the midst of which you might (through the diaphanous Cornea) see a round blackish spot, which is the pupil or apple of the eye, beset round with a greenish glistering circle, which is the Iris, (as vibrissant and glorious as a Cats eye) most admirable to behold" (1). The reader's eye, pressed against an imaginary microscope, discovers another eye. Like Marvell's "crystal mirror slick," which undergoes a seamless metamorphosis from reflective surface to magnifying lens, the flea's eyes present us with a brief experience of self-recognition (or misrecognition) before we disidentify. Power encourages us to tarry somewhere between admiration and disgust, praising the intricacy of the diminutive creature. "How critical is Nature in all her works!" he exclaims, "that to so small and contemptible an Animal hath given such an exquisite fabrick of the eye, even to the distinction of parts" (1). Here, we see what Daston and Park mean when they describe the power of the microscope to transform ordinary things into marvels. Yet Power soon discovers a greater source of wonder in mental vision: "We have heard it credibly reported [by Thomas Muffet] . . . that a Flea hath not onely drawn a gold Chain, but a golden Charriot also with all its harness and accouterments fixed to it, which did excellently set forth the Artifice of the Maker, and Strength of the Drawer" (3). Like Hooke's swift passage from the actual observation of a needle's sharp point, blunted by magnification, to an imaginary assembly of insects, Power's description of the flea generates further interest by leaving literal vision behind. Even if Muffet the naturalist is a "credibl[e]" source, the incident is far too strange for a good Baconian to accept on hearsay. (Perhaps because he finds it insufficiently guided by firsthand inspection, Hooke never cites Muffet's book on insects.) What Power's first observation most conspicuously accomplishes is a graceful transition from ocular to mental sight—and one that shows little concern for the dangers of credulity.

Eventually, Power comes close to an explicit reflection on mental (as opposed to ocular) vision. When, at the very end of the collection, he quotes from Bacon's *Sylva Sylvarum*, we notice that he has already cited this source without acknowledgment, describing the flea on the very first page as a "small and contemptible" creature (1). Approaching his conclusion, he writes: "The Eye of the Understanding, saith he [Bacon], is like the Eye of the Sense; for as you may see great Objects through small Cranies or Levels; so you may see great Axioms of Nature through small and contemptible Instances and Experiments" (82).[110] Power's collection is short on axioms, but it does participate in Bacon's fantasy of wondrous dilation from the minuscule to the enormous. However, it also runs the risk of contradicting Bacon's stated intention, implying a principle very much like the sin imputed to scholastic natural philosophy: an "anticipation of nature" that generalizes too quickly on the basis of individual cases. For Power, a small, observable thing is an opportunity for perceptual breadth. Moreover, the "pleasant Spectacle" of the individual object is an inducement to the pleasures of speculation. Power sees no reason to hold himself back when he wishes to make "a paradoxical and extravagant Quaere," even when the only kind of vision it concerns is strictly metaphorical: that, for instance, the animal spirits that mediate between mind and body (on Descartes's model) might very well serve as "the Soul's Vehicle and Habit" in the "other world" (71–72). Power seems to derive pleasure from the avowed fancifulness of his speculation that the lens grants us access to the supernatural. In his "Poem in Commendation of the Microscope," he writes:

> Nay then yow pretty sprit's & fairy Elues
> That houer in ye aire Looke to your selues.
> For with such prying Spectacles as these
> wee shall see yow in yr own essences.[111]

Here, Power indulges in the kind of claim Boyle would have set apart with parentheses.

Power's "small and contemptible Instances" are often similar to the very first one in the book: the eye of the flea. He goes on to discuss the eyes of bees, flies, butterflies, lice, spiders, and other creatures (eventually, he allows himself a digression on the human eye), dilating from detailed visual description to reflections on the pleasures of vision, which include the gratification of simply thinking about sight. Of special interest here is the anatomical variability of the eye: the visual gifts different creatures get to enjoy.

When Power corrects Browne's claim that snails do not have eyes at the ends of their horns, he observes: "Through a good *Microscope*, he may easily see his own errour, and Nature's most admirable variety in the plurality, paucity, and anomalous Situation of eyes, and the various fabric and motion of that excellent organ; as our Observations will more particularly inform him" (37). One might expect Power to confirm his argument with further discussion of the snail, but he has nothing else to say on this matter. Instead, he takes the opportunity of a single anatomical misjudgment to promise Browne (and the rest of us) further observations on vision in general. Nature, he marvels, fashions organs of sight with "admirable variety." In his appreciation for nature's multifariousness, he resembles Boyle the natural theologian. In these lines, however, Power's most powerful fantasy, like Marvell's, is an experience of hyperreceptivity—for which one can only envy other creatures.

The eyes staring back at us from Power's observations show us how poorly Nature has provided for us. Offering us an alternative to accounts like Harrison's, which emphasize the weakness of the senses (without technological assistance), Power makes this point in order to dispel it; the considerable power we actually do have is undamaged by what we lack.[112] Moreover, examples of visual power from other creatures are promises of future achievement for humankind. When, at the very end of the collection, Power finally presents us with a whole section explicitly devoted to "Anatomical Considerations about the Eye," which might be taken as the culmination of his foregoing digressions on vision, he encourages us to relish the wide and easy prospect we currently enjoy—no matter, with respect to other creatures, our anatomical poverty. It's worth quoting at length from Power's conclusion, which reveals the wavering line of his thought:

> The two Luminaries of our *Microcosm*, which see all other things, cannot see themselves, nor discover the excellencies of their own Fabrick: Nature, that excellent Mistress of Opticks, seems to have run through all the Conick Sections, in shaping and figuring its Parts; and Dioptrical Artists have almost ground both their Brain and Tools in pieces, to find out the Arches and Convexities of its prime parts, and are yet at a loss, to find their true Figurations, whereby to advance the Fabrick of their *Telescopes* and *Microscopes*: which practical part of Opticks is but yet in the rise; but if it run on as successfully as it has begun, our Posterity may come by Glasses to out-see the Sun, and Discover Bodies in the remote Universe, that lie in Vortexes, beyond the reach of the great Luminary. At present let us be content with what our *Microscope* demonstrates; and the former Observations, I am sure, will give all

ingenious persons great occasion, both to admire Nature's Anomaly in the Fabrick, as well as in the number of Eyes, which she has given to several Animals. (78)

Power points out how little we know, but he also explains how little we should be worried by our incomprehension. What's interesting is how both points receive emphasis throughout the passage. Our eyes, he says, "which see all other things, cannot see themselves"; they encompass the entirety of our little world (our *"Microcosm"*), but there's a blindness in the very seat of vision. The eye's anatomy, he goes on to explain, is too intricate for us to make out, and yet our capacity to observe the natural world advances at an ever-accelerating speed. "Dioptrical Artists have almost ground their Brain and Tools in pieces" in order to discover the inner workings of the eye, but we can look forward to the invention of "Glasses" with which we will "out-see the Sun, and Discover Bodies in the remote Universe, that lie in Vortexes, beyond the reach of the great Luminary." How *much* we will eventually observe, Power thus seems to say, and how *little* we presently do! Yet Power's double perspective is no puzzle; it expresses the attitude with which he responds to the inexorable fact of limited vision. He savors the power he has and the fact of its continual expansion, which promises to penetrate even the mysteries of Cartesian vortices—a forward motion premised on present insufficiency. To "see all other things" in the *macro*cosm would extinguish the pleasure of advancement. We are lucky, instead, to inhabit an ever-widening field of perception. At the beginning of the passage, eyes are "Luminaries," predicting their eventual competition with (and triumph over) the "great Luminary" at the center of the universe. Future success is not an excuse for present failure. What we already have is no small thing—except in the literal sense, of course, that it consists of a great number of very small things. "At present," he writes, "let us be content with what our *Microscope* demonstrates." While the engineers who construct optical instruments "have almost ground their Brain and Tools in pieces," author and reader ("all ingenious persons") might simply take "content[ed]" pleasure in the foregoing observations—and everything else the microscope has to offer. Turning our attention to the realm of what we can actually observe, we discover how much, even here, remains to be seen. Most interesting of all, Power brings the technical limitations of the microscope into focus by drawing a contrast with the eye itself; it's the mechanics of ordinary vision that "Dioptrical Artists" have failed to replicate. I have argued that the microscope owes some of its appeal to the conceptual simplicity of its design; using it is, in the imagination if not in fact, no more difficult than peering through it. Yet here Power reminds us that the most ingenious device of all is already installed in our faces: we use it (indeed, two of them) all the time—and without even trying. The lens is only an extension of an extraordinary gift he encourages us to understand in exactly those terms: an unearned affordance of beneficent Nature—which is to say, a gift from God. Trust in technical advancement assumes the endless enhancement of natural power: gift upon gift.

At last, when Power directs our attention back to the eyes of animals, he adopts the same rhetoric of diminishment and amplification, where the former implies the promise of the latter's continuation. He dilates on the undeniable weakness of human vision, which pales in comparison to the powers of other creatures—but he also uses our very perception of ocular variety as an illustration of humankind's extraordinary powers of sight (including our present capacity for technological enhancement). We are reminded of what we don't have—but also of how much more we have, thanks to the microscope, than we otherwise might.

> I could never find any Animal that was monocular, nor any that had a multiplicity of Eyes, except Spiders, which indeed are so fair and palpable that they are clearly to be seen by any man that wants not his own. And though Argus has been held as prodigious a fiction as *Polypheme*, and a plurality of Eyes in any Creature, as great a piece of monstrosity, as onely a single one; yet our glasses have refuted this Errour (as *Observat*. viii. and ix. will tell you) so that the Works of Nature are various, and the several wayes, and manifold Organizations of the Body, inscrutable; to discover the more mysterious Works of that divine Architectress; but especially, when she draws her self into so narrow a Shop, and works in the retiring room of so minute an Animal. (81–82)

Power continues with the rhetoric of grateful adequacy we saw in the previous lines, recasting the unknown as the sphere of potential discovery. We should not rule out what "prodigious fictions[s]" represent to us, he explains, since we have good evidence that some of them are true. We might doubt the existence of Argus (a giant with one hundred eyes) and Polyphemus (a giant with only one eye), but we have already discovered versions of the former under the microscope. Observations VIII and IX are Power's descriptions of spiders, one of which he wonderfully describes as "*Argus* his head being fix'd to *Arachne*'s shoulders" (12). Where Bacon transforms ancient myths into allegories of fact, Power calls on fact to revise myth.[113] The image of Arachne fails to acknowledge her ocular majesty; with the help of the microscope, Power corrects the oversight. If such wonders actually exist, why may we not discover others?

In keeping with the pattern we've observed, Power conveys a desire to see as well as—or even better than—Argus does. He is playful with metaphors of vision (almost as playful as Hooke is with his "point"), explaining that the many eyes of spiders "are clear to be seen by any man that wants not his own."[114] It is hard not to identify Power himself with poor Polyphemus, whose vision is restricted to a single eye.[115] When he looks through the microscope, Power likewise has only one. Yet this Polyphemus wishes he were Argus. The collection's purpose is to inhabit the dream of better vision but not to understand it as a hopeless wish. Instead, it's a forever-outstanding promise, but one Nature is forever in the midst of fulfilling: exactly the Baconian "apocalypse" Marvell's poem narrates.

Power is not alone in wanting more, but he is especially good at feeling good about it. Milton's Samson expresses a similar wish, but with a mournful emphasis on vulnerability. (He has suffered catastrophic defeat: he has been shorn of his strength and blinded.)

> Why was the sight
> To such a tender ball as th' eye confin'd?
> So obvious and so easie to be quench't,
> And not as feeling through all parts diffus'd,
> That she might look at will through every pore?[116]

Here, the emphasis falls on the ease with which human sight is extinguished. Indeed, blind Samson intensifies our sense of exposure and danger by speaking like Polyphemus (after Odysseus blinds him); he laments that "sight" is "confin'd" to "a tender ball" when he might have mentioned the second one. If sight were instead widely distributed across the body—if its seat in our two eyes were not "so obvious and so easie" to discern and thus to damage— perhaps then he would not have been reduced to this unhappy state. Yet the passage also conveys a desire to see again, as well as a less emphatic version of that wish: Oh, to see more, and better! Cavendish is again an interesting point of comparison, fulfilling Samson's desire by insisting that "every several Pore of the flesh is a sensitive organ, as well as the Eye, or the Ear."[117] Power's vision, we might say, acknowledges a desire like Samson's but with the exultation of a Cavendish (notwithstanding her distrust of optical technology). He rests satisfied with his small power, trusting the world to remain ever in excess of what he can see. Since, in *Experimental Philosophy*, vision is sometimes metaphorical, encompassing even the creatures of thought, what this conviction amounts to is the anticipation of the persistence of discoveries by every sensory organ—and even by the imagination. Recall Marvell's use of vision as a metaphorical container for perception in its several dimensions: images are foreground and background for a multisensory—and multifaculty—experience. The eye is the sensorium, with its inward-facing and outward-facing parts.

Like Boyle, who takes cognitive pleasure in whatever happens to cross his path, Power enjoys a succession of images that might in principle be infinitely extended. I concluded the last chapter by reflecting on the question of temporal succession: whereas God sees everything in the universe in a single glance, Boyle covers ground by letting his attention wander. In reflecting on the eyes of other creatures, Power draws a similar conclusion about human vision. In the following passage, he sets up a comparison that flatters the human eye by suggesting that there might be something lacking in the multidirectional eyes of spiders: "Since their Eye is perfectly fixed, and can move no wayes, it was requisite to lattice that Window, and supply the defect of its motion, with the multiplicity of its Apertures, that so they might see at once what we can but do at several times, our Eyes having the liberty and advantage to move every way (like Balls in Sockets) which theirs have not" (79). Suddenly, human weakness is an asset; the extraordinary "circumspect[ion]" of spiders makes up for the fixity of their eyeballs. We don't get to see everything at once (not even everything in our immediate surroundings), but we get to keep on looking. Where the spider enjoys a single but composite view, human beings experience the surrounding world in continual succession. Like Marvell, then, who discovers in reckless wandering the method by which a poem can take everything as its subject, Power enjoys the "liberty" of "mov[ing] every way." The "lattice" of the "Window" of human perception is the division of time into moments.

Conclusion: A Terrestrial Star

I bring this chapter to a close by gazing with new eyes at another of Marvell's poems, and by holding it up for comparison alongside a resonant passage in Power. Together, they crystallize those visual features of the observational mood to which the foregoing readings have drawn our attention. Consider first the lines with which Marvell brings *The Mower to the Glow-worms* (1651–52) to its voluptuous but melancholic conclusion:

Ye glow-worms, whose officious flame
To wand'ring mowers shows the way,
That in the night have lost their aim,
And after foolish fires do stray;

Your courteous lights in vain you waste,
Since Juliana here is come,
For she my mind hath so displaced
That I shall never find my home.[118]

The forlorn speaker recognizes himself in Nature's night lights, and yet he barely glimpses the reflection. The poem's strange achievement is to make such glancing attention the content of the speaker's recognition. The mower and the glow-worms share an inclination to look past each other. The poem's final words bring their reciprocal unconcern into focus. When the speaker mentions ghostly lights (*ignes fatui*) that lead "wand'ring mowers" into darkness, he offers little assurance that purportedly "courteous" glow-worms are any more dependable. "Officious" is exactly what glow-worms are not. The poem might thus be read as a humorless joke, deliberately mistaking random flashes of misdirection for gestures of eager solicitude. The mower delivers a bewildering punchline by discounting the possibility of a homeward journey, rejecting an offer the glow-worms never made.[119] It makes little tonal difference whether the destination is literal (the hearth) or metaphorical (self-possession); up until the final stanza, the mower feigns interest in an outcome for which he has in fact lost hope. When he at last informs flagrantly inattentive and uncomprehending creatures that their labor has been sadly "waste[d]," he brings the outlandishness of the monologue's premise to a cool crescendo. It dwells on a relationship that doesn't really exist, or just barely does.

If we understand the absence of emotional focus as Marvell's subject, we discard the sensible thesis that he has written a love poem. The mower attends to the quiet scene in which he happens to find himself rather than the disquieting object of his heart's desire. No matter the temptation to read attention's swerve as pointed deflection, his ready avowal of "displace[ment]" dissuades us from the view that he carefully protects himself from pangs of desire. Though he only mentions the disturbance in passing, he does not muffle his distress. Eros simply lacks the magnetism we expect from it. Perhaps erotic anxiety has given way to exhaustion—or the syncopated background thrum of a definitively broken heart.[120] What we know for sure is that passion's fire has little fuel within the poem's modest compass. If we foist a Petrarchan fire (even the cold kind) on the mower's reflections, we fail to apprehend the distinctiveness of a scenario in which speaker and addressee are at oddly smooth cross-purposes. If we pay as little mind to longing as the serenely disoriented speaker, however, we find ourselves disarmingly engaged by a scene of fleeting appearances in which no powerful emotional content accounts for our interest. In this, we resemble the mower. Immersion has the quality of unremarkable calm.

Though distant from the scene of trial (the subject of the next chapter), Marvell's glimmering nightscape perfectly captures the observational mood. Though simultaneity doesn't prove anything, *The Mower to the Glow-worms* dates from the same period as *Upon Appleton House*—inviting us to wonder whether the poet might indeed have a natural-philosophical context in mind when he sends the mower out into the meadows to observe the glowing creatures. The glow-worm is a kind of marvel, to be sure, but it's an ordinary enough feature of rural life—and it doesn't, Marvell tells us, mean much: it "shin[es] unto no higher end / Than to presage the grass's fall."[121] In other words: it reminds us quite simply of the turning of the seasons. Marvell ushers us into a field of cool awareness unsharpened by fixation or anticipatory excitement. Whether or not his interest in the New Science has left a trace in this poem (I do not know whether it has), we can take the mower's faded dejection as emblematic of an assumed intimacy between absent passion and open-ended receptivity. Attentive to flashing glow-worms for which he feels no powerful interpretive desire, the mower displays the everyday evenness of a mind engrossed but adrift, responsive to phenomena but unburdened by the responsibility of figuring anything out. Marvell doesn't narrate the passage from indifference to insight, but he pictures languorous absorption in directionless contemplation. Things come to light as passion fades—and at a pace no more predictable than the vagaries of feeling.

Power's collection of microscopic observations shows a similar interest in what's right before his eyes. In Observation XX, he waxes poetic on the adequacy of what happens to be available to human sight, and takes the glow-worm as his example. After removing the creature's eyes and inspecting them through the lens (anticipating the passage from Hooke that typifies the strain of calculating violence on which some of our best interpreters of microscopy have focused their attention), he wanders away from his role as meticulous observer and enters a scene of contemplation: "This is that Night-Animal with its Lanthorn in its tail; that creeping-Star, which seems to outshine those of the Firmament, and to outvye them too in this property especially; that whereas the Coelestial Lights are quite obscured by the interposition of a small cloud, this Terrestrial-Star is more enliven'd and enkindled thereby, whose pleasant fulgour no darkness is able to eclipse" (24). Like the episode in Boyle's *New Experiments* in which he savors the "little flash of Lightning" produced by phosphorescent wood, Power steps back from his apparatus to observe the creature's "pleasant" flame. In this case, of course, the spectacle takes place in memory or imagination. What interests him about the glow-worm is simply that it happens to make itself available to sight. The distinction he draws between the "Terrestrial-Star" and "Coelestial Lights" says almost nothing about either of them. Indeed, it's a bad comparison—or perhaps a knowingly loose one worthy of Boyle's *Reflections*. When he tells us that glow-worms are "enkindled" by the "small cloud" that "obscure[s]" the stars, all he means is that he is lucky enough to have them right there in front of him. (The stars would be "enliven'd" by the clouds too, if they weren't on the wrong side of them!) Yet Power's gratitude for what's immediately available to perception is actually quite strange, since in reality the glow-worm is out of sight. Power is at the microscope—not out under the stars. He stretches his legs in the expanses of imaginative digression; glow-worms appear to him there as well.

CHAPTER 4

The Paradise Without

John Milton in the Garden

In Milton's *Paradise Lost*, one of the distinct pleasures of life in Eden is labor. This might be taken to imply that innocence is utterly unlike the experience of the fallen world—that the prelapsarian condition entails a unique anthropology, distinguishing Adam and Eve from their postlapsarian descendants. A number of decisive interpretive consequences follow from this premise. Only for superhumans is work indistinguishable from play. If Adam and Eve have little in common with mere mortals, then they are inducements to false consciousness. Endlessly striving to resemble their "first parents," readers approximate Weber's Protestant ethic, holding fast to the idea that work is inherently virtuous—that suffering can be counted on to reap rewards.[1] Like good Stoics, they come to understand rectitude as the capacity to ignore the feeling of being burdened or even beaten down. The desire for moral standing functions as an anesthetic. Many interpretations of *Paradise Lost*, some of which I discuss below, follow this line of reasoning—though without my suspicious assessment of the value of pain.

In this chapter, I defend a much simpler explanation for the apparent mismatch between the work cut out for Adam and Eve and their experience of it. Paradisal labor is just not very demanding; hence its conduciveness to that feeling of easygoingness I call the observational mood. Adam and Eve's charge, tending the garden, never asks more from them than they are willing to give. Labor is easy for the same reason it sometimes can be in the fallen world; in fact, the affective dimensions of work join pre- and postlapsarian anthropologies rather than holding them apart. My view assumes that labor is inherent to wellbeing—that, indeed, human thriving depends on the surmounting of challenges. Adam and Eve's work takes something, but far less than everything, out of them. The demands of gardening nicely meet their dependable desire to exercise their faculties, preserving the effortless feeling of their daily routine. Labor in Milton's Paradise is that perfectly calibrated assignment that gives the body enough to do without making it hurt.

Adam and Eve's experience of gratified appetite comes into focus at the beginning of book 5, in a scene that serves as a useful point of departure for this chapter's investigation of the psychology and phenomenology of trial. The scene suggests that the concept—exemplified not only by the labor of tending paradisal plant life but also by the minding of God's prohibition against the Forbidden Fruit—is of more use to the interpreter when taken as closer in meaning to "experiment" or "exercise" than "evaluation" or "assessment." The trial of paradisal labor is an open-ended experience of receptivity before it's anything else, including an appraisal of moral fiber. Early-rising Adam "whisper[s]" to the still-sleeping Eve, asking her to join him in taking up their day's work.[2] His inviting words both evoke the observational mood and raise the possibility of Baconian resonances.

> Awake
> My fairest, my espous'd, my latest found,
> Heav'n's last best gift, my ever new delight,
> Awake, the morning shines, and the fresh field
> Calls us; we lose the prime, to mark how spring
> Our tended Plants, how blows the Citron Grove,
> What drops the Myrrh, and what the balmy Reed,
> How Nature paints her colors, how the Bee
> Sits on the Bloom extracting liquid sweet. (5.17–25)

Calling to mind the erotic physicality of the Song of Solomon by borrowing both its phrasing ("Rise up, my love, my fair one, and come away") and its sensuous language, Adam's seductive words are also goads to productivity (KJV, Song of Solomon 2:10).[3] The tenderness of morning pillow talk gives way to urgency, yet Adam's sense of haste—"we lose the prime"—reveals a different meaning from what it first suggests. Objects of concern dissolve into objects of inquisitive interest. When Adam voices a wish to see "how spring our tended Plants," he sounds very much like a dutiful gardener, but the activities he subsequently enumerates have less and less to do with obligation: from the ambiguous case of "balm[s]" that "drop" overnight from the myrrh and the reed, which possibly (but only possibly) need clearing and collecting, to the purely delectable (that is, unproductive) observation of nature's "colors," which he compares to the work of a painter, and finally to the enjoyment of the sight of a happy bee that drinks its fill from a luscious flower.[4] The final item on Adam's list might remind us of Marvell's description of the jaunty bee who sips nectar from a flower's "flask": a similarly gratifying object of interest.[5] Milton's image is a representation of the experience Adam anticipates, much as Marvell's buzzing avatar of cheer evokes the poet's indulgent carelessness; "extracting liquid sweet" is a good gloss of the shapeless enjoyment Adam dependably finds waiting for him in the garden. By the end of the speech, Adam has acknowledged the centrality of pleasure to innocent labor, raising the possibility that even the first enumerated tasks, checking in on "tended plants" and looking to see "how blows" (how blooms) the "Citron Grove," are likewise anticipated as opportunities for exploratory pleasure. Adam's eagerness upon waking Eve is just the readiness of someone who is about to undergo an experience he trusts will bring him pleasure. *Come on*, he says, *join me, look what nature has in store for us.*

In the end, then, work doesn't feel much like work. The language of erotic love gives way to the language of labor, but, almost as quickly, labor turns out to be an experience of investigation at once underfocused, undermanaged, and underplanned. Adam's catalogue of possible activities cannot be mistaken for a checklist: it's a seductive description of possible gratifications the day holds in store for them. In place of the focused desire one might expect from a lover when he beckons to his beloved, especially given the scriptural cue, Adam conveys an interest in the lives of plants and at least one animal (that self-gratifying bee). Labor takes shape as a rite of love—but the enjoyment of marital togetherness consists in the sharing of an experience of gentle receptivity to the world. What I most want to emphasize here is that readers also find an opportunity to taste it. After all, they can have no idea where this passage is going: from the art of love to the art of horticulture to the artlessness of just looking around. In this respect, form mirrors content: *not knowing what you have in front of you* sums up the condition readers share with Adam the investigator. The relaxation of fixed attention Adam displays *for* them is also, perhaps, the effect *on* them of Milton's characteristically elongated sentence, which cascades down the page in nine lines; they allow themselves to be carried along by a mellifluous voice with little sense of where it's headed. Critics of *Paradise Lost*, most notably Stanley Fish, have sensationalized the poem's disorientations, discouraging soft responses to the experience of going astray. For Fish, indeed, Milton pulls the rug out from underneath his readers in order to produce an experience of mortifying self-correction.[6] Yet losing one's footing need not entail shock or anxious perplexity. Sometimes, as in Adam's morning reverie, confusion belongs to an experience of floating, multidirectional interest in a world that reveals itself gradually through a sequence of unexpected disclosures. This is not to say that interpretations that link poetic reversals to shock or distress are incorrect but rather that the feeling of vexed incomprehension can't be taken for granted. This chapter explores one alternative possibility, the observational mood, in Milton's Eden, where it also shows up as content in descriptions (like this one) of the unfallen workday.

Where might Milton have gotten the idea of linking the Song of Solomon to the effortlessness of paradisal labor? The Vulgate's version of the Song is the source of the topos of the *hortus conclusus*, the enclosed garden that appears in medieval and Renaissance poetry and visual art as an image of perfect purity: not only of Paradise but also of the Virgin Mary and other repositories of inviolate divinity.[7] Exegetical tradition associates the hortus conclusus with Solomon's pleasure garden, where his wedding is said to have taken place, but it also allegorizes Solomon's love poetry as a description of the holy matrimony of Christ and His church.[8] Bacon frequently invokes Solomon as an avatar of wisdom, perhaps most memorably in the name of the research institute, Salomon's House, he imagines in his utopian fiction, *The New Atlantis* (1627). The connection is stronger than this: in the *Advancement of Learning*, he calls on Solomon to defend his interpretation of Adam and Eve's transgression, which is also a defense of open-ended inquiry. In a passage I have already discussed, Bacon argues that the first sin was not a desire for "the pure knowledge of nature and universality" but rather for "the proud knowledge of good and evil, with an intent in man to give law unto himself, and to depend no more on God's commandments" (3:264–65). Though Solomon worries that knowledge, wrongly pursued, can produce "contristation" and "anxiety," Bacon quotes the wise king to clarify how little selfless inquiry has to do with the puffed-up arrogance of moral presumption: "the eye is never satisfied with seeing, nor the ear with hearing; and if there be no fullness, then is the continent [the container] greater than the content."[9] He extends this point to "the mind of man," which he describes as "a mirror or glass, capable of the image of the universal world, and joyful to receive the impression thereof, as the eye joyeth to receive light"—an image I used in chapter 3 as a gloss of an experience of grateful awareness captured by Marvell on the grounds of Nun Appleton.[10] For readers who know the *Advancement*, Adam's inviting words might likewise recall "joy[ful]" receptivity, the never-quite-satisfied but less-than-desperate hunger for knowledge crystallized in Bacon's formula for the observational mood: "satisfaction and appetite are perpetually interchangeable."[11] The voluptuous change in scale occasioned by the appearance of the bee in Adam's speech might also recall the words of that sometime Baconian Thomas Browne, who enlists Solomon's support for the idea that the smallest creatures ("narrow Engines") reveal "the wisedome of [God's] hand": "Out of this ranke *Solomon* chose the object of his admiration; indeed what reason may not go to Schoole to the wisedome of Bees, Aunts, and Spiders? what wise hand teacheth them to doe what reason cannot teach us?"[12] This last idea, of an inborn facility for craftsmanship or skillful labor that cannot be taught, resonates with Milton's suggestion that Adam and Eve's very bodies express a precognitive appetite to work the garden—irrespective of any obligation. Whereas the church had long spiritualized the erotic physicality of the Song of Solomon, Milton rematerializes it—with, I suggest, a Baconian license. He sublimates eros in the sense that he vaporizes it: Adam and Eve's attention remains amorous, but it's released from the impassioned privacy of the married couple and distributed as gentle awareness across the perceptual field.

My description of Adam's beckoning speech clarifies my perspective on scholarly discussion of Milton's interest in science. Though she doesn't draw a connection with Adam's variation on the Song of Solomon, Picciotto has argued persuasively that Baconian intellectuals, Milton included, devote themselves to the project of "digging up" the hortus conclusus, working to achieve the epistemological privilege of innocence through the practice of experiment.[13] Rather than locate paradisal perfection behind high walls, placing their hope for redemption in Christ alone rather than in their human faculties, they engage in a sustained effort to "[bring] paradise to the world beyond the garden"—to restore their capacity for Adamic insight through self-discipline and mutual correction.[14] Picciotto goes on to argue that the poem's formal difficulty demands from readers the level of mental effort on which the practice of innocent knowledge production depends.[15] Though she argues that "labors of innocence" integrate *otium* and *negotium*, her emphasis on the importance of rising to the challenge of Adamic perfection ensures that her description of experimental practice *feels* much more like muscular activity than quiet contemplation.[16] For both intellectual-historical and formal reasons, Picciotto's account chimes with that of Karen L. Edwards, who not only demonstrates Milton's engagement with the new natural history, which lends him strategies for representing plant and animal life, but also locates the poem's "experimentalism" in its demand "for the reader's imaginative engagement in the process of making meaning"—in opposition to lazy confidence in traditional sources of knowledge.[17] Though I'm indebted to both readings, it should be clear by now that my interpretation cuts a different path. Both Picciotto and Edwards lend emphasis to difficulty, but Adam's speech offers a description of paradisal labor that dilates smoothly into the leisurely pleasure of wandering attention.[18] On this question of absent laboriousness, Catherine Gimelli Martin is illuminating.[19] In addition to establishing the ongoing importance of Bacon to Milton's thought, she draws attention to the investment of *Paradise Lost* in the Baconian fantasy of effortless work.[20] "Adam and Eve's prelapsarian labor," she explains, "is un-laborious and it serves as its own reward, yet, even after the loss of Eden, work is still not a covenant, a curse, nor even a duty, but rather a blessing in disguise—a Baconian cure for Adam's lost 'idyll' of contemplative bliss."[21] In this chapter, I make a case for the interpretive benefits of radicalizing the idea of "un-laborious" "labor." What if we take this idea seriously—neither as an impossible feature of a fantasized life in Eden nor as a vague because fallen approximation of innocence but rather as a true description of what the investigation of nature is like for the fallen no less than the unfallen? What if we ourselves answer Adam's call, accepting the invitation he extends to Eve—and so conduct an experiment in reading comparable to the one they conduct together in their daily passage through the garden?

My answer to this last question is that our willingness to develop an interpretation of *Paradise Lost* from within the observational mood unlocks a strange but compelling perspective on Milton's project. Some have argued that he refuses the comfort many of his contemporaries take in the topos of the "happy fall," which redescribes the tragic outcome of Adam and Eve's temptation as the basis for salvation in Christ, proposing instead that *Paradise Lost* enjoins its readers to do the hard work—in the here and now—of reclaiming innocence.[22] Though I embrace the better part of this view, especially when it takes the Baconian form that links the restoration of Paradise to the retrieval of epistemological privilege, I have already articulated doubts about how challenging scholars have made self-redemption out to be. To be sure, there is much to say on behalf of Milton's ethos of struggle. He shows a commitment, throughout his theological and political writings, to the necessity of ongoing vigilance against temptation.[23] Furthermore, the difficulty of regaining innocence follows plainly from the poem's narrative premise: the starkness of the distinction between Paradise and the fallen world. An effortless journey from *here* to *there* would undermine our sense of the Fall's momentousness, which Milton takes as his point of departure in the poem's opening phrase, announcing his intention to sing to us "Of Man's First Disobedience" (1.1). Yet I affirm that Milton frequently implies just such ease of passage—not definitively, and perhaps not always knowingly, but with extraordinary consequences for our experience of reading the poem. By making the observational mood characteristic of life in prelapsarian Eden—and by creating opportunities for readers to partake in it—Milton short-circuits the distinction between innocence and fallenness.[24] All of a sudden, readers find themselves spirited away to Paradise, blinking their eyes with relieved surprise. This is not to say that Milton denies the Fall's fatefulness; it's to distinguish between two discordant features of the poem and to draw attention to the one that has yet to receive much exegetical elaboration. The very fullness with which Milton realizes the psychology and phenomenology of innocent experience, including its pervasive atmosphere of effortlessness, undermines the difference between the pre- and postlapsarian.[25] If readers are intimidated by the prospect of returning to Paradise, the poem's success at transporting them there offers more than a little encouragement. Though my attention to epistemology precludes focused engagement with Milton's theology, it's worth considering that his Arminianism makes this proposal less surprising than it at first appears. Against a Pelagianism that places the responsibility for salvation in human hands or a Calvinism that understands election and damnation as determined ahead of time by God's decree, Arminius envisions a God who offers grace universally, requiring only that human beings choose freely to accept it—an idea that chimes with Milton's repeated suggestion that innocence is there for the taking. I do not imply the simple confirmation of my reading by Milton's theology—and not only because I lack the space to pursue the question. Salvation might be taken, from an Arminian perspective, as emphatically earned. At the very least, Milton's invitation to an eminently achievable return to Paradise resonates with a conception of salvation as the magnanimous extension of an offer—rather than as the heroic achievement of the faithful self or a decree from on high to which the agency of the self is irrelevant.[26]

One reason I began with Adam's prospective description of the pleasures of gardening is that this chapter mainly explores the emotional life of innocence by way of a negative example—by contrast, that is, with the experience of falling and of fallenness. By locating failure in prelapsarian Eden, I risk a perspective that suspends awareness of undeniable differences between paradisal life and the aftermath of transgression. Yet this is not to speak, as some critics have, of a "fall before the fall."[27] For me, the most coherent reading of the poem casts Paradise as beautifully forgiving insofar as *all* errors are permissible other than the eating of the Fruit; Eden tolerates an almost infinite number of mistakes that do not rise to the level of disobedience.[28] Indeed, I accept Adam's description of God's stipulation that he and Eve avoid the Tree of Knowledge as an "easy prohibition," seeing how they are otherwise provided with "choice / Unlimited of manifold delights"—rejecting the conviction of some scholars that stern admonish-ment ends up intensifying interest in the forbidden (4.433–35).[29] Paradise isn't booby-trapped: everything (or *almost* everything) is there for the taking.[30] Yet in exploring softer distinctions between clarity and error, between an atti-tude that savors and preserves innocence and an alternative that jeopardizes it, I end up examining attitudes and behaviors that look like versions of fallen experience within the precincts of the prelapsarian world. Since my interest is Milton's seductive pedagogy, the request he makes of readers to appreci-ate their observational moods, I see it fit to emphasize continuity between ordinary (fallen) experience and Adam and Eve's innocent psychology—even if this sometimes means losing sight of the difference between proximity to transgression and the event itself.[31] My reading is normative in the limited sense that it grants value to the observational mood, implying that one can at least learn not to talk oneself out of it (or to do so less often); perhaps, Milton suggests, steps can be taken to encourage it—but it feels nonetheless like a gratifying gift from God rather than a personal achievement. From this angle, the Fall might be taken for a diagnostic opportunity. Like a slow-motion replay of a competitive diver who breaks the water's surface with clumsy splashes, the poem's detailed narration of the Fall brings the defin-ing features of error into focus without necessarily lending emphasis to its significance for the future of humankind. How far readers wish to take this "convergent reading" of Paradise and the fallen world can only remain an open question. Recalling the words of the angel Abdiel, who observes that "God and Nature bid the same," readers might come to see the effect of a taste of the Forbidden Fruit as closer to a natural phenomenon than an event in an interpersonal drama (between God and humankind), assimilating the logic of transgression and punishment to that of cause and effect (6.176).[32] This is not to imply a claim about the secular—and not only because I am tempted to speak of the sacredness of natural law in *Paradise Lost*. More prosaically, such questions are beyond the scope of this chapter. My focus is the experience of Paradise—for Adam, Eve, and the poem's readers. The observational mood appears to them in flashes—fleeting opportunities to entertain the idea that innocence is not just available to them but that the promise of its achievability is already in the process of being fulfilled.

This chapter builds its argument by elaborating an anatomy of error. Eve, I suggest, rejects the premise that innocence is effortless, and so she need-lessly pursues an experience of difficulty. To understand the logic of this substitution, I explore the sequence of gratuitous mistakes Eve makes on her journey to the Forbidden Tree. In this respect, I follow Milton's own man-ner of proceeding: it's by narrating Eve's winding approach to the tempting

Fruit that he enumerates the defining features of the attitude that ultimately vanquishes innocence. Unlike Milton, however, I take for granted the toxicity of the view that blames Eve for Adam's fall—and, by extension, womankind for mankind's. My emphasis on atmosphere reveals aspects of the poem that resist the translation of the Eden story into a misogynist parable. First, Eve is—some of the time—a better example of the affective rhythms of innocent cognition than her husband; Adam is not predictably more successful in this regard. Second, by the end of the poem, he repeats the several mistakes she makes on her way to the Tree, demonstrating that she is not alone in her susceptibility; it's not just that he likewise falls but that he stumbles in the same way. Third, by dedramatizing the Fall, my reading casts doubt on the special burden of guilt apparently incurred by Eve's initial tasting of the Fruit. This is not what Milton says—but the more he persuades readers of the accessibility of Paradise, the less they find themselves worrying over their losses and about where to place blame for them.

Eve's distorted perspective comes most clearly into focus when, in book 9, she seeks to persuade Adam that they should tend the garden separately, parting ways in order to cover more ground. Their exchange introduces her conviction that work is incompatible with the default psychology of innocence. The very turns of the conversation, I suggest, allegorize the Fall itself, which Milton goes on to narrate through the drama of temptation that leads first one and then both of the protagonists to the "fatal fruit" (10.4). I proceed by describing the distortion of the innocent mind by way of three related motifs: difficulty, adversity, and intensity. My discussion tacks between the debate about the nature of labor, which I take as the epic's thematic center, and other moments in Milton's narrative—both within and beyond the scene of satanic seduction.

Labor and Laboriousness; or, Two Versions of Digression

In *How Milton Works* (2001), Fish offers a synthesis and critique of scholarly discussion of Adam and Eve's debate, inhabiting a perspective that illustrates the literary-critical aversion to easygoing unselfconsciousness I have diagnosed in this book. Fish rightly "insist[s] . . . that the scene be allowed the contingency that attends its dramatic unfolding, a contingency that disappears if it is always being understood in relation to an event of which it is supposedly the cause"—though I would add that the perception of a close relationship between Adam and Eve's disagreement and their eventual transgression need not imply that the latter controls the interpretation of the former.[33] I affirm the contingency of both events but nonetheless trace connections between them. Milton seems to have gone out of his way to convey the messiness of their argument; both Adam and Eve seem at different times to be "right" about what to do. There is, however, a simple opposition at the heart of the exchange; it just doesn't rest on the difference between correct and incorrect propositions about what it means to show obedience to God. What distinguishes Eve's position from Adam's is her distrust of the affective experience they get to enjoy by default. Fish's approach rules out this possibility. Citing Barbara Lewalski and J. M. Evans, he affirms that Adam and Eve "are their own chief crop," assimilating the work of gardening to self-cultivation.[34] On his account, attending to the world means attending to the self—to the exclusion, indeed, of any real interest in whatever is to be found out there in the world. Fish's commitment to self-cultivation and self-presentation rules out the possibility of nontheatricality. Returning now to an argument I discussed in the introduction, Fish disputes Marshall Grossman's observation that Adam and Eve seek the rhetorical upper hand by dramatizing themselves, turning their attention from the actual "substance" of their argument to the theatrical effects they produce in one another, and insists to the contrary that everything is theater in the context of their relationship.[35] To this view, my response is that a distinction between self-aware theatricality and unselfconsciousness need not imply a belief in the possibility of expressing some authentic interior self without any trace of mediation; it need only assume that one is not always attending to self-presentation with the same intensity.[36] Grossman's point is explicitly psychological rather than ontological: "the dispute takes place on the level of emotions—the argument is about how they feel."[37] Fish takes a similar approach in rejecting Mary Nyquist's perspective. Responding to her view that Adam and Eve ultimately lose control of their conversation, Fish argues that there is no such thing as lost control ("control can never be lost") because everything in Milton's cosmos is reducible to a binary distinction between commitment to God and other misbegotten commitments.[38] Subjective experience, I suggest, is not reducible to commitment, and there is no good reason to assume that selves are saturated by deliberateness. Indeed, the freedom to be less than fully deliberate about one's labor is the very thing at issue in the debate.

In Adam and Eve's exchange, Milton juxtaposes two conceptions of labor, each of which has a distinctive affective dimension, and two corresponding descriptions of deviation: the kind that wanders contentedly away from efficiency and the kind that enforces it by crossing greater expanses. Eve initiates the dispute by worrying over "wild" botanical growth. Though she and Adam are earnest gardeners, they can't keep pace with the lushness of Paradise. It isn't clear, however, that they actually need to do so. In this scene,

Eve presents her anxious response to the situation as a predictable, natural reaction—going as far as to personify nature itself as a malicious character in a counterfactual drama.

> . . . the work under our labor grows,
> Luxurious by restraint; what we by day
> Lop overgrown, or prune, or prop, or bind,
> One night or two with wanton growth derides
> Tending to wild. (9.208–12)

In these lines, Eve's concern is speed. As Evans puts it: "their pastoral retreat will be gradually engulfed by the jungle outside it."[39] Notice that Eve imagines botanical overabundance as deliberate cruelty: "with wanton growth," she says, the garden "derides" their efforts to keep it under control. Thus she imposes a moral evaluation on a natural phenomenon, misconstruing fecundity as nasty peals of laughter. Milton has already used the nominal form of this word, "derides," to characterize God's scorn for the doomed rebellion of the traitor angels. In book 5, as Satan gathers his forces in preparation for war, the Son of God tells His father:

> . . . thou thy foes
> Justly hast in derision, and secure
> Laughst at thir vain designs and tumults vain . . . (5.735–37)

Eve imagines herself and Adam in just this situation: "laugh[ed] at" because of their "vain designs" by someone whose power completely dwarfs their own. The most obvious "someone" she can imagine as a looming presence, hovering above or behind the foliage and imbuing it with a maddening capacity to sprout new leaves and branches, is this same God—perhaps here blasphemously misrecognized as standing in the same relation to Adam and Eve as He does to rebellious Satan. Yet I do not wish to translate Eve's misprision into the language of faith and disobedience; I want to establish what it tells us about the emotional life of innocent labor. Eve's description of the garden exteriorizes her defensive suspicion—as if it were a logical conclusion rather than a gratuitous affective response. The poem underscores the availability of other reactions—and of one alternative in particular. The passage with which I began, where Adam echoes the Song of Solomon, suggests that labor is a privilege rather than a burden: more than simply an opportunity to relish the body's powers, the specific commission God gives the first human beings is an opportunity to investigate the perceptual riches of the natural world. But Milton does not simply rely on our memory of earlier reflections on paradisal gardening to provide a sufficient counterweight to Eve's concern.

When she asks Adam to affirm her wish for increased productivity, both the form and the content of her speech undercut urgency:

> Thou therefore now advise
> Or hear what to my mind first thoughts present,
> Let us divide our labors, thou where choice
> Leads thee, or where most needs, whether to wind
> The Woodbine around this Arbor, or direct
> The clasping Ivy where to climb, while I
> In yonder Spring of Roses intermixt
> With Myrtle, find what to redress till Noon . . . (9.212–19)

Eve's "therefore" presents this proposal as a solution to the problem she has just diagnosed, but her speech conveys leisurely open-endedness. These are just her "first thoughts," she points out, and so she invites Adam to "advise" her otherwise. Unfolding a plan she has thereby located somewhere on the continuum between considered thought and mere whim, she instructs him to go occupy himself "where choice / Leads thee, or where most needs," as if the state of the garden might or might not confirm her perception of a problem desperately in need of solving. When she goes on to enumerate those tasks that might—but just might—capture his interest, she calls to mind his morning delectation of the day's potential: he will "wind" "Woodbine," "direct" "clasping Ivy," or pursue whatever unpredictably strikes his fancy. Similarly, when she tells him she will "find what to redress till Noon," she professes a surprising unawareness of anything in particular that requires her attention. The phrase also suggests that what she intends is to fill the available time rather than to bend the time to make room for the meeting of nonarbitrary goals. None of this sounds like a response to a crisis.

While Eve unconsciously returns to the observational mood, even as she seems to take her leave from it, Adam's response to Eve's complaint is an explicit affirmation of the sufficiency of what they already manage. He directly says what she implies by drawing a line from paradisal anthropology to God's intentions.

> For not to irksome toil, but to delight
> He made us, and delight to Reason join'd.
> These paths and Bowers doubt not but our joint hands
> Will keep from Wilderness with ease, as wide
> As we need walk . . . (9.242–46)

Adam's line of reasoning begins with his understanding of the psychology of labor—from which he then derives a conclusion about the task ahead.

"Delight," he reminds her, is the end for which God "made" them—which rules out the possibility that they actually do find themselves in a race against engulfment by a hostile environment. Indeed, Paradise would not be Paradise if God's provision were insufficient. It's not that "delight" is an alternative to "toil" but that "toil" should itself be "delight[ful]" rather than "irksome." With this expectation in mind, Adam trusts that whatever he and Eve accomplish will be more than enough. Without disavowing the necessity of work (the need for constant clearing of an encroaching wildness or, in Milton's parlance, "Wilderness"), Adam disputes the necessity of "doubt" and apprehension.[40] In Paradise, he affirms, labor really is as "eas[y]" as it feels. When "join'd" chimes with "joint," we notice that "Reason" harmonizes with "delight" much as Adam and Eve synchronize their labors, working side by side—and not only if we agree with those critics who understand Eve as an allegory of self-indulgent fancy over which Reason, personified as Adam, maintains order.[41] "Joint hands" signify collective labor, but they also call to mind an image of hands that are literally joined: a perfect illustration of a cooperative project that takes as much time as it needs. Christopher Ricks discusses the "poignancy" of Milton's return throughout the poem to the image of Adam and Eve "hand in hand," lending credence to my perception that the image flashes up again in the midst of this description of labor.[42] Unbothered by unused hands (two or more, depending on how we picture handholding), Adam confirms his inattention to efficiency. Taking one's time doesn't just mean working slowly; slowness follows from the variety of activities that count as work—including even as apparently unproductive a pastime as the reciprocal giving of comfort.

Adam explicitly defends outright truancy and distraction:

> . . . not so strictly hath our Lord impos'd
> Labor, as to debar us when we need
> Refreshment, whether food, or talk between,
> Food of the mind, or this sweet intercourse
> Of looks and smiles, for smiles from Reason flow,
> To brute deni'd, and are of Love the food,
> Love not the lowest end of human life. (9.235–41)

Adam celebrates a style of relaxed exertion that blends seamlessly into leisure. As the marriage of "Reason" and "delight," he concludes, labor shouldn't have to be difficult in the first place. Instead, productivity can be assumed as an effect of merely persistent—rather than strained or even constant—effort. One of the affordances of innocence is industry divested of imperative force: a pleasure to be enjoyed like any other. Labor is flexible enough to encompass flirtation and intimate contact. It makes space for "looks and smiles," which Milton describes as the "food" of "Love"—imagined in this passage not as Petrarchan longing or rapturous consummation but rather as an experience of "sweet intercourse" in which wordless expressions of affection play no less a role (and have no less to do with "Reason") than utterances. Eve says all of this very well—but disapprovingly:

> For while so near each other thus all day
> Our task we choose, what wonder if so near
> Looks intervene and smiles, or object new
> Casual discourse draw on, which intermits
> Our day's work brought to little, though begun
> Early, and th' hour of Supper comes unearn'd. (9.220–25)

In this scene, Adam and Eve have just awakened and joined the "Choir / Of Creatures" in "vocal Worship" of God; they have not yet begun their day's work. Repeating the phrase "so near," perhaps Eve gestures at the presently unfolding situation: How can labor be bona fide if it remains as pleasurably intimate as *this*? Caring for the garden feels disconcertingly like the preface to it. If they were sufficiently industrious, she suggests, work wouldn't feel so much like what they presently enjoy: ordinary conversation—or, as Eve puts it, "casual discourse," which responds to whatever happens to capture their attention. Labor unfolds in Paradise not as a single task but as a multidimensional experience. Eve's emergent desire is to do just one thing at a time— surely a more difficult practice than the casual rotation between one activity and another that defines the gentle temporality of the innocent workday.

Before elaborating on the role of labor as an opportunity for inquiry, I want briefly to return to the passage in which Eve first articulates her dissatisfaction. Having clarified the difference between Eve's misconstrual of paradisal labor and the poem's dominant understanding of it (as established by the poet, Adam, and, notwithstanding her emergent attitude, Eve herself), I suggest that her original assessment of the problem of unmanageable growth can be read against the grain: "The work under our labor grows, / Luxurious by restraint." Milton plainly indicates Eve's worry about the speed of the garden's growth, and there is little doubt most readers understand her observation as an expression of concern—but, to that subset of readers who have eagerly accepted Milton's invitation to the observational mood, her words might also be taken against her wishes to describe the good news of disburdened labor. "Work" signifies the object of Adam and Eve's ministrations: the disconcerting robustness of plant life that grows too quickly for its keepers, given the unavoidable "restraint" of finite human power. Yet also audible in "the work under our labor" is a reference to the project rather than the object (or objects) of gardening. In a thought-provoking reading of this passage, Kevis Goodman shows that Eve makes a distinction between "what ought to get done" and the "activity of doing it."[43] Where Goodman, looking to Marx, pursues the possibility that Eve describes an experience of alienation from the product of her labor (and also from her coworker, Adam), I observe an affective benefit in the autonomy of innocent activity with respect to its material results: a relationship of independence rather than estrangement. The task (of pruning, clearing, and so on) might be taken to "[grow] / Luxurious by restraint" in the sense that the evident impossibility of bringing the work to completion dissolves the burden Eve imagines she bears: having to keep pace with "wild" proliferation. Should Milton have entertained this counterreading of these lines, it wouldn't be the first time someone had identified Paradise with bounteous "luxury"— no matter the frequently negative connotations of that term in the period. Indeed, Nicholas Billingsley's exclamatory response to the botanical profusion of Paradise in his hexameral poem, *The Infancy of the World* (1658), would make a good caption to the exertions of Milton's Adam and Eve:

. . . how all things smile
And e'ne luxuriate! Oh delightful soile![44]

Adam and Eve take their sweet time with a job that never has to proceed any faster than they wish. After all, a task you can't possibly accomplish is easier than one that pushes you to the very limits of your abilities; it invites the happy resignation of doing what you can. Unlike under grinding capitalism, paradisal workers really don't have to get much done—not one jot past the threshold that separates pleasure from pain, satisfaction in the exercise of natural faculties from resentment at the necessity of overexertion.[45]

Eve is often an exemplar of paradisal easygoingness: reading her words against her present intentions means reading them in accordance with her previous intentions. In book 8, when Adam puts questions about the cosmos to his angelic visitor Raphael, Milton explains that Eve prefers to receive such information from Adam instead—precisely in order to savor the digressiveness she now disdains:

Her Husband the Relater she preferr'd
Before the Angel, and of him to ask
Chose rather: hee, she knew, would intermix
Grateful digressions, and solve high dispute
With conjugal Caresses, from his Lip
Not Words alone pleas'd her. (8.52–57)

What Eve now refuses (in their disagreement, which brings their idyll to an end) is exactly what she formerly treasured in natural-philosophical dialogue. In this respect, Eve exemplifies the casual effortlessness of innocence more successfully than Adam; in this earlier moment, she opts for the integration of rational inquiry and sensuous "digression" while Adam pursues "studious thoughts abstruse"—against which, eventually, Raphael has to caution him (8.40). At the beginning of book 9, however, she finds their labors inappropriately intermittent. From Adam's amenable "Lip," she once requested kisses just as eagerly as she solicited descriptions of the universe, but, somewhere along the way to the scene of contention (perhaps no earlier than the moment before she speaks), she loses her confidence in the virtues of waywardness. Labor, she decides, should hurt.[46]

One might object that Eve hasn't actually undergone a change; from her perspective, perhaps, easy speculation is well and good but actual work is taxing. Eve would then insist on the distinction between otium and negotium, serene contemplation and exertion. Yet Milton, especially when he stages Adam's conversation with Raphael, encourages us to identify the innocent work of gardening with the labors of understanding. Indeed, Adam's argument for the leisureliness of work, coming as it does immediately after this dialogue, can be read as a defense of the persistence of effortlessness as he passes from exploratory discussion to dilatory spadework—out from the "shady Bow'r" and into the heat of the garden (5.367). Since Adam did not passively absorb wisdom from Raphael's remarks on the cosmos but rather engaged him in vigorous conversation, the poem passes from one kind of intellectual labor to another—both of which make easygoingness an engine of success. There's a telling symmetry here: Just as "casual discourse" belongs to the experience of exploratory labor, exploratory labor is itself an aspect of Adam and Raphael's "casual discourse." Indeed, Adam's exchange with Raphael reveals to him the following precept, which directs our attention to the act of gardening:

> That which before us lies in daily life,
> Is the prime Wisdom. . . . (8.193–94)

Although Raphael confirms that Adam is right to pursue knowledge widely, he ultimately concludes that "things at hand" are the worthiest objects of his inquiring attention (8.199). For Adam and Eve, the garden in which they perform their daily labor is the most significant of such proximate "things." Bruce R. Smith has noted Milton's interest in producing and juxtaposing "curiously multiple" "vantage points" from which to observe Paradise, including, in book 4, the "*almost* panoptic view commanded by Adam."[47]

Indeed, Adam and Eve's perspective is always pushing outward to encompass new objects—but without arriving at Archimedean mastery. The casual but immersive exploration in which their work consists localizes without limiting the cosmos-spanning investigation Adam performs in conversation with Raphael. "That which before us lies" turns out to be an endlessly capacious realm of perceptual novelty. According to Joshua Scodel, there is an echo of this discovery at the very end of the epic: "The World was all before them" (12.646).[48] Thus the fallen "World" turns out to be just the sort of valuable object on which Adam and Eve had previously settled their investigative attention. It's no accident, then, that Milton, describing their departure from Paradise, compares the cherubim who "descend" on them to an "Ev'ning Mist" that "gathers ground fast at the Laborer's heel / Homeward returning" (12.628, 629, 631–32). Adam and Eve do not leave Paradise intent on heroism, shoulders squared in preparation for the endless task of remaking their lost home. Milton pictures them "wand'ring . . . and slow," which is also a good description of the "luxurious" "work" of innocence, and he underlines the point by conjuring forth the image of a laborer at rest—or nearly so, walking quietly "homeward" from the fields (12.648). This anonymous creature of metaphor remains in motion, suspended between work and leisure, perhaps still wet with perspiration but temporarily unburdened of duties.[49]

Milton does not rest content with the assertion of paradisal labor's effortlessness. He also develops a psychological portrait of careless pleasure. By the time readers get to Adam's debate with Eve, Raphael's visit to the garden has prepared them to understand how it matters for the emotional life of innocence. The very passage in which Adam absorbs the lesson about the importance of attending to the surrounding world (rather than the far-flung secrets of the universe) also describes the affective difficulty that accompanies more ambitious forms of inquiry. Like Bacon, who uses Solomon's positive appraisal of inquiry to insist on the limited scope of his worry about the anxiety that attends the quest for knowledge, Adam praises Raphael's pedagogy.

> How fully hast thou satisfi'd me, pure
> Intelligence of Heav'n, Angel serene,
> And freed from intricacies, taught to live
> The easiest way, nor with perplexing thoughts
> To interrupt the sweet of Life, from which
> God hath bid dwell far off all anxious cares,
> And not molest us, unless we ourselves
> Seek them with wand'ring thoughts, and notions vain. (8.180–87)

With Raphael's prompting, Adam identifies immersive attention to "things at hand" with "the easiest way" of life. "Anxious cares" are simply foreign to innocent experience, "unless we ourselves / Seek them out." One question this passage raises is the status of "wand'ring" in Milton's scheme, since Adam's critique of overanimated and overweening inquiry seems to cut against my defense of errancy. I return shortly to this theme; it's too important to Milton's conception of innocent labor to discuss in passing. For the moment, I observe that Adam's expression of "[full] . . . satisf[action]" both celebrates the perfection of his enjoyment and conveys an absent intensity. Such is the distinctiveness of prelapsarian pleasure. The experience Adam treasures is defined by stability in the sense of mere ongoingness: "the sweet of Life" can be counted on to persist unless needlessly "interrupt[ed]." Referring to Raphael as an "Angel serene" and "pure / Intelligence" only draws attention to what Adam's praise for the persistence of ease already implies: that "sweet" refers to purity, as it often does in early modern English, rather than, say, the intensity of sugary fruit.

On this matter, Milton is crystal clear. In chapter 1, I argued that Bacon's description of "learning" as an experience in which "satisfaction and appetite are perpetually interchangeable" need not refer to intellectual rapacity, as scholars have sometimes maintained, but might instead suggest the steady but various pleasure that attends ongoing investigation.[50] As, say, attention wanders (or, under sustained inspection, an object of inquiry takes on a new appearance), the inquiring mind is unperturbed without having to restrain itself: because it doesn't know what it wants, it always finds something that gratifies interest. "Of knowledge," Bacon writes, "there is no satiety."[51] Sounding much like an Epicurean himself (and, during the unfolding of his garden symposium with Raphael, looking very much like one), Adam describes the pleasures of talking philosophy in exactly these terms.

> For while I sit with thee, I seem in Heav'n,
> And sweeter thy discourse is to my ear
> Than Fruits of Palm-tree pleasantest to thirst
> And hunger both, from labor, at the hour
> Of sweet repast; they satiate, and soon fill,
> Though pleasant, but thy words with Grace Divine
> Imbu'd bring to their sweetness no satiety. (8.210–16)

Like, on Bacon's account, "learning" or "understanding" in general, Raphael's lessons "bring to their sweetness no satiety" in the specific sense that they generate steady contentment rather than wild hunger: emergent desire swiftly answered but just as quickly redirected—rather than, say, intense longing met by exultation before returning at full strength. Recall that the "looks and smiles" that punctuate the workday "are of Love the food"— another version of this experience of appetite without scarcity, gratification without intensity. As in Adam's equation of the "sweet of Life" and the "easiest way," what he here names "pleasantest" (or "sweeter" than "pleasantest") is not to be confused with what packs the biggest punch.[52] Though Milton speaks repeatedly of "sweet[ness]," in all three cases he points away from *strength* or *concentration* of flavor: the enjoyment to which he refers is a property of "discourse," of "repast" (which could mean either food or rest), and again of "words."[53] Adam uses his sense of taste as a point of departure for reflecting on the pleasures of conversation because he and Eve are persistently engaged in sensuous self-education; it's in this specific sense as well that his dialogue with Raphael about the cosmos is itself a version of the effortless labor of understanding they experience daily in the garden.[54] W. B. C. Watkins has written in a rapturously speculative vein about Milton's delight in poetry as an art both "aural" and "oral": "For words must be formed in the mouth and throat and propelled by breath before they can impinge on the ear; and though it may seem outlandish on first consideration, probably Milton derived as much pleasure from the actual formation of words on the tongue, the sense of muscular expansion of the diaphragm to expel air through vocal cords, as from the sound they make."[55] Watkins detects in Milton a conception of *breath* as an embodied medium of sensual enjoyment, integrating the diverse, local sensations of lungs, throat, mouth, and nose into a unified experience of pleasurable exhalation. Milton's description of the relationship between matter and spirit resonates with Watkins's account. In these lines, Milton takes plant life as a metaphor for the structure of the cosmos, ascending gradually from the heavy to the ethereal.

> So from the root
> Springs lighter the green stalk, from thence the leaves
> More airy, last the bright consummate flower
> Spirits odorous breathes . . . (5.479–82)

Here, Milton saves the verb for the end, and, with the adverb "last," calls our attention to how the process of sublimation ends. Following Watkins's lead, we might observe that form merges with content when we finally arrive at the word "breathes" and enjoy—on the tongue, lips, and palate—a soothing release of air. Though Milton goes on to depict self-transformation as a matter of ingestion and digestion, breathing is actually a much better metaphor than hunger (as far as fallen readers, for whom food can be scarce, are concerned) for an endless alternation between "satisfaction and appetite" that doesn't feel like desperation.

What's astonishing is how swiftly Eve takes her leave from so deep, per-sistent, and self-evident a pleasure as the gentle cultivation of knowledge. Her transformation is distinctly affective—a change of tone that makes all the difference. Milton conveys the meaningfulness of such an alteration, the powerful bearing of attitude on everything it touches, by presenting a single image of both innocent labor and its corruption. I'm thinking here of the topos of "wandering," which I flagged above as requiring further attention. Some readers have noticed that this section of the poem is a tale of "error"—in the specific sense that Eve "wanders" away from safety and thus exposes herself to danger.[56] Baconian echoes complicate this picture: Eve's disorientation is something like a perfect inversion of the style of care-less investigation in Boyle's meditations and Walton's leisurely dialogue—in Marvell's carefree promenade and Power's drifting journey across the micro-scopic world. The protagonists of Bacon's "instauration" remake the way-ward path, formerly associated with moral and epistemological failure, as an image of their collective aspiration to unveil the world's mysteries. Yet Eve understands her decision to drift away from Adam as a strangely indistinct alternative to exploratory rambling. She reimagines breadth as speed: the zigzag as a straight line. Another way to describe the attitude with which Eve wanders is to borrow Milton's remark on the Serpent's rhetorical prowess: it "[makes] intricate seem straight" (9.632). Recall that Adam uses this word, "intricac[y]," to describe the "anxious care" from which he is by nature free, but which he might foolishly, needlessly, and perilously choose.

As it happens, the Serpent can be understood as a revealing emblem of the wayward path's sudden transformation from a corollary of easygoingness into an impatient strategy for thoroughness. I complete my account of the poem's double interpretation of digression—as both unearned effortlessness and gratuitous difficulty—by rethinking the Serpent's role in book 9. One need not recall the vast literature of chivalric romance, in which vectors of desire send heroes down endlessly forking paths (or Spenser's representation of Error as a serpentine monster) to notice that Milton's Serpent is, in his very corporeal form, an image of indirection.[57] Yet nobody, as far as I can tell, has observed therein the contrary image of innocent freedom from con-cern. It's hard to look past the malevolent physicality of the Serpent when it takes disorienting shape as the "surging Maze" in which Eve loses her way (9.499). His "side-long" maneuvers are blatant representations of "fraudulent temptation," redoubling the already powerful sense of dramatic irony that attends almost any version of the story that delivers Eve to the foot of the Forbidden Tree (9.512, 531). One aim of my appraisal is to show that Milton draws a line from the most familiar of the Serpent's emblematic features, with their suggestion of duplicity, to the specific delusive fantasy that guides Eve onward as she makes her self-destructive way forward: the substitution of burdensomeness for efficiency. Yet more important (and much more surprising) than the satanic quality I observe in the Serpent's winding motion is another—indeed, prior—interpretation: the intimation of paradisal languor.

The "wanton wreath[s]" with which the Serpent means to "lure [Eve's] Eye" at first suggest a happy aimlessness (9.517, 518). The poem presents the serpentine "Labyrinth" as an image of unguarded immersion in the world, and thus Eve's journey through the "Maze" can be understood as a decisive repurposing of the path of innocent exploration. In his fleshly form, the Serpent makes immediate, multidirectional contact with the surrounding environment. Without having to do anything, he drapes himself expansively across the garden; such is the birthright of snakes. It's tempting to read Eve's encounter with him as psychomachia; perhaps it's her devotion to emotional suffering that renders the Serpent's disordered entanglement with Paradise threateningly "tortuous" (9.516). In any case, Satan takes up residence in the Serpent's body just in time to lend Eve assistance on the hasty journey to anxiety she has already begun. Milton's early description of the Serpent shares the uncertainty of an exegetical tradition that sometimes blames the lowly creature for inviting Satan's use.[58] Yet rather than simply invoke predictable associations, Milton draws a sharp contrast between Satan and the creature "in whose mazy folds" he "hide[s]" his "dark intent" (9.161–62). When Satan first comes upon the sleeping Serpent, Milton draws a distinction between malevolent intensity and reptilian languor:

> . . . through each Thicket Dank or Dry,
> Like a black mist low creeping, he held on
> His midnight search, where soonest he might find
> The Serpent: him fast sleeping soon he found
> In Labyrinth of many a round self-roll'd,
> His head in the midst, well stor'd with subtle wiles:
> Not yet in horrid Shade or dismal Den,
> Nor nocent yet, but on the grassy Herb
> Fearless unfear'd he slept . . . (9.179–87)

As soon as Milton observes that the Serpent's brain is "stor'd with subtle wiles," he goes on to subtract every predictable accessory of dangerous cunning. The Serpent, Milton explains, has yet to take up residence in "horrid . . . or dismal" hiding places, which would perfectly materialize his presumed sneakiness. When Milton remarks that he is not yet "nocent," harmlessness converges with unconcern in a single, compact phrase: "fearless unfear'd."

The chiasmus suggests that innocuousness mirrors sangfroid—and for good reason. Violence is an effect of fear; aggression, of defensiveness. The Serpent is not yet anyone's enemy—just as nobody is his. It's for this reason that he need not hide himself away in "Shade" or "Den" but rather gets to luxuriate "on the grassy Herb." The fact that Satan locates him without any trouble signals the vulnerability he embraces in unconcernedly unrolling himself on the lawn.

The Serpent's slumber, then, is not simply a convenient opportunity for demonic (satanic) possession; it's an image of absent vigilance that strangely epitomizes the pleasures of Paradise. The eagerness of Satan's quest for a fitting instrument finds its opposite in the sleeping snake. Indeed, when Satan chooses the creature that best serves his purposes, he conducts a "narrow search" (9.83). Once he decides that the Serpent is the "fittest Imp of fraud" (he is only described as "wily" from Satan's perspective, and even this loaded adjective might convey sheer aptitude rather than a predisposition to treachery), he does so once again, ranging over Paradise ("through each Thicket Dank or Dry") with a singleness of purpose Eve soon adopts herself. In another contrastive turn of phrase, Milton juxtaposes Satan's gimlet-eyed efficiency and the soundness of the Serpent's slumber: "him fast sleeping soon he found." Although sleep suggests an absence of awareness, it nicely illustrates the experience of relaxed self-forgetfulness Adam has already defended as an attribute of labor. In fact, my concession might be overstated, since Jane M. Petty has shown that Milton actually does treat "the twilight between sleep and consciousness" as a state of receptivity, "allow[ing] the senses to transmit messages of external stimuli to the sleeper's imagination."[59] In any case, sleep conveys the pleasure of slackness. As I discussed in chapter 2, Boyle makes a display of his easygoingness by presenting himself to his readers in states of sleepy inattention. If insomnia often appears in early modern literature as an image of unrelenting and overheated watchfulness (to which Lucifer, readers discover in book 5, himself falls victim), the Serpent hyperbolizes the opposite when he takes shape before us "in Labyrinth of many a round self-roll'd."[60] Indeed, the word "Labyrinth" encourages us to visualize a mess of undulations, at ease in unprotected extravagance—rather than the controlled neatness of a geometrical spiral or cylinder. Sleep amplifies our sense of the luxuriousness for which the Serpent's body is the very image.

When Satan imbues the Serpent with intensity of purpose, he mirrors Eve's unfolding self-transformation. No longer capable of easy abandon, the Serpent is neither "fearless" nor "unfear'd": he is a dangerous seducer with, at minimum, the fearfulness of the schemer who lacks confidence in success.[61]

When Satan speaks of "the dark intent I bring" to the "Serpent sleeping," he refers not only to the specific contents of his scheming mind (the desire for "Ambition and Revenge") but also to the very form of mental fixation (having any sharp "intent" at all) of which the Serpent's slumbering disarray is the inverse (9.161, 162, 168). Perhaps nothing better conveys the disarmingly fast transition between aimless and focused wandering than Milton's reinvention of a topos we recognize from Marvell's *The Mower to the Glow-worms*: the "wand'ring Fire," or ignis fatuus, that leads the unwary traveler into danger (9.634).[62] When Milton imagines the Serpent as a "fraud[ulent]" night light guiding Eve to her doom, he illustrates the treacherous equation between drifting and overeager footsteps.

> Lead then, said *Eve*. Hee leading swiftly roll'd
> In tangles, and made intricate seem straight,
> To mischief swift. Hope elevates, and joy
> Brightens his Crest, as when a wand'ring Fire,
> Compact of unctuous vapor, which the Night
> Condenses, and the cold invirons round,
> Kindl'd through agitation to a Flame,
> Which oft, they say, some evil Spirit attends,
> Hovering and blazing with delusive Light,
> Misleads th'amaz'd Night-wanderer from his way
> To Bogs and Mires, and oft through Pond or Pool,
> There swallow'd up and lost, from succor far.
> So glister'd the dire Snake, and into fraud
> Led *Eve* our credulous Mother, to the Tree
> Of prohibition . . . (9.631–45)

I suggested above that Eve herself "[makes] intricate seem straight" in the sense that she wanders off in the interest of efficiency. Milton's use of the phrase draws attention to the onward rush of satanic misdirection, insisting twice on the "swift[ness]" of the Serpent's "tangle[d]" "roll[ing]." Indeed, "straight" conveys directness, but it also implies speed. Borrowing a definition from the *OED*, Eve and the Serpent proceed "without delay."[63] Eve's original interest in covering more ground intensifies to the point of sheer intentness on achievement; she makes a beeline for (what seems to be) the ultimate payoff, a new course of action that only qualifies as "intrica[cy]" in the narrow sense that it represents a needless deviation from work—as it usually proceeds in Paradise. The Serpent knows exactly where he's going, and Eve knows just what she wants. Though the near-redundancy of a "Night-wanderer" "mis[led]" by phosphorescent "Flame" is commonplace (recall Marvell's drifting mower), Milton finds a more precise use for it; like the "wanderer" who chases the will o' wisp, Eve wanders away from wandering.

The simile highlights the continuity between Eve's initial wish to bring her work to completion and her present journey to the Forbidden Tree by drawing her perspective into close alignment with the Serpent's. Eve is less the dupe of a successful con artist than an eager victim of self-deception. Indeed, this is as true for Satan (in the guise of the Serpent) as it is for Eve; for him, triumph is flatly unavailable—in the long run, anyway. Like many of Milton's famously elaborate similes, this one disorients us with great success, drawing a comparison between the ignis fatuus and a productively undefined second term. Readers most likely understand the "glister[ing]" of the "dire Snake" as the right referent for the fatally "delusive Light," but this interpretation only comes into focus at the end of the passage. When Milton first proposes a comparison, we cannot tell whether the Serpent's passion or his inflamed "Crest" is the errant "Flame"; nor can we tell whether Eve (trailing the Serpent) or the Serpent (trailing his own malevolent desire) is the hapless "Night-wanderer." The Serpent follows the "Flame" of "Hope" and anticipatory "joy," while Eve follows the "bright[ness]" they together generate—as well as her own "leading" passion: much the same mixture of anticipation and premature triumph. Milton has prepared readers for the neatness of Eve and the Serpent's psychological convergence by having Eve command the Serpent to do what he previously suggested and now actually does: "Lead then, said *Eve*. Hee leading swiftly roll'd." Once again, the scene approaches psychomachia: Eve's intensity of focus materializes as a radiantly fleshy crown. Indeed, the conjuring forth of an "evil Spirit" by superstitious rumor ("they say" such spirits are responsible for will o' wisps) within the analogy encourages us to conjure Satan away. Once Milton translates the "lead[er]" of the expedition into metaphorical "Flame," he gives a full five lines to the production of fire from "unctuous vapor" before introducing the figure of the "Night-wanderer," thereby returning us to the unfolding narrative. Thus the simile also serves as a meditation on misbegotten passion: the poet explains at some length how "unctuous vapor" is "kindl'd through agitation" before it "blaz[es]" in the night. An unwillingness to take one's time is the source of passion's brightest and most delusive flame.

Most interesting of all, the simile also returns us to my opening discussion of the observational mood as an aspect of our experience as readers. Here, our initial confusion about what Milton is saying is neither prelude to violent correction ("Your mistake is evidence of your sinfulness") nor preparation for a feat of cognitive strength ("Your charge is to make sense of this difficult-to-visualize image") but rather an occasion for cognitive blurriness that needs no straightening out. When Milton's meaning finally comes into focus, clarity can only be experienced as quiet confirmation of what we already know. To be sure, there is much to discover amid the blur—I have already explored the semantic richness of the passage—but nothing of importance (no theological or natural-philosophical point, no significant feature of the narrative arc) rests on the return of clarity. Being wrong would have been just fine; indeed, being wrong is no different from being right. Whatever the will o' wisp signifies, it amounts to the same: the psychology and behavior of both Eve and the Serpent are in perfect alignment, and Milton's meaning is clear no matter how we originally made sense of the passage. At the level of form, Milton offers us an experience of languid digression; meanwhile, at the level of content, Eve and the Serpent electrify their journey with all the enthusiasm they can muster.

In Search of Adversity

In *Pseudodoxia Epidemica* (1646), Thomas Browne includes satanic deception among the causes of human error, of which he offers a catalogue much like Bacon's "idols of the mind." "To lead us farther into darknesse," he writes, "and quite to lose us in this maze of error, he would make men beleeve there is no such creature as himself . . . wherein . . . hee begets a security of himselfe and a carelesse eye unto the last remunerations."[64] Against his stated purpose, Milton places Adam and Eve in the inverse situation: attention to satanic danger is itself a "maze of error." William Empson writes: "Milton protests overmuch that there was no danger of sensuality at all among the pleasures of Eden. The Fall is due to carelessness, letting Reason slip for a moment, not living quite for ever as in the great Taskmaster's eye. It is odd to consider that the myth was probably invented to make the frightening and abstract question seem homely and understandable; Milton uses it to give every action a nightmare importance. To hold every instant before the searchlight of the conscious will."[65] When Empson declares this "terrific fancy" "too strange and too arid to be more than the official theme of such a poem," he is more revealing than he knows.[66] The observational mood demands a reappraisal of hypervigilance as the squandering of innocence. When Eve sets out to encounter danger, Milton takes up this theme. I've shown that Eve is skeptical about the value of labor when it feels insufficiently difficult. It makes good sense, then, that her quest for efficiency gives rise to an ambition to encounter resistance—one convenient form of which is straight-up enmity. Recall the Serpent's loss of cool: the collapse of his assurance of being "fearless unfear'd." Raymond Waddington has attributed to both Adam and Eve the "carelessness" for which the theological analogue, thinking now of Milton's Arminianism, is *securitas*—which, he explains, along with *desperatio* or despair, is a looming threat to "perseverance in grace."[67] This is certainly a useful account of Milton's convictions—but what *safety* Eve originally owed to carelessness (the absent *cura* of "security") she ultimately loses in her pursuit of efficiency, which is also a bid for heroism. Letting the emphasis fall on Eve's desire for approval rather than hardship itself (he compares her to a teacher's pet), David Quint has perceptively described the fantasy that leads to her corruption as the "wish" to be "a solitary, virtually self-sufficient being undefiled by worldly temptation and approved in faith by God."[68] This desire, I suggest, is a response to a fantasized privation. Conflict generates the sensation of struggle Eve finds wanting in work. Ultimately, the Forbidden Tree serves as an occasion on which to generate a level of frustration suggestive of—but not, in fact, evidence for—productivity.

The very manner of Adam and Eve's initial debate about the nature of labor signals the conceptual proximity of effort and adversity; their exchange grows labored in the specific sense that they both show increasing defensiveness. When Adam first responds to Eve's argument against leisureliness, he does so "mild[ly]"—anticipating the gentle dispassion the Son displays in *Paradise Regain'd* (1671) when he parries Satan's arguments rather than steeling himself against them (9.226). Even as Adam defends the rightness of a workday periodically interrupted by physical and intellectual "Refreshment" ("food," "sweet intercourse / Of looks and smiles") against Eve's ambition for speedy success, he nonetheless offers her praise: "Well hast thou motion'd, well thy thoughts imploy'd" (9.237–39, 229). For late modern readers, these lines probably feel patronizing, but the gentle expansiveness of Eve's initial proposal conveyed the same intention Adam here expresses: to create a space for easy disagreement. In this respect, he follows her lead. When he explains that God is not as "strict" an overseer as Eve assumes, he mimics God's abstention from over-"strict[ness]"—not only in conducting their daily tasks, but in debating the merits of innocent labor's casual style. Thus he recalls Bacon's objection to the schoolmen's "strictness of positions, which of necessity doth induce oppositions, and so questions and altercations."[69] Soon, however, Eve's counterargument falsifies Adam's view—not by challenging his premise that disagreement can be peaceful, but by intensifying their dispute to the point of sharp contention. By responding to Adam's "mild" manner with "sweet austere composure," she takes a small but decisive step away from quiet dissent, opting instead for a kindly but unyielding intransigence that presses the point.[70]

In addition to the prospect of greater efficiency, Eve makes a second argument about the benefits of working in different regions of the garden: exposure to danger creates an opportunity for her to prove her strength. It's as if the umbrage she takes at Adam's disagreement now materializes as an object of desire. Explaining that she knows very well that they have an enemy in Satan, about whom they have been warned, Eve defends adversity as an opportunity for "firmness."

> But that thou shouldst my firmness therefore doubt
> To God or thee, because we have a foe
> May tempt it, I expected not to hear. (9.279–81)

Her expression of surprise (perhaps tinged with disappointment) is a muffled accusation. She produces the struggle she now chooses to pursue, clashing with Adam in anticipation of a conflict with their mutual "foe." A few lines below, she again defends the "firm[ness]" of her "Faith and Love," turning the idea of rectitude over in her mind (9.286). As the conversation unfolds, the value of heroic resoluteness comes to serve as the central justification for her departure.

> And what is Faith, Love, Virtue unassay'd
> Alone, without exterior help sustain'd? (9.335–36)

Conversation having now given way to conflict, Eve celebrates the value of struggle and looks forward to proving her steadfastness. Coming disconcertingly close to the line of reasoning Milton adopts in his 1644 *Areopagitica* (and thereby illustrating the difficulty of seamlessly integrating the observational mood and the public positions the poet took during his lifetime), Eve insists on finding out what she's made of.[71] This is a "satanic" ambition in the etymological sense that she thinks her best self is an "adversary"—to Adam and to Satan (though *not* to the latter when she actually crosses his path), and, at the foot of the Tree, to God Himself.[72]

Adam answers Eve's reflection on their unfolding conflict (*"This* is what I want; *this* makes virtue possible") with his own: *"This* is already an injury—no matter the outcome." For Adam, the very fact of friction is a loss in itself, even when virtue is "firm." Having to be strong is already to suffer defeat. Adam explains:

> For hee who tempts, though in vain, at least asperses
> The tempted with dishonor foul, suppos'd
> Not incorruptible of Faith, not proof
> Against temptation: thou thyself with scorn

And anger wouldst resent the offer'd wrong,
Though ineffectual found: misdeem not then,
If such affront I labor to avert . . . (9.296–302)

If Eve were to prove as unmovable as she claims, Adam argues, she would nonetheless be harmed by Satan's perception of her weakness. Notwithstanding Milton's frequent celebration of besieged firmness (in his *Mask Presented at Ludlow Castle* [1634], the heroine embodies exactly this), the poet here observes (with Adam as his mouthpiece) that arduous trial and its attendant emotions ("scorn," "anger," and "resent[ment]") are inherently damaging.[73] Even when "ineffectual," the very gesture of "temptation" inflicts harm. To be sure, a good reader might conclude that Adam's perspective is simply incorrect: I am dishonored by someone else's bad intentions? Yet if the underlying source of their disagreement is a discrepancy in attitude, the point is less obscure: Eve's sense of besiegement would certainly compromise the experience of effortlessness Adam defends as their birthright. This disagreement calls attention to the limits of a prescriptive understanding of Milton's observational mood. When Adam says, "such affront I labor to avert," he acknowledges that what he fears is already unfolding before his eyes—and, more disturbingly, within his breast. It isn't entirely up to him to decide how he feels. He painfully "labor[s]" to persuade recalcitrant Eve against her quest for heroism: "So spake domestic *Adam* in his care" (9.318). While Milton affirms virtuous attention to the responsibility of the marital bond, the rising of tempers also suggests that something has already gone wrong.

Against the fantasy of self-burnishing (but in fact self-damaging) trial, Milton narrates defenselessness.[74] In this, he has something in common with the Marvell of *Upon Appleton House*, who disarms a fortress by rendering it botanical (who chooses to write a poem about an actually existing floral fortification—and to reflect elaborately on this feature of the site). For Milton, self-protection is a question of self-care—from which innocence is an experience of freedom. The point should not be overstated: it's hard to know how labor, speech, or any ordinary activity could proceed without some measure of self-awareness. There are certainly forms of activity that go by the name of "care" for which Milton offers a positive evaluation. Adam's distressing expression of "care" for Eve shows loyalty; it does not bespeak a failure in virtue, even if it does indicate that he has now joined her on the path that leads away from their habitual state of feeling. Notwithstanding Milton's emphasis on carelessness as an experience they inhabit by default, he identifies steps they can take to achieve or maintain it: not simple triggers for the observational mood but practices that are conducive to it. Collaboration is one example: if you know someone else is paying attention, you can afford regular lapses. Eve complains about the weakness persistent togetherness shows, but Adam treasures the advantages of mutual "mind[ing]":

Not then mistrust, but tender love enjoins,
That I should mind thee oft, and mind thou me.
Firm we subsist, yet possible to swerve,
Since Reason not impossibly may meet
Some specious object by the Foe subborn'd,
And fall into deception unaware,
Not keeping strictest watch, as she was warn'd. (9.357–63)

"Not keeping strictest watch": that's an eventuality Adam thinks ordinary. Letting down one's guard is just a regular feature of innocent experience. Yet this fact proves much less worrisome if Adam and Eve both know they can count on each other's mutual "mind[ing]." How likely is it that they will both simultaneously lose track of themselves precisely when they had better not? Like a perfect inversion of Joseph Hall's wish, which I discussed in chapter 2, to redouble vigilance by making his fellow men *"so many monitors . . . which shall point me to my own rules, and upbraid me with my aberrations,"* Adam's desire is to redouble the assurance of ease.[75] He takes the fallibility of Reason for granted; it's not as if he could simply decide ahead of time to guard carefully against all possible deceptions. Instead, he affirms the value of a situation in which distraction is no big deal. As I've shown, innocent labor proceeds by casual alternation: between gardening, talking, kissing, and otherwise passing the time.

Milton takes this argument a step further. Ultimately, he shows that even "mindless[ness]" is not the vice for which readers of *Paradise Lost* sometimes take it.[76] Absent the need for fearful circumspection, the mind is at liberty to wander. That's why mindlessness is a regular feature of innocent experience. Indeed, one example of the forgivingness of Paradise is Eve's sudden and fortunate absorption by "mindless" work—an unexpected reprieve from satanic entrapment that comes after her fateful decision to part company with Adam.[77] The habits of innocence die hard.

Each Flow'r of slender stalk, whose head though gay
Carnation, Purple, Azure, or speckt with Gold,
Hung drooping unsustain'd, them she upstays
Gently with Myrtle band, mindless the while,
Herself, though fairest unsupported Flow'r,
From her best prop so far, and storm so nigh. (9.428–33)

It's easy enough to read this description as most readers have: as the very picture of error—of going heedlessly astray. Yet Milton does not blame Eve's present danger on carelessness. He explicitly identifies Eve's "best prop," and it isn't circumspection. What she conspicuously lacks is another person who would make her "mindless[ness]" less dangerous—which is exactly the argument Adam made about the advantages of sticking together. Milton here gives proof of the seriousness with which he intends the premise of salutary effortlessness—for this is in fact a beautiful scene in which Eve has forgotten her will to "firmness" and found herself absorbed instead in the delicate pleasures of gardening: "upstay[ing]" those flowers she finds "drooping." Although Eve's desire for the satisfactions of conflict has already set her mind on the wrong path, she here enjoys temporary release from the grip of ambition: immersion in "mindless" attention to the variety of Paradise.

Milton lends credence to an admiring interpretation of Eve's "mindless[ness]" by making it an accidental shield from the Serpent's first effort at seduction. Far from setting her at a disadvantage, it actually postpones her entanglement with her foe.

> With tract oblique
> At first, as one who sought access, but fear'd
> To interrupt, side-long he works his way.
> As when a Ship by skillful Steersman wrought
> Nigh River's mouth or Foreland, where the Wind
> Veers oft, as oft so steers, and shifts her Sail;
> So varied hee, and of his tortuous Train
> Curl'd many a wanton wreath in sight of *Eve*,
> To lure her Eye; shee busied heard the sound
> Of rustling Leaves, but minded not, as us'd
> To such disport before her through the Field,
> From every Beast, more duteous at her call,
> Than at *Circean* call the Herd disguis'd. (9.510–22)

This is one of the very strangest of Milton's similes. More precisely, there are two similes here, but they are interlaced. The passage doesn't disorient us for the same reason Milton's epic similes usually do: by introducing elements completely extraneous (or seemingly so) to the relationship between the analogy's terms (think of the appearance of "the *Tuscan* artist" Galileo in the comparison of Satan's shield to the moon) (1.288). Here, Milton achieves the opposite effect: overprecision rather than superabundance of detail. He gives us an extra term of comparison (the jagged steering of the ship) that suits the initial two terms (the winding Serpent and the would-be interlocutor) perfectly well, so that all three compose a harmonious set. The intentness of explanation conveyed by the supplementary image mirrors the avidity they together describe: the Serpent's fixation on capturing Eve's interest. Thus the poem tightens with the emotional pressure of Satan's eager bid for her attention, but Milton releases her from that turn of events, explaining that she "minded not" his "wanton" "lure." "Mindless[ness]," then, is all that stands between Eve and the fateful conversation with the Serpent that soon reactivates her desire for emotional intensity. In Eve's disregard, freedom from error survives a moment longer.

Perhaps nothing better illustrates the delusive pleasure of antagonism than Eve's attitude toward the Forbidden Tree. When she finds herself in its presence, she stages a scene of heightened emotional tension in which longing faces off against obligation, producing exactly the experience of frustration for which she has hoped. Strained self-management leads directly to transgression. When Nathanael Culverwel—an admirer of Bacon's, if not exactly a Baconian—complains in his *Spiritual Opticks* (1652) about the agitated oversophistication of the schoolmen, he might just as well be thinking of Eve: "'Tis their grand imployment to tie a knot, and then see if they can undo it; to frame an enemie, and then triumph over him; to make an objection, and then answer it if they can: there are speculations enough, but if you see through them, it will be very darkly."[78] In just this way, Eve disguises an encounter with herself as an encounter with the world, obscuring what Paradise invites her to behold.

Milton foregrounds the epistemological dimension of Eve's manufactured dilemma by observing that the Serpent's counterfeit versions of "Reason" and "Truth" continue to "[ring]" in her ears. Suspending herself between competing imperatives, she finds an opportunity for struggle. Her experience of "longing" fixation is perfectly contrary to the cool immersiveness of Adam's exploratory conversation with Raphael.

> Fixt on the Fruit she gaz'd, which to behold
> Might tempt alone, and in her ears the sound
> Yet rung of his persuasive words, impregn'd
> With Reason, to her seeming, and with Truth;
> Meanwhile the hour of Noon drew on, and wak'd
> An eager appetite, rais'd by the smell
> So savory of that Fruit, which with desire,
> Inclinable now grown to touch or taste,
> Solicited her longing eye . . . (9.735–43)

In relishing her proximity to catastrophe, she "longs" for what she knows she mustn't do. Once her eyes are "fixt on the Fruit," she is faced with the

problem of desire's suppression. The internal drama of self-discipline produces everything she finds lacking in innocent labor: if she weren't exerting
herself, she would already be eagerly eating—and yet the overcoming of the
prohibition will itself be experienced as a triumphant feat. Earlier, the Serpent argues that Eve's violation of God's law would be proof of "dauntless
virtue"—and at last she rises to the (supposed) challenge (9.694). In narrating
his own experience of tasting the Fruit, the Serpent has set a good (that is,
bad) example for her. He felt a "sharp desire" for it, he explains, and so he
made use of his special capacity for climbing to obtain what other creatures
could not—anticipating, in this (fictional) scene of intense competition, the
postlapsarian moment of "fierce antipathy" in which "Beast . . . with Beast
gan war, and Fowl with Fowl, / and Fish with Fish" (9.584; 10.710–11):

> . . . Round the Tree
> All other Beasts that saw, with like desire
> Longing and envying stood, but could not reach. (9.591–93)

Milton elongates Eve's parallel experience of breathless suspense. Like a cup
filled to the brim, she savors emotional surface tension before it breaks and
spills over.

As Milton describes Eve's "fix[ation]" on the Fruit, he simultaneously
recalls the very different world in which she might have remained. Though
he mentions several lines of influence that bear significantly on her decision
to reach for the Tree (the Serpent's "persuasive words," as well as the Fruit's
olfactory, visual, and tactile appeal), the structure of the sentence directs
us to the "hour of Noon" as the precipitating cause of her growing "appetite." It's the grammatical subject of the phrase, and thus Milton grants it
responsibility for "wak[ing]" the "appetite" that carries her toward the decision to transgress. The "sound" in her ears, the "smell" in her nostrils, and
the "desir[able]" sight of the Fruit are only enabling conditions for an all-too-
ordinary wish for something to eat. What explains this strange hierarchy of
causes? Why does Milton underscore the banality of lunchtime hunger by
juxtaposing it with the extraordinary performance of satanic persuasion and
the allure of the Fruit? Why does he locate special initiatory power in what
least distinguishes this climactic event from any old day? My answer is that
Milton mentions the "hour of Noon" in order to place Eve's present psychological state into stark relief against the backdrop of the default rhythms
of innocence, reintroducing the routineness of ordinary desire just as Eve
exults in the near-magical appeal of the Fruit.[79] Against the high drama of
the seduction plot, Milton calls our attention to the regularity of emotional
and corporeal vicissitude—of which hunger is a good example.

More pointedly, the poem has already identified the "hour of noon" as the appropriate time for "repose," creating an implicit contrast between Eve's excitement and the reliable effortlessness of innocence. When Eve first parts ways with Adam, she promises

> To be return'd by Noon amid the Bow'r,
> And all things in best order to invite
> Noontide repast, or Afternoon's repose. (9.401–3)

Immediately thereafter, Milton repeats these words, lamenting that Eve "never" again "[finds] either sweet repast, or sound repose" (9.406–7). (He expresses this sentiment after the Fall as well: overcome with shame after postlapsarian sex, she and Adam cover themselves with leaves but are still "not at rest or ease of mind" [9.1120].) Now, as Eve confronts temptation, she dramatizes her situation by rejecting the experience of relaxation she would ordinarily enjoy at this very moment.[80]

When Eve explains, in a final self-justifying speech, why she decides to taste the Fruit, this other world disappears entirely. Her words perfectly epitomize error; they demonstrate the role adversity plays as a persuasive simulation of virtuous effort.

> Great are thy Virtues, doubtless, best of Fruits,
> Though kept from Man, and worthy to be admir'd,
> Whose taste, too long forborne, at first assay
> Gave elocution to the mute, and taught
> The Tongue not made for Speech to speak thy praise:
> Thy praise hee also who forbids thy use,
> Conceals not from us, naming thee the Tree
> Of Knowledge, knowledge both of good and evil;
> Forbids us then to taste, but his forbidding
> Commends thee more, while it infers the good
> By thee communicated, and our want:
> For good unknown, sure is not had, or had
> And yet unknown, is as not had at all. (9.745–57)

Eve's argument is as "tortuous" as the Serpent's "Train" (9.516). After imply-ing a counterfactual narrative of sustained "for[bearance]," Eve's language reaches a crescendo of technical and definitional crunch, turning words from their senses with sheer rhetorical force. To "forbid" is to "commend," she argues, recasting prohibition as the conferral of interest. Knowledge, she goes on to explain, is no different from possession. She argues for the utter valuelessness of "good" unless it is "known" in the sense of being held fast. Yet Milton pictures innocent understanding as perennial in its unfold-ing: an experience of pleasurable receptivity uncontaminated by the pres-sure of having to draw conclusions. Like Walton and Boyle's celebration of the pastoral pleasure of enjoying what one doesn't actually own, Milton affirms an experience of intellection that doesn't need to seize hold of any-thing. Eve's exchange of open-endedness for certainty offers us an additional insight about the experience of intellectual grasping. She renders knowledge proprietary by moralizing it. Given the high stakes it necessarily implies, "good" is a thing you want to make definitively your own, which is not nec-essarily true, say, for knowledge of botanical specimens. Thus the *agon* of the Eden story—innocence versus guilt, obedience versus trespass—can be understood as an artifact of the attitude of distrust Eve embraces on her journey away from Paradise. Innocence, this scene confirms, is usually an experience of premoral contentment in which, as Richard Strier has argued, careful adjudicatory activity has no place.[81] Eve's corruption takes shape as grave decision making. Casting God as the Tree's "forbid[ding]" watchman, she clamorously opens the way with labored reasoning to the triumphant pleasure of breaking His firm law and wresting Truth from Nature.

Adam's decision to eat the Fruit, though narrated with less detail than Eve's, obeys the pattern I have delineated in her journey to the Tree. Adam resembles Eve in adopting a stance of muscular heroism. (I will return to Adam's fall below.) Scholars have pondered the alternatives Adam might have considered to following Eve's example: asking for God's pardon on Eve's behalf is one interesting possibility.[82] As John Leonard has observed, "It is impossible to say what would have happened had Adam made the Son's choice [by offering to suffer God's punishment in her place]."[83] I affirm the availability of courses of action other than the one he takes, but I wish also to point out that his decision to taste the Fruit is a precise (if abbreviated) repeti-tion of Eve's choice. Although his self-sacrifice might seem a moving expres-sion of love, his profession of unshakable "Faith" in the marital bond would sound to an ear like Montaigne's (or, I have been arguing, Milton's) like an excess of Stoic virtue: "with thee," he says, "certain my resolution is to Die" (9.906–7). He seems to have decided without cause that the present scenario is one in which something painful has to be undergone. In this sense, virtu-ous sacrifice is a euphemism for corruption: a fall from undeserved (freely given) peace to the endless trials of moral seriousness. If I seem to be coming down hard on Adam and Eve, one welcome consequence of the connection Milton establishes in this moment between error and moral self-seriousness is that it discourages the interpretation of his didacticism as plainly censori-ous. His teaching might instead be understood (much like Bacon's "idols of the mind") as a set of instructions for avoiding delusion rather than a stern warning against defying God's wishes. The distinction is real without being stark: a difference in emphasis, but not therefore an unimportant one. Guidelines for success are not solemn commandments.

The Trouble with Sharpness

Fallenness, like the experience of suspicion that leads to it, is defined by an appetite for intensities predicated on contrast. In exploring this issue, I draw inspiration from Martin's argument that, in their native innocence, "Adam and Eve experience the emotional equivalent of Milton's thornless roses: mutable and mutual sensations stripped of sharp grief or lasting pain."[84] In Eve's desire for emotional difficulty (and, as a means to it, conflict), she insists on feeling the difference between work and leisure. In Adam's description of the pleasures of philosophical dialogue, meanwhile, as well as in Eve's earlier appreciation for gentle intellectual exploration, Milton presents an alternative to her emergent eye for disparity. Before falling, both Adam and Eve savor an experience of "sweetness" defined by gentle continuity. In what follows, I clarify this distinction by describing two related dimensions of falling and fallen experience in which the overvaluation of emotional intensity is unmistakable. First, Milton aligns the desire for sharpness with epistemological error—most emphatically in Eve's conversation with the Serpent. Second, he narrates the emotional highs and lows of postlapsarian grief, especially as it corrupts sexual pleasure, as versions of this same delusion. The Fall is not only a metaphor for the passage from one attitude to the other; it's also an opportunity for readers to perceive the difference without sensationalizing it.

One last time, Adam and Eve's debate is the right point of departure. As Adam makes a final attempt to dissuade Eve from her departure, Milton observes that he argues "fervently"; the adverb conveys a change—not only from his usual state of calm but also from the experience of mounting "care" from which it emerges. For Adam, anxious "ferv[or]" is an experience of overheated agitation (9.342). This moment can also be read sympathetically as evidence of Adam's sense of responsibility for Eve: the mutual obligation of marriage. Yet when Eve denies his request and walks away, she leaves him with a surplus of effervescently terrible feeling: the sensation of fruitless importunity. "Her long with ardent look his Eye pursu'd," Milton writes, presaging by just a couple hundred lines the scene in which the Fruit "with desire . . . solicited [Eve's] longing eye" (9.397, 741–43). Adam's frustrated desire is much like hers: "ard[or]" directs his attention to an incandescently fixed point. In this way, their final prelapsarian conversation comes to an end in just the same way their innocence ends: on a conspicuous emotional high.

The temptation scene associates such longing with a failure of understanding. For Milton, outsized desire conveys a mistaken belief in the importance of the difference between the *now* of dissatisfaction and the *then* of fulfillment. In Eve's conversation with the Serpent, Milton establishes the natural-philosophical coordinates of her quest for knowledge—and the distortive effects of her impulse to narrate cognitive triumph. Thinking back now to my discussion of "th'amaz'd Night-wanderer," the poet's turn of phrase can be recognized as a homophone with an etymological claim to kinship. Trailing the ignis fatuus, this figural representation of Eve is also "the mazed Night-wanderer." Given that Milton locates Eve in the Serpent's "Labyrinth," readers are likely to notice the bewilderment implicit in "amaze[ment]."[85] More importantly, Milton's choice of words links the convergence of confusion and inflated passion to vain epistemological ambition. One cannot generalize about the value Milton places on terms (like this one) with a distinctive philosophical lineage: "wonder," for example, is the poet's word for both Eve's enthrallment by the Serpent and for Adam's virtuous response to Raphael's discourse on creation.[86] Without leaning on the assumption of an unchanging terminology, I suggest that Milton shares Bacon's suspicion of the kind of philosophical passion (which sometimes goes by the name of "wonder") that produces gaping awe. As I discussed in chapter 1, Bacon suspends an Aristotelian emphasis on beginnings and endings ("wonder" and satisfaction) in favor of the middle space of open-ended investigation. For both Eve and the Serpent, by contrast, the drama of understanding is exactly as it shouldn't be: a high-stakes coup de théâtre in which paralytic amazement is suddenly extinguished by certain knowledge. Just as Eve misunderstands innocent labor as the absence of work, she confuses an unfolding process of investigation with a failure to learn; blind to shades of difference, she flattens partial achievement into loss. Incidentally, Milton's much-discussed allusion to "the Glass / Of *Galileo*," which "less assur'd, observes / Imagin'd Lands and Regions in the Moon" can be understood as yet another affirmation of gradualism: Galileo's "glass" is "less assur'd" because he knows better than to take it for a guarantor of conclusive truth (5.261–63). Eve, however, wants answers—and she wants them now. The Serpent presents her with a false choice between "amazement" in the sense of fearful incomprehension and the supreme gratification of total understanding. Every good question she might ask about the Tree, which would aim (in Baconian fashion) for local or preliminary explanations,

dissolves in the presence of an irresistible choice between unbearable ignorance and exultant knowledge: to eat or not to eat.[87]

As Eve speaks with the Serpent, they both invoke "wonder" in order to dispel
a welter of possible lines of inquiry, confirming false assumptions by burying
them as premises in seemingly open questions.[88] Recall that Eve, in her debate
with Adam, casually expresses the seeming obviousness of a fact by rejecting
the possibility of "wonder": "what wonder if so near / Looks intervene and
smiles." Of course, she implies, gardening "so near" each other is unwise. Yet by
the time she completes the thought with the complaint that their "day's work"
is "brought to little" because they enjoy themselves too much, she shows herself to be a bad judge of obviousness. When the Serpent first addresses her, he
offers a similarly preemptive answer to a question by rejecting "wonder" about
his unprecedented powers of speech: Nothing to see here!

> Wonder not, sovran Mistress, if perhaps
> Thou canst, who are sole Wonder . . . (9.532–33)

What the Serpent explicitly explains away is his "approach" as well as his
"gaze"—rather than his speech (9.535). The question he imagines her asking
is: "How dare you approach me?" To this, his anticipatory answer is something like: "Because you are magnificently beautiful." Yet the appearance of
a serpent that opens its mouth and says, "Wonder not," is much more likely
to evoke the question Eve does in fact pose:

> What may this mean? Language of Man pronounc't
> By Tongue of Brute, and human sense exprest? (9.553–54)

The Serpent tells Eve not to ask after his interest in her when what she actually asks after is his capacity for speech. Thus he epitomizes the problem I
have just described: he offers an answer to a question *other than the one she
asks*—and other than any of the good questions she might put to him. To be
sure, the Serpent means to flatter: *she* is the "sole Wonder" of the garden and
much more deserving of attention than a lowly snake. Yet, in paying this
compliment, he models the relationship with the world he hopes Eve will
emulate: "ravishment" (which might, as with the "fatal fruit," take the form
of a specific desire to seize hold of knowledge) rather than promiscuous
interest (9.541). The Serpent eventually tells Eve that the Fruit itself enables
"Speculations high or deep," thereby locating the experience of understanding *after* that of knowledge acquisition (9.602). Knowledge takes shape as a
thing to have and use (as it also does in Eve's sophistical self-justification at
the moment she reaches for the Fruit) rather than as the suspended telos of
unfolding understanding.

I cannot accept Edwards's view that Eve's initial question about the Serpent's speech suggests a good Baconian response—that she only subsequently goes astray.[89] The problem is not with the content of the question but rather, once more, with Eve's attitude about her investigation. Milton informs us that she is "not unamaz'd" (9.552). If we translate her question into a statement about her state of mind, the result is less "I want to understand this better" than "How extraordinary!" For Aristotle, of course, the latter would be the right pathway to the former—but Eve's astonishment does not give way to inquiry; it invites a solution that dazzles no less than the initial mystery. She says as much in asking for an explanation:

Redouble then this miracle, and say
How cam'st thou speakable of mute . . . (9.562–63)

The observational mood would preclude such helpless pleading for "miracle[s]." Even a pre-Baconian natural philosopher would expect understanding to extinguish the feeling of "mirac[ulousness]"—rather than redouble it. By this late stage in the game, Eve has become an avid connoisseur of excitement.

The scene of temptation foregrounds the epistemological consequences of the desire for emotional extremes, whereas the scene of postlapsarian grief shows how resolutely one might take up residence in that delusive oscillation. Milton describes fallenness as the awful fulfillment of the fantasy that guided Eve to transgression in the first place: a redoubled pleasure that also redoubles pain. Even setting aside the "price" of such enjoyment (the unhappiness it necessarily entails), pleasure's magnification is exactly that: not improvement but enlargement. It's only better or more satisfying on the false and ultimately self-disproving premise that quantity is quality. Just as Eve's embrace of overheated desire in the presence of the Tree is a fall into fervent but solemn moralizing—she insists on a sharp distinction between right and wrong, along with an equally clear line between having the requisite knowledge for making such distinctions and remaining in ignorance—the psychology of fallenness adheres to the logic of contrastive juxtaposition.

Adam and Eve discover intensified pleasure in the superlative flavor of the Fruit, which they both describe by way of eager distinction making. Here is Eve's account of the Fruit's flavor—followed by Adam's:

. . . what of sweet before
Hath toucht my sense, flat seems to this, and harsh. (9.986–87)

. . . if such pleasure be
In things to us forbidden, it might be wish'd,
For this one Tree had been forbidden ten. (9.1024–26)

Notice how Eve elides the difference between soft pleasure and actual pain—the "flat" and the "harsh." All at once, she recasts the pleasure of innocence as privation of her present state of exaltation. Testing one thing against another, she discovers the capacity to make bodily sensation a means of evaluative judgment—rather than, as in her native state, simply enjoying whatever ordinary pleasures garden tending affords. Similarly, Adam describes a "pleasure" the "such[ness]" of which distinguishes it from everything he has previously known; crediting "forbidden[ness]" itself with the intensification of enjoyment, he recapitulates Eve's discovery of the gratifications of transgression. The gradations of pleasure inherent to innocence give way to a sense of disparity Satan has already expressed, albeit with an emphasis on pain:

> . . . the more I see
> Pleasures about me, so much more I feel
> Torment within me, as from the hateful siege
> Of contraries . . . (9.119–22)

This description of inner life resonates with Eve's experience of suspense before the Fruit: both scenes convey the intensity of the sensation of lacking something. From within the observational mood, fulfillment is readily available without imagining an outside against which to compare it.

It should come as no surprise, then, that Milton describes fallenness as a commitment to anxious calculations about better and worse, right and wrong. What Adam and Eve ultimately lose is an experience of contentment in which such questions have no place. When Eve first tells Adam of her decision to eat the Fruit, she tries (and fails) to perform such happiness.

> Thus *Eve* with Count'nance blithe her story told;
> But in her Cheek distemper flushing glow'd.
> On th'other side, *Adam*, soon as he heard
> The fatal Trespass done by *Eve* amaz'd,
> Astonied stood and Blank, while horror chill
> Ran through his veins . . . (9.886–91)

Fallen Eve displays a visibly counterfeit merriment, her "blithe[ness]" belied by the "distemper" in her "Cheek."[90] She is overwhelmed by certainty of her corruption, but she wishes to hide it. For his part, Adam experiences something like the enervating, mind-numbing "amaz[ement]" she felt when she first heard the Serpent's words—but in terribly exaggerated form. "Blank[ness]" and "horror" attend an experience of cognitive weakness in which Adam's interrogative attitude gives way to wretched certainty. In perceiving the truth, and taking it as an irreversible harbinger of doom, Adam discovers deep suffering. Though such emotions might be taken to convey a properly realist response to Eve's unwise decision, Adam in fact draws a hasty conclusion where he might have wondered about what it means—and what to do next. To recall Adam's morning invitation, that alternative is one of the conditions of the observational mood: *not knowing what you have in front of you.*

Adam and Eve's postcoital gloom shows a similar incuriosity: "Confounded long they sat, as struck'n mute" (9.1064). Perhaps nothing better illustrates the bitter fruits of fallenness than the distortion of sexuality from "adoration pure" to "contagious Fire" (4.737; 9.1036). After Milton's description of innocent sex, he offers the following observation:

> . . . Sleep on,
> Blest pair; and O yet happiest if ye seek
> No happier state, and know to know no more. (4.773–75)

After Adam and Eve experience *fallen* sex, which is already moralized when Milton describes it as "the solace of thir sin," Milton identifies sleep as exactly the time during which the "fumes" of the Fruit wreak havoc in their bodies—and so they eventually awake "as from unrest" (9.1044, 1050, 1052). The conversion of repose to "unrest" is a perfect gloss of the point I've been arguing. Pleasure has been made to *work*; in enjoyment, they struggle hopelessly to compensate for moral failure. Notice too that Milton's description of prelapsarian sexual satisfaction is itself a meditation on "knowledge." "Know to know no more" sounds like a warning against knowledge seeking, but the context suggests—in keeping with Bacon's interpretation of Adam and Eve's transgression as an attempt at moral self-legislation rather than an effort to understand the world—that Milton cautions against the specific desire to "know" a "happier state." He approves the wish to "know" as much as possible about the one they presently inhabit. Recalling the sequence in which Satan exploits the Serpent's pleasurable abandon in sleep, Adam and Eve stop enjoying sleep as luxury and begin to understand it as liability: exposure to harm.

After postlapsarian sex, Adam and Eve experience "Shame," which collapses intense suffering and extreme self-consciousness—two experiences of which innocence knows little (9.1097). Adam and Eve recently inhabited a world of gentle pleasure and immersive "mindless[ness]," but now they are plagued by a feeling of failed self-minding that redoubles their feeling of self-protectiveness. Shame is the conversion of fallen inquisitiveness (acquisitiveness) into the language of eros. As a consequence of getting exactly the thing you want ("burn[ing]" "Lust" cleaves to its object), sexual shame can only mean disappointment (9.1015). As she approaches the poem's narrative climax, Eve comes to inhabit a philosophical attitude that generates burning questions and jubilant answers, the destitution of ignorance and plenitude of certainty. This perspective then proves typical of the postlapsarian world.

Conclusion: The Paradise Without

Milton commends to us a state of emotional gentleness we already know from the ordinary experience of paying attention to something other than ourselves. For Adam and Eve, the mind wanders pleasurably until it pleadingly errs, igniting intense desire and equally intense self-awareness. Even love, which sounds like it should be focused, is best realized as adjacency: the sharing of exploratory time rather than reciprocally returned and steady gazes. *Looking around* is different from looking *at* (something) or looking *as* (someone), which are versions of the same (impassioned) experience. One reason it's useful to put it this way—in psychological rather than theological terms—is that it renders mobile an experience we might otherwise assume is particular to Eden. If I'm right that Milton grants his readers moments of enjoyment in which nothing separates the *here* of reading from the *there* of paradisal labor, we might benefit in conclusion from further reflection on our willingness to entertain the idea of the ready availability of innocent experience. Perhaps Paradise is a name for anywhere you don't know very well—anywhere you can only come to know through wide-ranging exploration. Perhaps, moreover, when the observational mood reveals unfamiliar aspects of a place you *do* know well, it *becomes* a place you don't.

One reason for the success of Milton's invitation to the observational mood is that it feels like an opportunity for affective ordinariness. Rather than encourage readers to achieve an otherworldly state of perfection through imaginative projection, Milton identifies a familiar mood as itself paradisal bliss. That's why grasping Adam and Eve's observational mood is less a question of mediation (of, say, visualization) than of recognition. The minimal conditions for it—soft emotion and a wandering mind—ensure its ready availability (if in a great variety of different forms). We know how to answer Milton's request that we partake in his protagonists' easygoingness, or at least we know that we have plenty of occasions on which to do so: if not now, then sometime soon. The observational mood comes and goes.

I completed this chapter's anatomy of error by arguing for Adam and Eve's fallen appetite for stark distinctions. That point supplies us with an additional reason to ponder meaningful similarities between Paradise and the fallen world. Have we too often ruled out the possibility of convergence because we are fixated on what cannot be reconciled? To accept the softness of the disparity is itself to approximate the innocent mind. Paradise looks infinitely far away when our states of mind most resemble the postlapsarian despair of Adam and Eve.

Yet there are problems with my emphasis on sameness. Why would Milton go to the trouble of enjoining us to reclaim Paradise if he understood it as eminently available all along? If innocence is well within our reach, moreover, why does it remain so rare an achievement? (Recall Milton's well-known vision of human history as the serial resilience of "one just man" in the face of general depravity, which is difficult to square with the view that innocence is effortless [11.818].[91]) To these queries, the first response has to be my refrain that Milton cannot be expected to be perfectly consistent in every line of this expansive poem: sometimes he really does emphasize the difficulty of remaking Paradise. My second answer is also a repetition of a point I already discussed: we can affirm the obvious difference between innocent psychology, which I have suggested is freely available to us, and the prelapsarian condition, with its endlessly elastic workdays, natural plenitude, and freedom from the threat of death, which isn't already ours for the taking. My third and last answer is the most compelling, since it elaborates my Baconian-inflected perspective on the poem. At the risk of stating the obvious: ease is not the same as probability. We know very well that a thing can be both doable and unlikely. For Milton, humankind has made a lamentable habit of making things much harder than they need to be.[92] Perhaps when we place our trust in exertion, as the fallen are wont to do, we adopt a self-assured complacency that assumes consciousness of effort determines value. Perhaps, indeed, it's easier to think things difficult (and therefore not have to accomplish them) than to think them easy (and divest oneself of every last excuse): such is the laziness that captures our imaginations under the guise of laboriousness. To be sure, the opposite assumption, when formulated as a bald generalization (a blanket trust in effortlessness), is no more persuasive—but a softer version of it (detachment from what is reassuring about the feeling of effort) serves as an antidote to the habit of mistaking difficulty for meaningfulness. Like other critics of the old learning, which Bacon caricatured in the *Advancement* as "fierce" futility, Milton directs our attention to the tranquil pleasure of exploration.

Drawing a connection between this disarmingly easy shift of perspective and the undeniable arduousness of the projects to which Milton committed himself (such as the program of political reform crushed—or stalled—by the Restoration), I suggest the poem can be understood as an earnest attempt at seduction. It makes the first step of, say, ideological commitment—rightness of attitude—utterly inviting, and then leaves the rest to our resultant clarity of vision. David Norbrook writes: "One of the consolations the poem offers is that even if the new order created by the English republicans was unstable and short-lived, at least it lasted longer than God's. As eventually published, *Paradise Lost* had no occasion to seize; but it could still transmit to future generations the republican principles that were waiting in their Chaos."[93] Another of its gifts to the future is an invitation to effortless demystification: an entryway to an experience of "advancement" not limited to, but certainly inclusive of, political change. One goal of this book has been to show that Baconianism can be much more welcoming than scholars have noticed: that, alongside arduous heroism, it offers us an effortless but not therefore useless experience of enjoyment. One aim of this chapter was to show that *Paradise Lost* fulfills its version of this affective promise—if only in miniature. In moments, the observational mood is ours.

Critics have not given this dimension of Milton's epic much thought. One sensible reason for this oversight is the motif of sublimity that pervades (and sometimes controls) the scholarship. I am not the first to express reservations about the priority given to this aspect of Milton's poetry. Ricks, for instance, observes that Milton's "Grand Style is as remarkable for its accurate delicacy as for its power," cautioning us against a "dangerously exclusive" focus on intensity.[94] What are the consequences of such "dangerously exclusive" attention? Samuel Johnson's case for a poetics of "astonish[ment]" sets the agenda for subsequent appraisals. What interests me here is not his belief in the importance of sublimity to *Paradise Lost* (which seems to me incontestable) but rather his suggestion that this quality is "natural" to Milton's voice. In the following passage, this assertion brings us back to the very question of artificial "grace" with which this book began: the refusal of scholars to take seriously the idea that artlessness might not be a calculated performance. Johnson writes: "[Milton] had accustomed his imagination to unrestrained indulgence, and his conceptions therefore were extensive. The characteristic quality of his poem is sublimity. He sometimes descends to the elegant, but his element is the great. He can occasionally invest himself with grace; but his natural port is gigantic loftiness. He can please when pleasure is required; but it is his peculiar power to astonish."[95] Johnson acknowledges that Milton can be "elegant," "grace[ful]," and "pleas[ing]," but he subordinates these features of the poem to "great[ness]," "gigantic loftiness," and "astonish[ment]." The former are deliberate deviations from the default "sublimity" of Milton's imagination; thus he "*descends* to the elegant," "*invest[s]* himself with grace," and only "please[s]" "when pleasure *is required*" (emphasis added).

Just like Montaigne's interpreters, Johnson attributes qualities like "grace" and "elegance" to strenuous cultivation, attributing even the simplest effects of Milton's poetry to artfulness. He epitomizes a habit I would like us to unlearn: of casting effortless simplicity as an optical illusion. As Milton's chronicle of humankind's appetite for error has it, self-conscious artfulness can itself be a put-on: a thing we insist on because we are afraid of self-forgetfulness. In Milton's Paradise, however, the world offers itself up as the proper object (an inexhaustible set of objects) for innocent attention—irrespective of what it means for the self-image of the inquirer. At the end of the poem, the angel Michael promises fallen Adam that if he cultivates virtue in himself, he will "not be loath / To leave this Paradise, but shalt possess / A paradise within" (12.585–87). Like readers of Milton's poem, however, Adam and Eve have already known the mild but delicious happiness of lapsed self-awareness. Milton has presented Adam, Eve, and us with an alternative to Michael's emphasis on interiority: a Paradise without.

 # POSTSCRIPT

Here is one possible ending to this story. With the rise of Newtonian physics, science at last straightens its course.[1] The amateur hands the baton to the professional. The successors of the virtuosi redouble their commitment to rigor and develop a program for its enforcement. As Robin Valenza explains, the eighteenth century sees a simultaneous narrowing and refinement of expertise: the movement from a Renaissance paradigm of multidisciplinary involvement to the "modern" alternative of field entrenchment, and the promotion of technical languages in some fields but not in others—in physics, for example, but not in moral philosophy.[2] Confined both conceptually (by specialization) and linguistically (by jargon) to narrower lanes of investigation, the scientist has much less room to wander. We might think here of Hooke's famous rivalry with Newton, who prefers mathematical clarity to the mess of Baconian trials.[3] As for dispassion, Newton renders it "otherworldly": in place of the peculiar marriage of ordinary emotional peace and perceptual chaos I investigate in this book, the magus-priest of nature performs beatific serenity.[4] In *The Prelude* (1850), William Wordsworth, reminiscing about his time at Cambridge, describes a statue of Newton in Trinity College Chapel that confers oracular strangeness on him:

And from my pillow, looking forth by light
Of moon or favouring stars, I could behold

The antechapel where the statue stood
Of Newton with his prism and silent face,
The marble index of a mind for ever
Voyaging through strange seas of Thought, alone.[5]

The solitude of Newton's "mind," his distance from ordinary human community, suggests that there was already something marmoreal about his persona.

There are other stories to tell about the future of science: all roads do not lead to Newton. There are also other stories entirely: all roads do not lead to natural philosophy. Many of the genres we now call "literary" remain hospitable to—indeed, generative of—observational moods. Looking ahead from my vantage point in the seventeenth century, I can identify poems, essays, and novels that prioritize open-ended receptivity over self-reflection, the welter of phenomena over the shapeliness of the text: the Romantic era is the obvious (but not the only) place to go.[6] The orientation of literature toward a world that far exceeds its holding capacity—that distinctive quality of "outwardness" does not depend on any particular crossdisciplinary alliance. The intimacy of literature and science at the threshold of modernity simply calls it to our attention. Yet what comes into view in this book is not just a state of feeling but a distinctive relationship between emotional life and intellectual labor. The observational mood deforms the Baconian project from within—and yet many of its participants present themselves as grateful for rather than thwarted by affective interference. That looseness of purpose—an almost unlimited willingness to see things out—is distinctive of early modernity's scientific imagination. Is there a place to go in our own late modern culture to explore that same gentle chaos—not any old case of carelessness but drifting disobedience to rigor?

One possible point of comparison is my own field of literary studies. I bring this book to a close with a brief discussion of the observational mood as it might figure (and already has) in the practice of textual analysis. What I have to say privileges what happens far downstream from the dams and pumping stations of literary theory; I intend it as a query or suggestion that finds validation or receives needed qualification in actual experiences of reading. One efficient way to locate my approach among extant forms of literary criticism is to hold this book apart from the quest for a "postcritical" method that currently animates much scholarship in the humanities.[7] Unhappy with the routineness of the critic's adversarial relationship to the text—a suspicion of surface appearances sometimes attributed to the influence of Marxism, psychoanalysis, and poststructuralism—literary scholars have expressed frustration with that attitude and have gone in search of alternatives. The perspective I explore in these pages casts doubt on some of the reasons that have been given for such a reorientation of interpretive practice, even as it looks very much like a postcritical—or perhaps, to speak historically, "precritical"—inquiry. Outside the Baconian context, "carelessness" might sound like nothing other than the trusting embrace of surfaces: an inducement not to bother with decryption. Yet my protagonists are unimpressed by appearances; they are committed to the unearthing of secrets. One of the strange lessons of the Baconian moment—one from which the debate over postcritique shows we can still benefit—is that "idols of the mind" only sometimes call for smashing; other times, the observational mood is just the solvent we need to rid ourselves of them. When our active love rather than our passive indifference breathes life into a lie, seeing the world differently might mean unclenching the fist of our conviction. In its unknowingness, the observational mood is an escape from the tyranny of "gut instinct" and the sedimented "knowledge" of ingrained habit. By implicitly construing objects of inquiry as still-unfolding mysteries, it dependably forestalls their precomprehension. That is why I understand this book as closer in spirit to the target of postcritical polemicists than it is to their counterproposals. Though Paul Ricoeur's phrase, "the hermeneutics of suspicion," advertises its vagueness by yoking together Marx, Nietzsche, and Freud, the very looseness of the concept permits my affirmation that both this book's argument and its objects of study fit the bill—but without the appropriate emotional atmosphere.[8] Baconianism envisions the world as an endless series of hallucinations that need to be dispelled. With the benefit of our knowledge of the observational mood, we are now prepared to see how little we can assume about what it actually feels like to see through them. Bacon imagines his intellectual foes as animals cruelly deprived of daylight; they're "fierce with dark keeping" because of the inevitable frustration of seeking understanding without taking the trouble of inquiring into things.[9] In his view, then, it's exactly a hermeneutics of exposure that keeps the peace. That's why it's no paradox to affirm that careless disregard might grant effortless access not only to nature's secrets but also to whatever other ones are kept from view—be they psychological, social, or political. Some of the most compelling postcritical theories make the mistake of abstracting psychology from method—as if the former followed naturally or automatically from the latter.[10] For instance, Eve Kosofsky Sedgwick's brilliant critique of "paranoid reading," which has inspired much postcritical agitation, adopts a vocabulary that risks transforming procedure into pathology, when in fact there's nothing mysterious or compulsive about the practices of our best "paranoiacs"—who, upon closer inspection, turn out not to be "paranoiacs" after all, having described in some detail the forms of ideological subterfuge for which they seek out effective methods of exposure.[11] As I argued in the introduction, the connection between emotion and hermeneutic technique has to be established in every case. "Critique" encompasses too many different interpretive practices for it to entail a single psychological profile. Indeed, even a seemingly stable method can be employed variously, accommodating different attitudes, styles of execution, and horizons of expectation.

The debate over critique is only one instance of a general pattern. In literary studies, metareflection on interpretive practice often functions rhetorically as the ultimate expression of intellectual power. The observational mood, I suggest, deprivileges such theories, rendering their incompleteness conspicuous. We might say that it encourages a less "top-heavy" criticism: one that retains the flexibility to examine its underlying assumptions but also, more importantly, that understands the value of forgetting all about first premises—if only for a time. In this book, the content of the literary work depends on the various invitations it extends, eliciting underdetermined effects. Drawing readers into a sequence of experiences it doesn't fully anticipate, the work is an open structure that *leads* the way without *being* the way. When we inhabit the observational mood, we find ourselves pondering whatever turns up in the practice of interpretation—however apparently beside the point. This is not in the least to turn against hermeneutic self-consciousness but rather to confer value on the alternation, amply illustrated in this book, between methodological ambition and the pleasure of letting it go. Nor is it to suggest that we haven't already been doing a good job as literary scholars; indeed, there's something of this willingness to wander in many of our most influential critical practices—from the lingering of the historicist's attention on the sheer profusion of local detail to the sensitivity of the deconstructionist to the interpretive byways of *différance*. (I am comfortable using the language of "close reading" to describe such receptivity, as long as it does not imply the work's organic unity or "pattern of resolved stresses"; such disavowals of mess are anathema to carelessness.[12]) In his reflections on "the neutral," Roland Barthes speaks eloquently of an experience set free from norms of comprehension in which one might "accept the predicate [of a sentence] as nothing more than a moment: a time"—translating into grammatical terms the kind of free-floating observation (and style of reading) I explore in this book.[13] Like the natural world as it appears in Baconian writing, the "world" we access in literary works (in whatever works we examine with the exploratory attention encouraged by training in literary studies) does not always wait for us to prepare ourselves with motives, premises, or agendas before it begins to reveal itself to us. As scholars, however, we tend not to acknowledge the wayward experience that often precedes our interpretive rationale. What might happen to the experience of literary criticism if we embraced this aspect of what we do by professing our willingness to untether our perceptions from methodological machinery, teleological confidence, and even (following Boyle) logical entailment? To do so is not to sever all ties with procedures of sense making that have often served us well; it is only to grant the objects of our study temporary freedom from our ambitions for them. In early modernity, literature and science share a propensity to savor the strangeness of novelties that retain their interest regardless of how they came into view; across the field of Baconian inquiry, the moment-to-moment discontinuity of both literary and scientific writing detaches the present (if not final) value of a given phenomenon from the question of the rightness of the procedure that originally brought it to light. One of the unlikely conclusions I cannot help but draw from the readings that compose this book is that we can learn to be less scientific about our research and thus more alive to our discoveries by spending more time in the company of the first modern scientists.

Notes

Introduction

1. This introduction draws inspiration from Anne-Lise François's description of "a mode of recessive action that takes itself away as it occurs . . . locat[ing] fulfill-ment not in narrative fruition but in grace, understood both as simplicity or slight-ness of formal means and as a freedom from work, including both the work of self-concealment and self-presentation" (*Open Secrets: The Literature of Uncounted Experience* [Stanford: Stanford University Press, 2008], xvi). Though her ambition is a theory of action while my interest is the experience of perception, I have learned a great deal from her sui generis account of literary casualness. Two additional distinctions will go some way in clarifying the extent to which this book resonates with her project. First, my focus is the "fruition" of happily wasted (or "recessive") action: the rendering productive of the apparently squandered, or at least the expec-tation that carelessness will ultimately count. Second, I understand the feeling of unexpended energy as interestingly compatible with, though disruptive of, the most sophisticated of "formal means." Throughout this book, what looks complicated feels "simple." For further reflection on developments in literary theory outside the historical field, including the debate over the legacy of critique and suspicion in which François participates, see the postscript.

2. *OED*, s.v. "observation," n.

3. A wonderfully expansive collection of "histories of scientific observation" shows the distinction between my approach and emergent scholarly norms. "Like experiment," explain Lorraine Daston and Elizabeth Lunbeck in their introductory essay, "observation is a highly contrived and disciplined form of experience that requires training of the body and mind, material props, techniques of description and visualization, networks of communication and transmission, canons of evidence, and specialized forms of reasoning" ("Introduction: Observation Observed," in *Histories of Scientific Observation*, ed. Lorraine Daston and Elizabeth Lunbeck [Chicago: Uni-versity of Chicago Press, 2011], 3). My interest, however, is what is *not* "contrived" in observation. (I explain my perspective on this question below in my discussion of Steven Shapin.) One contribution to this volume that addresses my question of feeling as knowing is Elizabeth Lunbeck, "Empathy as Psychoanalytic Mode of Observation: Between Sentiment and Science," 255–75. For the connotations of "observation" and "experiment" in early modernity (and an account of their eventual convergence), see, in the same collection, Gianna Pomata, "Observation Rising: Birth of an Epistemic Genre, 1500–1650," 45–80.

4. This book tends to oppose the observational mood to concepts like objectivity and scientific dispassion. Another option is to defend them from misunderstanding on the grounds that we have failed to appreciate their amenability to disorder. Like

all concepts that have received relentless critique, objectivity and dispassion bear a heavy semantic burden: we know too well what they're supposed to mean. My adoption of a new vocabulary evades the problem of fighting an uphill rhetorical battle every time I invoke my key terms. Insofar as any understanding of these familiar concepts posits an origin story in the seventeenth century, however, *Light without Heat* should be understood as an argument for the unrecognized importance of drift and disorientation to the practices both paradigms underwrite. In understanding the place of dispassion in early modern philosophy, I have benefited in particular from Susan James, *Passion and Action: The Emotions in Seventeenth-Century Philosophy* (Oxford: Oxford University Press, 1997). With special attention to Descartes and Spinoza, she offers a nuanced description of the reliance of "dispassionate *Scientia*" on the opposition between distinctively intellectual emotions and potentially disruptive passions (183–207). My account does not sustain the tight association she observes in Descartes and Spinoza between pleasurable emotions that conduce to knowledge and the experience of reasoning in particular: the observational mood invites trains of associative thought and errant perceptions that cannot be neatly located under the rubric of the rational. I give sustained attention to conceptual cousins of scientific dispassion in the body of this introduction. On the notion of objectivity, see Lorraine Daston, "Baconian Facts, Academic Civility, and the Prehistory of Objectivity," in *Rethinking Objectivity*, ed. Allan Megill (Durham: Duke University Press, 1994). For a revisionary account of the emergence of this concept in the seventeenth century (and one with which I am in conversation throughout this book), see Joanna Picciotto, *Labors of Innocence in Early Modern England* (Cambridge: Harvard University Press, 2010). For a focused discussion of Bacon's role in this development, see Julie Robin Solomon, *Objectivity in the Making: Francis Bacon and the Politics of Inquiry* (Baltimore: Johns Hopkins University Press, 1998). For an interpretation that tilts modern, see Lorraine Daston and Peter Galison, *Objectivity* (New York: Zone Books, 2007).

5. Across the humanities, our descriptive language for unfeeling tends misleadingly to extremes, privileging the (exaggerated) fruits of untiring self-cultivation. Thus we understand the dream of an escape from turbulence much better than the unexceptional fact that some experiences carry little in the way of an emotional charge. Our critical vocabularies are disposed to capture flatness (widespread critical interest in boredom is one example), but they tend to fall short when mere emotional softness buoys awareness. I do not think I overstate my case when I observe how difficult it is even to utter the word "indifferent" without unconsciously (or semiconsciously) flinching at the impact of unbendable hardness. Yet cooler climes of emotional experience need not suggest severity, deficiency, or godlike invulnerability. The point is not surprising in itself; what is surprising is how little we have attended to the subject. For theories of boredom, see Elizabeth S. Goodstein, *Experience without Qualities: Boredom and Modernity* (Stanford: Stanford University Press, 2005); Adam Phillips, "On Being Bored," in *On Kissing, Tickling, and Being Bored: Psychoanalytic Essays on the Unexamined Life* (Cambridge: Harvard University Press, 1994), 68–78; and Patricia Meyer Spacks, *Boredom: The Literary History of a State of Mind* (Chicago: University of Chicago Press, 1995).

6. See my discussion of the figure of echo in the conclusion to chapter 1.

7. See "The Passions of Inquiry," in Lorraine Daston and Katharine Park, *Wonders and the Order of Nature, 1150–1750* (New York: Zone Books, 1998), 303–28.

8. My formulation resonates with Bruce R. Smith's imaginative meditation on the color green, which he uses as a lens to explore the "continuity between intellect and other ways of knowing" in early modernity (*The Key of Green* [Chicago: University of Chicago Press, 2009], 40). He takes Andrew Marvell's "green thought in a green shade" as a point of departure; I suggest (in chapter 3) that one of Marvell's other poems from the same period, *Upon Appleton House* (1651), is exemplary of the observational mood (11).

9. Between the sixteenth and eighteenth centuries, Daston and Park argue, "once-frivolous curiosity took on the virtuous trappings of hard work" (305).

10. I distinguish wonder (*admiratio*) and its cousins from the observational mood, but I have benefited considerably from scholarship that focuses on that passion. Mary Baine Campbell describes a philosophical emotion that "embraces surprise" and "enjoys the excess and alteration which generate it," while she associates "indifference" with "repression" (*Wonder and Science: Imagining Worlds in Early Modern Europe* [Ithaca: Cornell University Press, 1999], 3, 6). Daston and Park emphasize wonder's role as preliminary to Baconian purposefulness: "Wonder caught the attention; curiosity riveted it" (311). Their description of an experience of "musing admiration" that precedes "startled wonder" resonates with my discussion (303). On the original Aristotelian position, Jonathan Lear is illuminating: "Although philosophy begins in wonder, it ends in lack of wonder. . . . The desire to know achieves its deepest satisfaction in the philosopher who understands the principles and causes of the world" (*Aristotle: The Desire to Understand* [Cambridge: Cambridge University Press, 1988], 6). Interestingly, Aristotle's original formulations emphasize nonpragmatic desire: "It is owing to their wonder that men both now begin and at first began to philosophize; they wondered originally at the obvious difficulties, then advanced little by little and stated difficulties about the greater matters, e.g., about the phenomena of the moon and those of the sun and the stars, and about the genesis of the universe. And a man who is puzzled and wonders thinks himself ignorant (whence even the lover of myth is in a sense a lover of wisdom, for myth is composed of wonders); therefore since they philosophized in order to escape from ignorance, evidently they were pursuing science in order to know, and not for any utilitarian end" (Aristotle, *Metaphysics*, 1.2, 982b12–22, trans. W. D. Ross, in Aristotle, *The Complete Works of Aristotle: Revised Oxford Translation*, ed. Jonathan Barnes [Princeton: Princeton University Press, 1984], 2:1554).

11. Daston and Park, *Wonders and the Order of Nature*, 305. "Curiosity," they go on to explain, "had . . . become a highly refined form of consumerism, mimicking the luxury trade in its objects and its dynamic of insatiability" (310).

12. "Thinking without feeling" is what the observational mood feels like, relatively speaking, rather than what it is: a description of an experience rather than its ontology.

13. The persistence with which the quotidian experience of carelessness—as contingent but perceptually advantageous, as an accident that elicits gratitude—has eluded philosophical discussion deserves further reflection. Heidegger deserves credit for elevating care to a central place in philosophical discourse, but his account of *Sorge* encourages us to take for granted the arduousness of understanding: it conveys either "anxious effort" or "dedication." See especially his discussion of the fable of Care from Hyginus (*Being and Time*, trans. Joan Stambaugh, rev. Dennis J.

Schmidt [Albany: State University of New York Press, 2010], 191). On this matter, Heidegger cites Konrad Burdach, "Faust und die Sorge," *Deutsche Vierteljahrsschrift für Literaturwissenschaft und Geiesesgeschichte* I (1923), 1. For Heidegger's discussion of Burdach, see 191–92, 190n5, 191n7. When we confuse descriptions of emotional life with theories of subjectivity, I suggest, we tend to confirm the idea that difficulty is a basic feature of cognition, our perspective shifting from immersion in the world to self-delimitation (the shoring up of subjectivity). I intend no easy dig at Heidegger; my point is only that his ontological orientation precludes attention to undermotivated interest in the contents of perception (for him, they are interesting insofar as they redirect us to the primordial question of *Dasein*). In much the same way, many theories of emotion leave unasked the question of the mere contents of impassioned experience by attending instead to the kind of thing a human subject is. The observational mood suggests easy reconciliation to the fading of subjectivity—not as the ecstatic "death of the subject" or the transgression of bounded individuality but rather as the ordinary attenuation of a thing's (the self's) visibility when some other thing (the world) catches our attention. As critics go searching in emotional life for alternatives to the "linguistic turn" in the humanities, for which they blame the subject's dissolution (in language), and when they do the opposite (defending dissolution from its detractors), they tend to disregard everything emotions can do other than tell us about ourselves (or lack of selves). I take two books, each published in 2001, as exemplary illustrations, staking out near-opposite perspectives on the same polemical ground. Arriving at a conveniently allegorical moment (the new millennium as a fresh departure from the reign of High Theory), they develop starkly incompatible lines of retrospective interpretation, offering conjectures about feeling in order to make contrary cases about subjectivity. Philip Fisher wishes to remain above the fray (he tends not to name his adversaries), but it would be hard not to understand his defense of "vehement passions" that express "militancy about a perimeter of the self" as an argument on behalf of the subject's coherence in the face of deconstructive danger (*The Vehement Passions* [Princeton: Princeton University Press, 2001], 248). One of his striking insights is that almost every philosophy of emotion takes a single passion as a "template" for all the others—so that, for instance, the Stoic conception of feeling as a disturbing "interruption of the self" is an elaboration of a primary fixation on fear (7, 61). He follows the pattern himself, elevating anger to a position of privilege. I adopt a similar strategy, taking the observational mood as an invitation to think widely about the disorderly relationship between states of feeling and the forms of engagement they make possible. For my purposes, emotion raises the question of the means by which we gain access to whatever lies beyond the "perimeter" of self-understanding. Because my "template" is an invitation to rethink the relationship between emotions and the actions they animate, I refrain from making pronouncements about the nature of the self. For Fisher, by contrast, the paradigm of anger translates emotional experience into an occasion for the confirmation of unified self-identity. The implication seems to be that diffuse emotion would be an experience of self-dispersal. While this sounds amenable to my perspective (I do believe that carelessness, in the literature of early modernity, often attends an experience of drifting disarray), the enlistment of emotion in an argument about the coherence of subjectivity is exactly the interpretive practice from which I deliberately withdraw. Faced with an intellectual climate he thinks noxious for self-identity, Fisher presents us with a theory of passion as red-hot evidence for subjectivity's survival. My second example is Rei Terada's *Feeling in Theory: Emotion after the "Death of the Subject"* (Cambridge: Harvard University Press, 2001). Her allegiance is to deconstruction, and so she develops a theory that symmetrically counters Fisher's. Defending disunity, she argues against the possibility of the subject's self-sheltering "perimeter" and confirms emotion's status as evidence for the "death of the subject." Among her discoveries is that both Jacques Derrida and Paul de Man, far from disabling our capacity to describe what we feel (one common charge against them) or perhaps even to feel anything at all (recall Fredric Jameson's announcement of the "waning of affect" in postmodernity), were themselves philosophers of emotion (Jameson, "Postmodernism, or The Cultural Logic of Late Capitalism," *New Left Review* 146 [1984]: 53–92). In good deconstructive fashion, she describes impassioned states as experiences of mediation, which is why "pathos," usually a term for identification with someone else's feelings, is at the center of her account. For Terada, then, *je est un autre*: every emotion counts as "pathos" because the self is never transparently available to itself. I don't accept a possible corollary of this argument: that mediation implies primordial suffering. Terada argues that "intellectual difficulty" is the most basic feature of emotional life in general. If this is a strictly technical claim, referring to the interpretive obstacles necessarily encountered by thought, then I raise no objection to it—but, absent clarity on this issue, it risks translating the impossibility of seamless understanding into the impossibility of nonevasive contentment. "Logically," she explains, "the first emotion *is* cognitive difficulty" (31). Having a thought cannot mean having a care; if it did, easy unfeeling could only imply disavowal: the repression of cognitive effort. One might accept Terada's perspective (I am sympathetic to it) and still wish to deny the equation of emotion's inherent vicariousness (a claim about the *structure* of self-awareness) with an experience of laborious exertion (a claim about the *feeling* of self-awareness). Emotional peace is no more (or less) indicative of self-identity than it is of self-difference. See my discussion of Terada's more recent thinking in note 25.

14. "Careless" can be synonymous with "carefree," as in Charles's language of pastoral fantasy in *As You Like It* (1599–1600): "They say he [Duke Senior] is already in the Forest of Arden, and a many merry men with him; and there they live like the old Robin Hood of England. They say many young gentlemen flock to him every day, and fleet the time carelessly as they did in the golden world (William Shakespeare, *As You Like It*, in *The Complete Pelican Shakespeare*, gen. ed. Stephen Orgel and A. R. Braunmuller [New York: Penguin, 2002], act 1, scene 1, lines 109–13). My general preference for the term "careless" reflects my interest in foregrounding the unsought quality of the observational mood—as well as the overstepping of procedural rules it encourages. A "carefree" life sounds like the achievement of a fantasy—one to which people might deliberately "flock"—and it resonates with hoped-for therapeutic (Stoic, Epicurean, skeptical) outcomes foreign to the observational mood.

15. One might also draw a distinction between carelessness and the concept of "disinterestedness" that later comes to play a central role in debates around aesthetic theory. There is a rich vein of philosophical reflection on this theme, but suffice it for the purposes of this introduction to affirm that the observational mood is *interested*—though it demands a distinctive vocabulary of free-floating or generalized interest. Unlike disinterestedness, the observational mood does not rule out the

sensual appeal of objects of contemplation; nor does it preclude reflection on the uses to which they might be put. Paul Guyer usefully sums up Kant's line of thinking by observing that "judgments of taste are disinterested" if they "arise solely from the contemplation of their objects without regard to any purposes that can be fulfilled or interests that can be served by their existence" (Guyer, editor's introduction to *Critique of the Power of Judgment*, by Immanuel Kant, trans. and ed. Paul Guyer [Cambridge: Cambridge University Press, 2000]). Such stipulations do not apply to the observational mood, the inherently undisciplined quality of which invites trains of thought that might lead anywhere—to, for instance, utility.

16. For the reception of Lucretius in early modernity, see Jonathan Goldberg, *The Seeds of Things: Theorizing Sexuality and Materiality in Renaissance Representations* (New York: Fordham University Press, 2009); Stephen Greenblatt, *The Swerve: How the World Became Modern* (New York: W. W. Norton, 2011); Ada Palmer, *Reading Lucretius in the Renaissance* (Cambridge: Harvard University Press, 2014); and Catherine Wilson, *Epicureanism at the Origins of Modernity* (Oxford: Clarendon Press, 2008). For his searching exploration of literary, in addition to historical, issues, I am especially indebted to Gerard Passannante, *The Lucretian Renaissance: Philology and the Afterlife of Tradition* (Chicago: University of Chicago Press, 2011). For subtle attention to the resonance between different philosophical traditions, including their shared commitment to ataraxia, see Reid Barbour, *English Epicures and Stoics: Ancient Legacies in Early Stuart Culture* (Amherst: University of Massachusetts Press, 1998).

17. On Virgil's relationship to Lucretius, as evidenced by these lines, see Monica R. Gale, *Virgil on the Nature of Things: The* Georgics, Lucretius *and the Didactic Tradition* (Cambridge: Cambridge University Press, 2000), 9–11. For the passage in question, see Virgil, *Georgics*, ed. R.A.B. Mynors (Oxford: Clarendon Press, 1990), 2.490–94. I have borrowed the English translation from Gale (9).

18. James I. Porter, "Lucretius and the Poetics of Void," in *Le jardin romain: Épicurisme et poésie à Rome. Mélanges offerts à Mayotte Bollack*, ed. Annick Monet (Villeneuve d'Ascq: Presses de l'Université Charles-de-Gaulle, 2003), 225.

19. Sextus Empiricus, *Outlines of Skepticism*, ed. Julia Annas and Jonathan Barnes, 2nd ed. (Cambridge: Cambridge University Press, 2000), 3.

20. On exploratory Baconianism, see, for instance, Peter Harrison, *The Fall of Man and the Foundations of Science* (Cambridge: Cambridge University Press, 2008), 208.

21. Sextus, *Outlines of Skepticism*, 12.

22. Sextus, *Outlines of Skepticism*, 11.

23. Hans Ulrich Gumbrecht associates mood or *Stimmung* with metaphors of music and weather (*Atmosphere, Mood, Stimmung: On a Hidden Potential of Literature* [Stanford: Stanford University Press, 2012], 4). He speaks of "the present of the past in substance," by which he refers to the persistence of mood *in* or *as* literary form (14). My own approach is to identify the formal occasions literary works present for readers to inhabit the atmospheres they describe: form, then, as solicitation or opportunity. Perhaps the most focused extant study of "mood" (as opposed to emotion, passion, affect, etc.) is Thomas Pfau, *Romantic Moods: Paranoia, Trauma, and Melancholy, 1790–1840* (Baltimore: Johns Hopkins University Press, 2005). I have benefited in particular from his sophisticated defense of feeling as "a latent evaluative grid" for experience (13). Pfau doesn't adopt the vocabulary of "mood," as I do, in order

to emphasize slightness; it would, for instance, be counterintuitive to conceptualize "trauma," one of his organizing rubrics, as fleeting or minimal. One intriguing point of comparison is Leo Spitzer's account of the drawing of a distinction in the nineteenth century between "milieu" and "ambiance"—between, that is, the "factual situation" denoted by "milieu" and the "quality" or "abstract" principle conveyed by "ambiance," which implies continuity in time: not simply "a sensation for a moment" but a "substance" or even "eternality" implying "a spiritual climate or atmosphere, emanating from, hovering over, a milieu" ("Milieu and Ambiance," *Philosophy and Phenomenological Research* 3, no. 1 [1942]: 187–88). The states of feeling I discuss in this book, though they emerge with some—but only some—regularity from a specific set of environments (pastoral landscapes, for instance) would be far more predictable than they are if each one simply belonged to this or that locus as its "essence" (187). Also invitingly resonant is Ludwik Fleck's pioneering work in the sociology of knowledge, in which he argues that "collective mood" is a precondition for scientific understanding: it "produces the readiness for an identically directed perception" ("The Problem of Epistemology [1936]," in *Cognition and Fact—Materials on Ludwik Fleck*, ed. R. S. Cohen and T. Schnelle [Dordrecht: D. Reidel, 1986], 101). This book likewise tracks a shared atmosphere, but it's one imagined much of the time as privately wandering thought—an issue I discuss in chapter 2 as an experience of solitary togetherness. For a thoughtful reflection on Fleck's most significant work, *Genesis and Development of a Scientific Fact* (1935), which argues for his relevance to ongoing debates about constructivist epistemologies, see Barbara Herrnstein Smith, *Scandalous Knowledge: Science, Truth and the Human* (Durham: Duke University Press, 2005), 46–84.

24. The most influential argument to this effect is Brian Massumi, *Parables for the Virtual: Movement, Affect, Sensation* (Durham: Duke University Press, 2002).

25. Where might we turn for a theory of affect that looks beyond the question of selfhood? Let me offer two examples. In *Looking Away: Phenomenality and Dissatisfaction, Kant to Adorno* (Cambridge: Harvard University Press, 2009), Rei Terada offers a searching account of modern philosophy's "phenomenophilia": its appetite for fleeting appearances, understood as opportunities for escape (however temporary) from factuality. (As indicated by my discussion in note 13 of *Feeling in Theory*, her career interestingly encompasses different approaches to theorizing emotion.) Because the transitory phenomenon doesn't belong to a shared world of concreteness, it offers relief from the coercion of the ordinary imperative to concede the reality of the given. Although the observational mood traverses the borders between thought, appearance, wish, and memory, enabling discoveries that are not, strictly speaking, phenomenal, Terada's formulations resonate with my own: "The luxuriousness of lingering comes from the lifted obligation to declare oneself [as endorsing or denying the factual]" (15). Yet *experimental* "phenomenophilia," if I can adapt her term to my subject, understands "lingering" as the engine of the process of knowledge production from which we construct a shared world of facts. One explanation for our difference of perspective is that Terada's account is squarely post-Kantian. "By normalizing appearance (Erscheinung) and requiring its acceptance," she writes, "Kant unwittingly encourages fantasies of aberrant perception that might escape his strictures and hence his recommended path to world-acceptance" (6). She goes as far as to say that Kant's First Critique itself "triggers" the dialectical process she has in

mind. Her focus on a small set of canonical literary and philosophical figures lends credence to the claim; Coleridge, Nietzsche, and Adorno really are thinking about Kant. What the precritical moment gives us, however, is a world in which the maker of facts does not speak from the far side of epistemological confidence; certainty is a vague telos that authorizes the dissolution of other coercive certainties. Thus the ephemeral is less an alternative to the collective world than the means of its transformation: errant observations revise our sense of the given. As I discuss in chapter 2, the protagonists of scientific modernity report back to each other from private reverie and exploration. What they share is not a set of reified claims but an unfolding experience of receptivity. Similarly attentive to the adhesiveness of passion (its capacity either to bind or to release us from the difficult situations in which we find ourselves) is the work of Lauren Berlant. When she raises the question of feeling, she tends to inhabit the three-dimensionality of the world it animates— pursuing a theory of the scene, the attachment, or the case. Consider, for instance, *Cruel Optimism* (Durham: Duke University Press, 2011), her investigation of what happens (affectively, socially, and politically) when "something you desire is actually an obstacle to your flourishing" (227). For Berlant, emotional life is less a distinct conceptual problem than a means of access to an expansive world: affective attachment implies an object, a set of promises the object makes, and a fantasized situation that sustains the credibility of (forever?) outstanding promises. With Berlant's model in mind, one way to take up the question of the observational mood is to inhabit the receptive moment prior to attachment, which we might also understand *as attachment*—but to a vague object like "the world." That investment might be no stronger than the mere entertainment of thought. Berlant is more attentive than I can be, given my emphasis on the experience of surrender, to the postures and performances with which people secure, reshape, or suspend their attachments. She makes a persuasive case that we can imagine forms of self-fashioning that aren't sovereign decrees or enactments of discipline. "Viscera have been taught and are teachable," she writes, and yet they are "barely known" (159). Thus she describes the strategies with which people work to change their relationships to objects without therefore casting themselves as sovereign agents of self-transformation. I have drawn on her account in my effort to understand experiences that lose track of wishes for self-betterment.

26. Gregory David de Rocher makes this point about Laurent Joubert: "Although a distinction could be made at the time between humors (*humeurs*) and spirits (*esprits*), Joubert uses the terms synonymously as stylistic variants. The same is true for the terms *emocions, passions,* and *affeccions*" (Joubert, *Treatise on Laughter,* trans. and ed. Gregory David de Rocher [Tuscaloosa: Alabama University Press, 1990], xiii). Similarly, Joseph Glanvill speaks of "emotion," "passion," and "appetite" as if these terms amounted to the same thing: "'Tis clear from experience, that, though many of our *volitions* are *motions* from the *Passion,* yet some of our *Determinations* are from the *Understanding* and *immaterial* Faculties. And sometimes we set our *Wills* to determine in things that are purely *indifferent,* to make tryal of our *Liberty;* when we find not the least provocation or incitement to the action from any *emotion* of the *body.* And indeed to suppose every action of the Will to depend upon a previous *Appetite* or *Passion,* is to destroy our *Liberty,* and to inferr a *Stoical Fatality* with all the dangerous consequences of that Doctrine" (*Scepsis scientifica* [London: 1665], 29).

27. Linda M. G. Zerilli, "The Turn to Affect and the Problem of Judgment," *New Literary History* 46 (2015): 261–86.

28. Rei Terada, *Feeling in Theory*, 4–5. I share her view that such terms "differ most valuably in connotation" (4).

29. I'm thinking in particular of Pfau's account of emotion as "an intrinsically evaluative experience" (*Romantic Moods*, 12).

30. Francis Bacon, *The Advancement of Learning*, in *The Works of Francis Bacon*, ed. J. Spedding, R. L. Ellis, and D. D. Heath (London: Longmans, 1861–1874), 3:294.

31. This book has benefited from scholarship on the history of attention. In particular, I'm indebted to Lily Gurton-Wachter's account of poetry's response to the "militarization of attention" during the Napoleonic Wars (*Watchwords: Romanticism and the Poetics of Attention* [Stanford: Stanford University Press, 2016]). In seeking to understand the problem of witnessing war (the French Wars of Religion and the English Civil War), I have learned a great deal from her nuanced analysis of the interrelation of wartime psychology and poetic form. For a philosophical account of distraction that seeks to distinguish it from a simple absence of attention (he rejects what he calls the "attention theory of distraction"), see Paul North, *The Problem of Distraction* (Stanford: Stanford University Press, 2012), 5. I'm sympathetic to his interest in the "intermittent interruption" and "unintentional withdrawal of thinking," though my own theme is almost never *not-thinking* but rather something like *thinking sideways*—or "occasionally," to borrow Boyle's terminology (which I discuss in chapter 2) (6, 7).

32. My emphasis on the ongoingness of careless investigation resonates with Sianne Ngai's account of the "merely interesting" as a late modern aesthetic category (*Our Aesthetic Categories: Zany, Cute, Interesting* [Cambridge: Harvard University Press, 2012], 1–52, 110–73). Indeed, to read her discussion of the interesting with the observational mood in mind is to discover a number of striking continuities: not only ongoingness but also open-mindedness, receptivity, ordinariness, and coolness, as well as intimate connections to realism and to science. Of particular interest is the following observation: "Provoked by the absence of a name for the difference that one is nonetheless registering, the experience of the interesting is essentially a feeling of not-yet-knowing" (132). Were I to attempt a history that traversed the ground between early and late modernity, a project I only adumbrate in the postscript, Ngai's account would point the way. The most obvious difference between my theme and the "merely interesting" is the importance of languor and the delectation of effortlessness to the observational mood—which, as we have seen, is also what distinguishes it from curiosity.

33. Harry Berger Jr., *The Absence of Grace: Sprezzatura and Suspicion in Two Renaissance Courtesy Books* (Stanford: Stanford University Press, 2000); Frank Whigham, *Ambition and Privilege: The Social Tropes of Elizabethan Courtesy Theory* (Berkeley: University of California Press, 1984). I discuss other scholarly treatments of sprezzatura in chapter 1.

34. Baldassare Castiglione, *Il Libro del Cortegiano*, ed. Bruno Maier (Torino: Unione Tipografico-Editrice Torinese, 1964), 127. For the wide range of meanings encompassed by Castiglione's "grazia," see David Quint, "Courtier, Prince, Lady: The Design of the *Book of the Courtier*," *Italian Quarterly* 37, nos. 143–46 (2000): 186.

In this introduction, I use the term simply to convey the distinctive qualities of sprez-zatura (ease, artlessness) without the implication that they are necessarily simulated. It's worth stating plainly that I intend no intervention in the interpretation of Castiglione; my purpose is to explain what gets ignored in early modern studies due to the magnetic pull of sprezzatura and related vocabularies of self-cultivation and self-display.

35. For a theory of "care" that ranges across ancient, medieval, and modern sources with extraordinary erudition, see John T. Hamilton, *Security: Politics, Humanity, and the Philology of Care* (Princeton: Princeton University Press, 2013). One might expect that a history of care would also be a history of carelessness (the etymology of *security* promises as much), but Hamilton is interested in the paradox whereby the achievement of mental peace requires fretful self-care: "How," he asks, "can one ever be without care without care?" (6).

36. Stanley Fish, *How Milton Works* (Cambridge: Harvard University Press, 2001), 541–43.

37. Marshall Grossman, *"Authors to Themselves": Milton and the Revelation of History* (Cambridge: Cambridge University Press, 1987), 139. For his discussion of the debate in Paradise, see 138–43.

38. Fish, *How Milton Works*, 541–42.

39. Reflecting on poststructuralist premises, Rita Felski observes: "Suspicion... must be directed not only at 'nature' as an object, ideal, or value but also at 'naturalness,' as that quality possessed by any style of thought that fails to draw attention to its own contingency, that yields to the lure of the accepted, the obvious, the familiar" (*The Limits of Critique* [Chicago: University of Chicago Press, 2015], 71). I discuss the call for alternatives to critique in the postscript.

40. Consider the three "functions" Greenblatt attributes to the literary work, all of which, taken together, convey to us the author's vexed situation in a sociopolitical frame: literature, he explains, is (1) "a manifestation of the concrete behavior of its particular author"; (2) "the expression of the codes by which behavior is shaped"; and (3) "a reflection upon those codes" (Stephen Greenblatt, *Renaissance Self-Fashioning from More to Shakespeare* [Chicago: University of Chicago Press, 1980], 4). Insofar as the work tells us something about the world, then, it's one we find most meaningful insofar as it enables or constrains the capacity for action. The interest of the literary critic is importantly though not exclusively characterological: the discernment of the author's expressive or dissimulating face, if I can put it this way, as it responds to changing circumstances.

41. We might also distinguish my paradigm from Stanley Fish's influential argument about seventeenth-century literature: that many of its greatest works are "self-consuming" (*Self-Consuming Artifacts: The Experience of Seventeenth-Century Literature* [Berkeley: University of California Press, 1972]). Fish describes a text that produces a change in the reader after which she, having now undergone the desired change, no longer has any use for textual medicine. By contrast, the literature of the observational mood invites the reader to navigate her own course.

42. Gail Kern Paster, *The Body Embarrassed: Drama and the Disciplines of Shame in Early Modern England* (Ithaca: Cornell University Press, 1993), 8. See also her *Humoring the Body: Emotions and the Shakespearean Stage* (Chicago: University of Chicago Press, 2004).

43. Michael Schoenfeldt, *Bodies and Selves in Early Modern England* (Cambridge: Cambridge University Press, 1999), 15, 39.

44. The point holds for any theory of emotion that envisions indifference as an achievement of self-discipline. Consider, for instance, William J. Bouwsma's classic account of the period's intellectual history ("The Two Faces of Humanism: Stoicism and Augustinianism in Renaissance Thought," in *Itinerarium Italicum: The Profile of the Italian Renaissance in the Mirror of Its European Transformations. Dedicated to Paul Oskar Kristeller on the Occasion of His 70th Birthday* [Leiden: E. J. Brill, 1975]). He pictures Seneca and Augustine as humanism's "two faces," charting a polarity between self-discipline (prefiguring Schoenfeldt, *Bodies and Selves*) and emotional vulnerability (prefiguring Paster, *Body Embarrassed*). While Stoicism elevates the brain over the heart, he explains, Augustinian humanism encourages an impassioned encounter with scripture. For the Stoic, virtue requires reason's sovereignty over the body and its passions; indeed, it's the "divine spark" of reason lodged within us that grants us access to God (10). For the Augustinian humanist, on the other hand, "the will is not . . . an obedient servant of . . . reason; it has energies and impulses of its own, and man is a far more mysterious animal than the philosophers are inclined to admit" (11). Bouwsma argues that the Augustinian tradition rejects the rule of reason and encourages interest in "the affective life of the whole man" (51). Interestingly, the scheme divides features of emotional life that my protagonists often experience as seamlessly coherent. The observational mood is soft-hearted and undisciplined, and yet it's conducive to rational thought.

45. Simon Schaffer and Steven Shapin, *Leviathan and the Air-Pump: Hobbes, Boyle, and the Experimental Life* (Princeton: Princeton University Press, 1985). We can crystallize the difference between my argument and theirs by distinguishing between the *discovery* of "a calm space" (my interest) and "the creation and preservation" of one (theirs) (76). Where Shapin and Schaffer see "calm" as an opportunity for natural philosophers to "heal their divisions, collectively agree upon the foundations of knowledge, and thereby establish their credit in Restoration culture," I argue for the importance of *perceptual* effects (76). I also take my distance from Shapin and Schaffer's account of "virtual witnessing" as "the production in a *reader's* mind of such an image of an experimental scene as obviates the necessity for either direct witness or replication" (60). My interest is the extent to which reading itself can be understood as an experience of "witness" *as* "direct" as watching an event unfold.

46. Steven Shapin, *A Social History of Truth: Civility and Science in Seventeenth-Century England* (Chicago: University of Chicago Press, 1994), 120. For more on scientific sprezzatura, see Jay Tribbi, "Cooking (with) Clio and Cleo: Eloquence and Experiment in Seventeenth-Century Florence," *Journal of the History of Ideas* 52, no. 3 (1991): 417–39.

47. Michael Fried's account of eighteenth-century paintings that owe their success to the rejection of "theatricality"—that refuse, in other words, to acknowledge the beholder's presence—chimes with my description of the observational mood (*Absorption and Theatricality: Painting and Beholder in the Age of Diderot* [Chicago: University of Chicago Press, 1980]). When a painted figure is absorbed in something (like blowing a bubble or reading a book), Fried argues, she demonstrates her obliviousness to the beholder. Though such figures might or might not be taken to inhabit the observational mood, what they definitely can't do is grant us access to it. Indeed, the vector of attention the work sends away from the beholder is a strategy for hooking her interest. In the cases I explore in this book, by contrast,

unselfconsciousness is just one possible consequence of exploring a scene—and so the reader is left the freedom to wander away from the author's interests. That's why the absence of self-vigilance is crucial for me but not for Fried. Although the observational mode invites an experience of feeling unobserved (or simply not feeling like anyone is watching), which is exactly what Fried's engrossed figures display to the beholder, this (absent) feeling need not imply intensity of focus; in the observational mood, it often suffuses an experience of mental wandering. By implicitly arguing for absorption as theatricality, Fried ultimately returns us to the problem of sprezzatura: Fried's version of absorption belongs to the tradition of "art-concealing art." Walter Benn Michaels articulates this last point as a claim about the convergence of (radicalized) absorption and theatricality ("Neoliberal Aesthetics: Fried, Rancière and the Form of the Photograph," January 25, 2011, http://nonsite.org/article/neoliberal-aesthetics-fried-ranciere-and-the-form-of-the-photograph).

48. In the philosophy of science, the distinction has often been used to rule out the analysis of discovery. One influential formulation is Karl Popper's: "The initial stage, the act of conceiving or inventing a theory, seems to me neither to call for logical analysis nor to be susceptible of it. The question of how it happens that a new idea occurs to man—whether it is a musical theme, a dramatic conflict, or a scientific theory—may be of great interest to empirical psychology; but it is irrelevant to the logical analysis of scientific knowledge" (*The Logic of Scientific Discovery* [London: Routledge, 2002], 7). My own interest here is less the philosopher's technical distinction than the ordinary difference between thinking something up and demonstrating it to be the case. My premise is that literary criticism is better suited than "logical analysis" to understanding the generation of "new idea[s]," and I find it instructive that Popper sees "music," "drama," and "science" as similar creatures of thought.

49. For a critique of the discovery/justification distinction, see Paul Feyerabend, *Against Method*, 4th ed. (London: Verso, 2010), 150–61. For my interpretation, which explores a state of feeling that can pass seamlessly between different scientific and literary practices, his metaphor for the continuity of discovery and justification is apt: "A river may be subdivided by national boundaries but this does not make it a discontinuous entity" (150). In writing this book, I have drawn inspiration from his defense of epistemological anarchism.

50. Much of Bruno Latour's career speaks to this question, but see especially his *We Have Never Been Modern*, trans. Catherine Porter (Cambridge: Harvard University Press, 1993). The difficulty of a Latourian reading of the early modern case is that the first experimental natural philosophers are explicit about the artifactuality of knowledge. This is not to say that they qualify their claims to truth the way Latour might, but his polemic is more powerful in a culture (our own) that too often takes for granted the pristine naturalness of the scientific fact. Latour's description of his object of study is fitting for the early modern scene: "Quasi-objects are much more social, much more fabricated, much more collective than the 'hard' parts of nature, but they are in no way the arbitrary receptacles of a full-fledged society. On the other hand they are much more real, nonhuman and objective than those shapeless screens on which society—for unknown reasons—needed to be 'projected'" (55). Indeed, no one has ever been more knowingly "unmodern," even as they reject much of the wisdom of previous generations, than the Baconian natural philosophers who unlearn as much as they can in the interest of attaining Adamic innocence (see Harrison, *Fall*

of Man, and Picciotto, *Labors of Innocence*). In her subtle account of a shared theory of "making" that links literature and science in early modernity, Elizabeth Spiller beats me to my point about how little time we have to wait (no time at all, if we date the beginnings of modern science to the turn of the seventeenth century) for scientists to show an awareness that knowledge is *made*: "Gilbert, Harvey, Kepler, and Galileo do not simply do what Latour says but, indeed, articulate their own versions of his arguments" (*Science, Reading, and Renaissance Literature: The Art of Making Knowledge, 1580–1670* [Cambridge: Cambridge University Press, 2004], 9).

51. Shapin, *Social History of Truth*, 16.

52. For Shapin's claim that "distrust is something which takes place on the *margins* of trusting systems" and for his rejection of the possibility of a "practicing scientist" who embraces "thoroughgoing skepticism," see *Social History of Truth*, 19; see also 17–22. On the contingency of distrust on trust, Shapin cites Barry Barnes, *About Science* (Oxford: Blackwell, 1985), 59–63. Barnes's discussion of the "Mpemba effect" is indeed instructive about late modern scientific practice, but it does not speak to the importance of doubt—of some ongoing sense of distance from certainty—to the experience of receptivity this book sets out to understand.

53. Shapin, *Social History of Truth*, 10. As Montaigne puts it, including himself in the judgment: "Anyone who does not feel sufficiently strong in memory should not meddle with lying" (*The Complete Essays of Montaigne*, trans. Donald Frame [Stanford: Stanford University Press, 1948], 23).

54. For the philosophical importance of "maker's knowledge," see Antonio Pérez-Ramos, *Francis Bacon's Idea of Science and the Maker's Knowledge Tradition* (Oxford: Oxford University Press, 1989). For a thoughtful discussion of experimentalist interest in artisanal knowledge, and the resistance to it of artisans themselves, see Picciotto, *Labors of Innocence*, 178–87. For more on the "history of trades," see Walter E. Houghton Jr., "The History of Trades: Its Relation to Seventeenth-Century Thought: As Seen in Bacon, Petty, Evelyn, and Boyle," *Journal of the History of Ideas* 2, no. 1 (1941): 33–60, and Kathleen H. Ochs, "The Royal Society of London's History of Trades Programme: An Early Episode in Applied Science," *Notes and Records of the Royal Society of London* 39, no. 2 (1985): 129–58. For a nuanced account of scientific interest in utility, including the "history of trades," see Michael Hunter, *Science and Society in Restoration England* (Cambridge: Cambridge University Press, 1981), 87–112.

55. Hans Blumenberg, *Paradigms for a Metaphorology*, trans. Robert Savage (Ithaca: Cornell University Press), 6–12, 21.

56. Blumenberg, *Paradigms for a Metaphorology*, 22. Savage notes that the original has "splendid isolation" in English (22n32); my assumption is that the reference is to political isolationism (with the case of nineteenth-century Britain in mind), casting "truth's" association with "strain" as a strategic alliance.

57. The gloss has become common wisdom, but see *OED*, s.v., "fact," n.

58. Hans Blumenberg, *The Legitimacy of the Modern Age*, trans. Robert M. Wallace (Cambridge: MIT Press, 1993), 397.

59. Blumenberg, *Legitimacy of the Modern Age*, 397. The phrase comes from his discussion of Descartes.

60. Blumenberg, *Legitimacy of the Modern Age*, 380.

61. Blumenberg, *Legitimacy of the Modern Age*, 385, 383.

62. John Aubrey attributes the remark to William Harvey (*Brief Lives*, ed. Andrew Clark [Oxford: Oxford University Press, 1898], 299).

63. I've abbreviated Latour's formulation: "Nothing is, by itself, either reducible or irreducible to anything else" (*The Pasteurization of France*, trans. Alan Sheridan and John Law [Cambridge: Harvard University Press, 1988], 158).

64. Here are Daston and Park on the advent of modern epistemology as the replacement of ease with labor: "Augustinian curiosity had also been of the flickering sort [as Descartes's version would be], but slack and aimless, in the guise of distraction; Aquinas had gone so far as to liken this wandering curiosity to sloth, so undirected was its affect. In contrast, early modern curiosity replaced the earlier dynamic of self-dissipating passivity with one of self-disciplined activity, all faculties marshaled and bent to the quest" (108).

65. Stephen Gaukroger writes: "The choice . . . is . . . between the active or practical and the contemplative life, where philosophers and to a lesser extent poets had traditionally fallen in the latter category, although neither poesy nor philosophy were strictly incompatible with the former. The explicit shift to the defence of the active or practical life, however, does place new requirements on these activities, for their practitioners now had to show that they were able to live up to the aims of the active or practical life. What Bacon effectively does is to transform philosophy into something that comes within the realm of *negotium*. This is completely at odds with the Platonic conception of philosophy" (*Francis Bacon and the Transformation of Early-Modern Philosophy* [Cambridge: Cambridge University Press, 2001], 55). He goes on to observe that Bacon's view sometimes "looks like a reconciliation of *negotium* and *otium*, and, in context, such a reconciliation is a way of making the same point, that is, of undermining the purely contemplative view of philosophy" (55n50). This argument risks suggesting that negotium + otium = negotium. One of my wagers in this book is that the immersive experience of literary reading invites us to understand and even inhabit, rather than loosely affirm, the coordination of serenity and effort, which would otherwise remain an empty paradox or, as here, the subordination of one concept to the other. Picciotto is perhaps the only scholar who has pursued this question with the subtlety of the literary critic, but her argument also privileges negotium; the most important payoff of the integration of action and contemplation is the reinvention and expansion of the former. "In urging a union of the active and contemplative lives," she writes, "experimentalists were widening the meaning of the 'active life' beyond the martial, beyond the political, to embrace the arts that shaped the material conditions of human existence" (*Labors of Innocence*, 130).

66. I'm thinking in particular of Carolyn Merchant, who offers an illuminating critique of the ideology of the New Science, but not one that should be taken as adequately descriptive of Bacon's works or legacy. See her *The Death of Nature: Women, Ecology and the Scientific Revolution* (New York: HarperCollins, 1980), which I discuss in chapter 1. Bacon's reputation as a philosopher remains tarnished, even in the larger culture, as evidenced by the following example, which I take more or less at random: "Science in its wider modern sense, as Francis Bacon's quest for the relief of man's estate through the technological conquest of nature by reducing all matter and energy to a resource, has brought us to the point that the United Nations' Intergovernmental Panel on Climate Change now tells us we have 15 years to halt global warming or the results will be catastrophic and irreversible" (Gregory Fried, "The

King Is Dead: Heidegger's *Black Notebooks*," *Los Angeles Review of Books*, September 13, 2014, http://lareviewofbooks.org/review/king-dead-heideggers-black-notebooks).

67. Hans Blumenberg, *Care Crosses the River*, trans. Paul Fleming (Stanford: Stanford University Press, 2010), 14.

68. Blumenberg, *Care Crosses the River*, 14.

69. I exchange "consciousness" for "mind," which would thus not be confined to explicit awareness, and I drop Blumenberg's proprietary metaphor.

70. The first phrase, quoted from Voltaire, appears on the very first page of Max Horkheimer and Theodor Adorno's most famous work. See their *Dialectic of Enlightenment: Philosophical Fragments*, trans. Edmund Jephcott, ed. Gunzelin Schmid Noerr (Stanford: Stanford University Press, 2002). For more on the second phrase, see Max Horkheimer, *Critique of Instrumental Reason*, trans. Matthew J. O'Connell et al. (London: Verso, 2012), especially vii–x, but also 1–33 and 136–58.

71. Max Horkheimer, *Eclipse of Reason* (New York: Continuum, 2004), 4.

72. Horkheimer and Adorno, *Dialectic of Enlightenment*, 2.

73. Horkheimer and Adorno, *Dialectic of Enlightenment*, 2.

74. "We have no doubt," write Horkheimer and Adorno, "and herein lies our petitio principii—that freedom in society is inseparable from enlightenment thinking" (xvi). Yet they also affirm, in the same breath, that such thinking "contains the germ of the regression which is taking place everywhere today" (xvi).

75. Horkheimer, *Eclipse of Reason*, 38.

76. Harrison, *Fall of Man*. For the importance of biblical hermeneutics to the rise of science, conceived as the restoration of innocence, see his *The Bible, Protestantism and the Rise of Natural Science* (Cambridge: Cambridge University Press, 1998). See also Charles Webster, *The Great Instauration: Science, Medicine, and Reform, 1626–1660* (Oxford: Peter Lang, 2002). I discuss millenarian aspects of Baconianism (Webster's interest) in chapter 3.

77. Harrison, *Fall of Man*, 8.

78. For a wonderfully revisionary account of the scientific revolution as a set of anxiogenic paradoxes, which offers a philosophical rather than literary vocabulary for restaging Picciotto's conflict between "fiction" and "instrument," see Ofer Gal and Raz Chen-Morris, *Baroque Science* (Chicago: University of Chicago Press, 2013). Here, the contest takes place between cold neo-Stoic withdrawal and impassioned empirical investigation. "One has to choose," they write, "between the vigorous inquisitiveness of the savant and the tranquil wisdom of the scholar; neo-Stoicism was not a viable moral option to complement the New Science—it was a competing, alternative intellectual program" (272). I affirm the acuity of the observation, but only insofar as it applies, to borrow their vocabulary, to a coherent philosophical "program." If the only states of feeling available to early modern scientists were inextricable from systematic theories, a stark choice would indeed have to be made. Yet the apparent illogic of the actual complementarity of pensive indolence and relentless pursuit is exactly my interest.

79. *OED*, s.v. "nonchalance," n. Between 1500 and 1700, Early English Books Online yields a single instance of "nonchalance" (and no instances of the adjectival form) in English-language sources other than French dictionaries.

80. This has been a subject of general interest, but see especially Michel Foucault, *The History of Sexuality*, vol. 2: *The Use of Pleasure*, trans. Robert Hurley (New York:

Vintage Books, 1990); Michel Foucault, *The History of Sexuality,* vol. 3: *The Care of the Self* (New York: Vintage Books, 1988); Pierre Hadot, *Philosophy as a Way of Life: Spiritual Exercises from Socrates to Foucault,* ed. Arnold Davidson, trans. Michael Chase (Oxford: Blackwell, 1995).

81. For a history of early modern science focused on exactly this question of self-cultivation, see Matthew L. Jones, *The Good Life in the Scientific Revolution: Descartes, Pascal, Leibniz, and the Cultivation of Virtue* (Chicago: University of Chicago Press, 2006).

82. Stanley Cavell, *The Claim of Reason: Wittgenstein, Skepticism, Morality, and Tragedy* (Oxford: Oxford University Press, 1999), 19.

83. Raymond Williams, *Marxism and Literature* (Oxford: Oxford University Press, 1977), 134.

84. Williams, *Marxism and Literature,* 132.

85. Kathleen Stewart, "The Point of Precision," *Representations* 135 (Summer 2016): 38, 40.

86. To offer a single example of each, which together conveniently illustrate the slow swing of the pendulum from one approach to the other, we might think here of Marjorie Hope Nicolson, *The Breaking of the Circle: Studies in the Effect of the "New Science" upon Seventeenth-Century Poetry,* rev. ed. (New York: Columbia University Press, 1950), and Frédérique Aït-Touati, *Fictions of the Cosmos: Science and Literature in the Seventeenth Century,* trans. Susan Emanuel (Chicago: University of Chicago Press, 2011).

87. I have not been able to follow Henry S. Turner's stimulating suggestion that scholars might "expand concepts of form beyond linguistic and textualist models" to encompass networked relationships between all manner of physical and immaterial entities—though it's true that I often trace connections between reading in the strict sense and other forms of interpretation (such as visual inspection). See his "Lessons from Literature for the Historian of Science (and Vice Versa): Reflections on 'Form,'" *Isis* 101 (2010): 582, 580.

88. Some of my casting decisions have been painful. Thomas Browne is almost a personification of gentleness. "I could never divide my selfe from any man upon the difference of an opinion," he writes, "or be angry with his judgement for not agreeing with mee in that, from which perhaps within a few dayes I should dissent my selfe" (*Religio Medici,* in *The Works of Thomas Browne,* ed. Geoffrey Keynes [Chicago: University of Chicago Press, 1964], 1:15).

1. "Nonchalance" and the Making of Knowledge

1. Horkheimer and Adorno, *Dialectic of Enlightenment,* 2.

2. Bacon's "method," Carolyn Merchant writes, "so readily applicable when nature is denoted by the female gender, degraded and made possible the exploitation of the natural environment" (Merchant, *Death of Nature,* 169). Peter Pesic argues that Bacon's use of the term *violence* "did not have the abusive sense it has in modern usage" ("Francis Bacon, Violence, and the Motion of Liberty: The Aristotelian Background," *Journal of the History of Ideas* 75, no. 1 [2014]: 69–70). An argument about terminology, however, doesn't settle the thematic question. For the many entries in the contentious debate around Bacon's "torture of nature," see Pesic, 70n3.

3. For Merchant's discussion of Bacon's use of the Proteus image, see *Death of Nature,* 169, 171.

4. Francis Bacon, *The wisdome of the ancients*, trans. Arthur Gorges (London, 1619), 69.

5. Scholarly inattention to affective inertia (I borrow this metaphor of restful directionality from physics) makes good sense; for the ancient traditions that capture Bacon's attention, thinking about the passions means learning how to avoid, control, and manage them. The Epicurean's bid for imperturbability (ataraxia), the Pyrrhonian skeptic's suspension of judgment (on which he likewise pins hopes for blissful ataraxia), the neo-Aristotelian's program of even-keeled moderation (*mediocritas*)—conceptions of equilibrium like these are avowedly prospective when they aren't frankly asymptotic (a near-impossible outcome calls for endless striving). The point is especially clear in the case of northern Humanism's great touchstone of self-supervision, Stoic indifference (*apatheia*), with its many shades of meaning from "cheerfulness" (*euthumia*) to Seneca's translation of the concept into neutral-sounding "tranquility" (*tranquillitas*). For these interpretations of apatheia and of the Stoic tradition, see Richard Sorabji, *Emotion and Peace of Mind: From Stoic Agitation to Christian Temptation* (Oxford: Oxford University Press), 182. His nuanced interpretation of Stoicism saves it from caricature. Even with our sympathy, however, Stoicism remains an ethos of self-control; its therapeutic aim is not conducive to the observational mood. The point is subtle, since even as important a text of the Stoic revival as Justus Lipsius's *De Constantia* (1584) distances itself from a hyperbolic insistence on discipline by distinguishing "constancy" (*constantia*) from "obstinacy" (*pertinacia*). See Justus Lipsius, *Concerning Constancy*, ed. and trans. R. V. Young (Tempe: Arizona Center for Medieval and Renaissance Studies, 2011), 26–29.

6. Bacon, *Cogitationes de natura rerum*, in *Works*, 3:25.

7. Ronald Levao, "Francis Bacon and the Mobility of Science," *Representations* 40 (1992): 15, 16. Another example is Charles Whitney, who argues that Bacon "strains" the "wonderful ambiguity" of *instauratio* as both conservative "reform" and spontaneous "revolution" "to the breaking point" (*Francis Bacon and Modernity* [New Haven: Yale University Press, 1986], 15). Joshua Scodel's sensitive description of Bacon's oscillation between means and extremes offers a different approach to Levao's point about alternation (*Excess and the Mean in Early Modern English Literature* [Princeton: Princeton University Press, 2002], 48–76). When he notes "an audaciously offhand parenthesis" in which Bacon acknowledges the sufficiency of "good Thoughts" to God even as he insists on the importance of "good Works," he touches on my theme: Bacon's habit of producing seismic effects with casual turns of phrase (72).

8. John C. Briggs, *Francis Bacon and the Rhetoric of Nature* (Cambridge: Harvard University Press, 1989), vii.

9. Peter Dear points out that Bacon never actually uses the word "method" to describe the procedure he recommends in the *Novum Organum*, but does not object to our habit of doing so. See *Revolutionizing the Sciences: European Knowledge and Its Ambitions, 1500–1700*, 2nd ed. (Princeton: Princeton University Press, 2009), 137–38.

10. Guido Giglioni, "From the Woods of Experience to the Open Fields of Metaphysics: Bacon's Notion of *Sylva*," *Renaissance Studies* 28, no. 2 (2014): 243, 242. Intriguingly, Giglioni remarks that "Bacon was . . . convinced that lack of order— an attitude, as it were, of *sprezzatura* and *neglegentia diligens* in the way information was being collected—could favour the active engagement of the interlocutor, the reader, or the collaborator"—but he does not elaborate the point (243). His

attention, here and elsewhere, to self-conscious strategy indicates the difference between our perspectives: "Lingering in the forest of experience is no lackadaisical stroll in the well-tended gardens of curiosity, but is based on strong ontological commitments, with clear therapeutic ends in view: to lose one's way in the forest of particulars is the prerequisite to finding the right path to an intellect purged from the spectres of imagination" (256). Still, his account is the best precedent of which I am aware for my interpretation of Bacon's oeuvre. Another scholar who attends to what in Bacon is not reducible to his disciplinary ethos is Catherine Wilson: "We need to remark from the outset," she writes, "that the Baconian ideal of scientific knowledge embraced a number of competing models. Both pure, undistorted, direct reception and the elaborated, indirect, tortured 'vexing' of nature furnish Bacon with metaphors for the acquisition of knowledge" ("Visual Surface and Visual Symbol: The Microscope and the Occult in Early Modern Science," *Journal of the History of Ideas* 49 [January–March 1988]: 95).

11. In this chapter, I pay special attention to interpretations that link Bacon's interest in Epicureanism to Montaigne's: Kenneth Alan Hovey, "'Montaigny Saith Prettily': Bacon's French and the Essay," *PMLA* 106, no. 1 (1991): 71–82; and Passannante, *Lucretian Renaissance*, 120–53.

12. For Bacon's interest in Epicureanism, see Reid Barbour, "Bacon, Atomism, and Imposture: The True and the Useful in History, Myth and Theory," in *Francis Bacon and the Refiguring of Modern Thought: Essays to Commemorate* The Advancement of Learning *(1605–2005)*, ed. Julie Robin Solomon and Catherine Gimelli Martin (Aldershot: Ashgate, 2005), 17–44; and Passannante, *Lucretian Renaissance*, 120–53.

13. For the influence of Montaigne's awareness of "la faiblesse de la raison humaine abandonée à ses seules forces" on Bacon's *Novum Organum*, see Pierre Villey, *Montaigne et François Bacon* (Paris: Revue de la Renaissance, 1913), 109.

14. I do not follow Villey in seeking empirical proof of influence: "En somme, je ne trouve dans l'apologie de la science [Bacon's *De augmentis scientiarum*] ni une opposition complète au point de vue de Montaigne, ni des réminiscences qui soient de nature à prouver qu'en écrivant Bacon avait le texte des *Essais* présent à l'esprit" (*Montaigne et François Bacon*, 62). He is intent on making a sharp distinction: "Toute cette conception d'une science rigide que la philosophe anglais, pénétré qu'il est des méthodes des sciences physiques, prétend imposer aux études morales, tout cela est absolument étranger à Montaigne" (75).

15. Bacon, *Of the Dignity and Advancement of Learning*, in *Works*, 4:421.

16. On Montaigne's Socrates, see Timothy Hampton, *Writing from History: The Rhetoric of Exemplarity in Renaissance Literature* (Ithaca: Cornell University Press, 1990), 134–95; Eric MacPhail, "Montaigne and the Trial of Socrates," *Bibliothèque d'Humanisme et Renaissance* 63, no. 3 (2001), 457–75; Joshua Scodel, "The Affirmation of Paradox: A Reading of Montaigne's 'De la Phisionomie' (III: 12)," *Yale French Studies* 64 (1983): 209–37.

17. Michel de Montaigne, *Essais*, ed. Alexandre Micha (Paris: Garnier-Flammarion, 1969), 2:220. All citations of Montaigne's *Essais* are from this edition. For the English, I follow *The Complete Essays of Montaigne*, trans. Donald Frame (Stanford: Stanford University Press, 1948), but here I slightly alter the translation to capture Montaigne's diction more precisely (see 359). Through the end of this chapter, I henceforth cite these sources parenthetically. When citing both the original and the translation,

I give the page number for the English edition followed by volume and page number for the French.

18. David Quint, *Montaigne and the Quality of Mercy: Ethical and Political Themes in the Essais* (Princeton: Princeton University Press, 1998), ix. He argues persuasively that flexibility is a central ethical value in the *Essais*.

19. I thank Andrea Gadberry for her reflections on polyptoton in Montaigne.

20. For Montaigne's "nonchalance" as an appropriation from Castiglione, see Felicity Green, *Montaigne and the Life of Freedom* (Cambridge: Cambridge University Press, 2012), 141–83; Marcel Tetel, "The Humanistic Situation: Montaigne and Castiglione," *Sixteenth Century Journal* 10, no. 3 (1979): 69–84; and Quint, *Quality of Mercy* (see note 21). I do not share Green's view that Montaigne's project in the *Essais* is to secure a personal space of liberty, imagined as the achievement of self-possession. "Carelessness," she writes, "is about being unaffected . . . in the sense that one's will is preserved in a state of indifference and equanimity" (147). She faults Quint for suggesting that "nonchalance" is "a repudiation of autonomy and independence," which exaggerates his view—but incidentally approximates the one I elaborate in this chapter (155). Green explains that Montaigne's "powerlessness . . . is the mark of an unacquired, unmastered goodness, owed to no one but himself"; she transforms weakness into self-mastery (163).

21. Quint writes, "[Montaigne's] claim to an easy native goodness as opposed to a virtue that requires struggle can thus be read as an expression of aristocratic hauteur and *sprezzatura* comparable to the disdain for pedants and professional writers against whose carefully structured and argued works he pointedly opposes the apparently—and one must, of course, emphasize 'apparently'—impromptu and dilettantish jottings of the *Essais*. The very style of the essays that Montaigne describes with two adjectives that translate the idea of Castiglione's *sprezzatura*—'desdaigneux' and 'mesprisant'—mimes a kind of natural effortlessness that, in turn, proclaims the nobility of the writer" (60). I wish to suspend his compulsory "apparently."

22. Green is an exception (*Life of Freedom*, 141–83).

23. For a meditation on the "accidental" in Montaigne, see Ann Hartle, *Michel de Montaigne: Accidental Philosopher* (Cambridge: Cambridge University Press, 2007), especially 33–38. I do not share her view that "accidental philosophy *ends* in wonder, not at the rare and extraordinary but at the most familiar," but only because this word carries philosophical baggage; wonder tends to imply either a purposeful, investigative posture or an attitude of gaping awe, and it gives too much emotional coherence to a book that conspicuously wanders (38). I affirm her conclusion about the "content" of wonder in Montaigne: "What is did not have to be at all" (38).

24. See especially Quint, *Quality of Mercy*, 42–74.

25. Quint, *Quality of Mercy*, 43.

26. Quint, *Quality of Mercy*, xiv, xv.

27. I accept Quint's wonderful suggestion that the "easygoing morality of yielding that Montaigne advocates may be the ground and condition, rather than the result, of his skepticism" (*Quality of Mercy*, x).

28. Wayne A. Rebhorn observes: "Castiglione . . . use[s] the words *natura* and *naturalmente* to describe forms of behavior opposed to affectation and related to the notions of purity and simplicity" (*Courtly Performances: Masking and Festivity in Castiglione's* Book of the Courtier [Detroit: Wayne State University Press, 1978], 43).

29. Thus he finds an unlikely ally in Adorno, who is likewise an innovator in fragmentary forms. See his *Minima moralia: Reflections on a Damaged Life* (New York: Verso, 2005), and, for his reflection on the genre, "The Essay as Form," in *Notes to Literature*, vol. 1, ed. Rolf Tiedemann, trans. Shierry Weber Nicholson (New York: Columbia University Press, 1991).

30. Interestingly, when Montaigne *disavows* "nonchalance," he treats it as a verb: a thing you can do rather than a state you simply inhabit. In "Des menteurs," responding to the charge that he is a bad friend because he forgets his obligations, he writes, "Certainly I may easily forget; but careless about [mais de mettre à nonchalloir] the charge with which my friend has entrusted me, that I am not [je ne le fay pas]" (1:22, 71; emphasis added). Montaigne signals that he does not self-consciously *adopt* a careless style of behavior; he just so happens to behave this way.

31. Catherine Belsey, "Iago the Essayist," in *Shakespeare in Theory and Practice* (Edinburgh: Edinburgh University Press, 2008), 157–71.

32. William Shakespeare, *Othello*, in *Complete Pelican Shakespeare*, act 3, scene 3, lines 118–24.

33. On Montaigne the diplomat, see Timothy Hampton, *Fictions of Embassy: Literature and Diplomacy in Early Modern Europe* (Ithaca: Cornell University Press, 2009), 62–72.

34. One example is Hovey, "'Montaigny Saith Prettily.'" For Renaissance interpretations of *honestum*, see Hampton, *Fictions of Embassy*, 46–47 and, for the specific case of Montaigne, 63.

35. Cicero writes: "The principle with which we are now dealing is that one which is called Expediency [utile]. The usage of this word has been corrupted and has gradually come to the point where, separating moral rectitude from expediency, it is accepted that a thing may be morally right without being expedient, and expedient without being morally right. No more pernicious doctrine than this could be introduced into human life" (Cicero, *On Duties*, trans. Walter Miller [Cambridge: Harvard University Press, 1913], 177, 176). He makes the following observation about the consequences of dishonorable action: "But if he [who wrongs his fellow men] believes that, while such a course could be avoided, the other alternatives are much worse—namely, death, poverty, pain—he is mistaken in thinking that any ills affecting either his person or his property are more serious than those affecting his soul" (293).

36. The tradition is epitomized by François Mitterand's official portrait of 1981, in which the twinkly-eyed president of the French Republic looks up at the camera from his reading of Montaigne's *Essais* before a backdrop of leather-bound volumes. The Montaigne we meet by way of my reading is too much the moral provocateur to serve this purpose.

37. Hartle comments that "the character of Epaminondas is largely his [Montaigne's] own invention" (*Accidental Philosopher*, 82). Quint argues that he is a "*post facto* projection of the moral qualities that the essayist discovers to inhere *naturally* in himself as a kind of instinct he has retained unchanged from his nurse and infancy. . . . the essayist implicitly holds himself up as an ethical model" (*Quality of Mercy*, 64). I'm partial to Hampton's argument that Montaigne takes La Boétie as his model for Epaminondas, which would explain his interest in softening an image of Stoic hardness: "Montaigne's humanist formation was refined through his friendship

with Etienne de la Boétie, whose commitment to Stoic models of virtue rings throughout the early essays, and whose death was a blow from which Montaigne seems never
to have recovered. Montaigne felt that La Boétie was the single figure of his age who
rivaled the ancients. His depictions of La Boétie are peppered with commonplaces
about ancient heroism. For example in 'De la praesumption' he describes de la Boétie
as 'un'ame pleine.' This is the same phrase he reserves for one of the few ancients
whose valor he praises without reserve, Epaminondas, described as 'pleine par tout
et pareille,' whose homeland of Beotia recalls La Boétie's own demesne. Of the three
greatest men in history, described in 'Des plus excellens homes,' the first was a poet,
Homer, and the second was a general, Alexander. The third and greatest, Epaminondas, was both a general and an orator, both a virtuous actor and a virtuous writer—the
double of La Boétie, 'un'ame à la vieille marque,' whose untimely death is even more
devastating than the loss (also lamented by Montaigne) of Plutarch's biography of
Epaminondas" (*Writing from History*, 138–39). Despite Montaigne's hyperbolic praise
of La Boétie, I detect a discordant note.

38. We also learn that Scipio "write[s] comedies," and that he does so "nonchalantly" (*nonchalamment*) (3:851, 321).

39. To be sure, Epaminondas shows mercy to those he conquers; in this respect,
he deserves his reputation for gentle goodness. Yet his anomalous goodness doesn't
quell the strangeness of Montaigne's admiration, in this context, for martial might.

40. In "De l'experience," Montaigne writes, "I take pleasure in seeing an army
general, at the foot of a breach that he means to attack presently, lending himself
wholly and freely to his dinner and his conversation, among his friends" (3:851).

41. See, for example, "De l'experience," where Montaigne writes, "L'affirmation
et l'opiniastreté sont signes exprez de bestize" (3:286). We are not surprised, later in
the same essay, to find him describing himself as follows: "La meilleure de mes com-
plexions corporelles c'est d'estre flexible et peu opiniastre" (3:294).

42. Lucretius, *De rerum natura*, trans. W. H. D. Rouse, rev. ed. Martin Ferguson Smith,
Loeb Classical Library (Cambridge: Harvard University Press, 1982), book 2, lines 1–2.

43. For more on Montaigne's *Schadenfreude* and its comic dimensions, see David
Carroll Simon, "The Anatomy of *Schadenfreude*; or, Montaigne's Laughter," *Critical
Inquiry* 43 (Winter 2017): 250–80.

44. Lucretius, *De rerum natura*, book 2, line 3.

45. Frame inserts "likely" to convey Montaigne's "être pour."

46. Hans Blumenberg, *Shipwreck with Spectator: Paradigm of a Metaphor for Existence*
(Cambridge: MIT Press, 1997), 18. I do not accept Blumenberg's reading of Montaigne
as the skeptic "who stands unimperiled on the solid ground of the shore" (17).

47. Though he doesn't share my focus on Montaigne's coordination of affective
cool and epistemological success, my interpretation is informed by Timothy Hampton's suggestion that Montaignian individuality consists in a "jaunty," "improvisatory" responsiveness to the "contingencies of the moment" ("Difficult Engagements:
Private Passion and Public Service in Montaigne's *Essais*," in *Politics and the Passions,
1500–1800*, ed. Victoria Kahn, Neil Saccamano, and Daniela Coli [Princeton: Princeton University Press, 2006], 45).

48. My emphasis on contingency and changeability sets me on a different track
from Richard Strier, who argues for the constancy of selfhood in Montaigne, but I
agree with Strier's main point: that Montaigne thinks the ambition to be other than

you are is foolish. For Montaigne, such a transformation will not be an achievement of the self. Strier's unusual and suggestive claim that Montaigne approximates a Protestant position when he concedes that the self is subject to transformation (just not through force of will) actually allows for the most dramatic alterations in character. Even if we conclude that the kind of self-betrayal I discuss in this chapter is less a *change of* than a *deviation from* underlying character, my sense of Montaigne's moral quandary would remain the same. See Richard Strier, *The Unrepentant Renaissance: From Petrarch to Shakespeare to Milton* (Chicago: University of Chicago Press, 2011), 208–29.

49. Aristotle, *Metaphysics*, 1.1, 980b23, trans. W. D. Ross, in Aristotle, *Complete Works*, 2:1552.

50. In "Des Coches," when Montaigne does in fact speak of causes, he professes a willingness to "pile up" multiple explanations: "It is very easy to demonstrate that great authors, when they write about causes, adduce not only those they think are true but also those they do not believe in, provided they have some originality [*invention*] and beauty. They speak truly and usefully enough if they speak ingeniously. We cannot make sure of the master cause; we pile up several of them, to see if by chance it will be found among them" (3:685, 113). He cites Lucretius as an ally in advocating this procedure. See Lucretius, *De rerum natura*, book 6, lines 703–4.

51. *Essays of Michael, seigneur de Montaigne I three books, with marginal notes and quotations of the cited authors, and an account of the author's life; new rendered into English by Charles Cotton, Esq.* (London, 1685), 499; *Essays, Translated by Charles Cotton, with some account of the life of Montaigne, notes and a translation of all the letters known to be extant*, ed. William Carew Hazlitt (London, 1877), 3:392.

52. For this and related topoi ("making something of nothing" and "anything of anything"), see Gerard Passannante, *Catastrophizing: Materialism and the Making of Disaster* (Chicago: University of Chicago Press, forthcoming).

53. See *OED*, s.v. "convenient," adj.

54. Bacon, *Advancement of Learning*, in *Works*, 3:122; hereafter cited parenthetically by page number through the end of this chapter.

55. Passannante, *Lucretian Renaissance*, 130. He goes on to argue persuasively that the concerns of "Des Coches" animate this section of the *Advancement*. See also *Montaigne's Annotated Copy of Lucretius: A Transcription and Study of the Manuscript, Notes, and Pen-Marks*, ed. M. A. Screech (Geneva: Droz, 1998).

56. Hovey, "Montaigny Saith Prettily," 75–76.

57. Hovey, "Montaigny Saith Prettily," 75–76.

58. Bacon, *Speeches of the Philosopher, The Captain, The Councillor, and the Squire*, in *Works*, 8:383; "Of Truth," in *Works*, 6:378. For Bacon's revisions of the passage, see Hovey, "Montaigny Saith Prettily," 76; and Passannante, *Lucretian Renaissance*, 130.

59. For the history of the image, see Blumenberg, *Shipwreck with Spectator*.

60. Bacon, *De Augmentis scientiarum*, in *Works*, 2:171.

61. Daston and Park, *Wonders and the Order of Nature*, 307.

62. Bacon's interest is *both* the resurgence *and* the satisfaction of desire. The history of philosophy is better stocked with affirmations of the importance of quelling it. Consider, for instance, Seneca's approving account of the Epicurean garden: "Go to his Garden and read the motto carved there: 'Stranger, here you will do well to tarry; here our highest good is pleasure.' The care-taker of that abode, a kindly host,

will be ready for you; he will welcome you with barley-meal and serve you water also in abundance, with these words: 'Have you not been well entertained?' 'This garden,' he says, 'does not whet your appetite; it quenches it. Nor does it make you more thirsty with every drink; it slakes the thirst by a natural cure—a cure that demands no fee. This is the 'pleasure' in which I have grown old" (Seneca the Younger, *Epistles 1–65*, trans. Richard M. Gummere, Loeb Classical Library [Cambridge: Harvard University Press, 1917], 147). Bacon is just as committed to the reappearance of fledgling desire as he is to its moderating gratification.

63. See *OED*, s.v. "addict," v.

64. On the one hand, this reading distinguishes the observational mood from the "merely interesting," which, Ngai explains, Schlegel associates with "'restless' striving" (*Our Aesthetic Categories*, 122). On the other hand, Ngai's description of "a feeling so indeterminate that it can even be hard to say whether it counts as satisfaction or dissatisfaction" speaks directly to what is at issue in Bacon's account of learning (135). Though my emphasis falls on gratification, it is necessarily slight: a minimal pleasure.

65. Daston and Park, *Wonders and the Order of Nature*, 307.

66. Francis Bacon, *Refutation of Philosophies*, in *The Philosophy of Francis Bacon*, ed. Benjamin Farrington (Chicago: University of Chicago Press, 1964), 106.

67. Bacon, *Refutation of Philosophies*, 106.

68. See the account of Bacon's "charitable extremism" in Scodel, *Excess and the Mean*, 48–76, esp. 70–72.

69. See the introduction for reflections on the place of wonder in early modern natural philosophy.

70. Bacon, *Sylva Sylvarum*, in *Works*, 2:570.

71. For more on "gawking wonder," including a discussion of Bacon's wariness of it, see Daston and Park, *Wonders and the Order of Nature*, 316–28. For Bacon, they explain, "in small doses wonder whetted the edge of curiosity, but in larger amounts it both betokened and prolonged ignorance"; they also point out that Bacon's Restoration successors increasingly emphasize the latter (321). See also the discussion of this passage in Susan James, *Passion and Action*, 188.

72. See *OED*, s.v. "spatiate," v.

73. Brian Vickers glosses and discusses other appearances of "Lumen siccum optima anima" in Bacon's works (Bacon, *Francis Bacon: The Major Works*, ed. Brian Vickers [Oxford: Oxford University Press, 1996], 588n125).

74. In a similarly unpresumptuous mood, he explains that "time seemeth to be of the nature of a river or stream, which carrieth down to us that which is light and blown up, and sinketh and drowneth that which is weighty and solid" (131).

75. Bacon, "Of Fortune," in *Works*, 6:473.

76. Castiglione, *Il Libro del Cortegiano*, 129. Rebhorn's gloss is useful for its attention to the possibility of simple rather than forced or performative indifference: "Here [in a discussion of "graceful dancing"] the count uses *desinvoltura*, a synonym for sprezzatura that refers specifically to the ease and nonchalance of physical movements, and he most instructively modifies this synonym with the participle *sprezzata*, thus emphasizing the latter's quite different connotations. He does not describe mere indifference, the freedom from complexities and entanglements that *desinvoltura* suggests, but a scornful indifference" (34).

77. Bacon, "Of Fortune," in *Works*, 6:472.

78. See *OED*, s.v. "disinvoltura," n. For a discussion of both disinvoltura and sprez-zatura in connection with Cicero's *neglegentia diligens* and other proximate concepts, see Peter Burke, *The Fortunes of the Courtier: The European Reception of Castiglione's Cortegiano* (University Park: Pennsylvania State University Press, 1996), 31. *Desembol-tura* seems to have been a Spanish word before it was an Italian one (Burke, *Fortunes of the Courtier*, 31). For the period resonances of this word and its place in Boscán's translation, including its suggestion of "looseness" (*soltura*), see Margherita Mor-reale, "*Desenvoltura, Suelto y Soltura* en Boscán," *Revista de Filologia Española* 38 (1954): 257–64. Intriguingly, given the direction of my argument, she notes that the term can imply the danger of too much freedom: "La *desenvoltura* . . . lleva en sí el peligro de llegar a libertades inaceptables" (258). Sprezzatura's first English rendering as "reck-lessness" likewise implies risk: "But I, imagynyng with my self oftentymes how this grace commeth, leaving a part such as have it from above, fynd one rule that is most general whych in thys part (me thynk) taketh place in al thynges belongyng to man in worde or deede above all other. And that is to eschew as much as a man may, & as a sharp and dangerous rock, *Affectation* or curiosity & to speak a new word to use in every thyng a certain Reckelesness, to cover art withal, & seeme whatsoever he doth & sayeth to do wythout pain, & (as it were) not mynding it" (*The Courtyer of Count Baldessar Castilio*, trans. Thomas Hoby [London, 1561], cii).

79. I borrow Brian Vickers's translation of this phrase (*The Major Works*, 762n420).

80. Bacon, "Of Fortune," in *Works*, 6:473.

81. Bacon, "Of Fortune," in *Works*, 6:473.

82. Bacon, "Of Fortune," in *Works*, 6:472. Jessica Wolfe points out that Bacon "imagines machinery as the apotheosis of *disinvoltura* in that it both mimics and assists in the effortless display crucial to success at court" (*Machinery, Humanism, and Renaissance Literature* [Cambridge: Cambridge University Press, 2004], 25–29). She also describes Bacon's comparison of his method to a "hoisting machine," "alluding, per-haps, to the 1585 competition instigated by Pope Sixtus V to design a machine capable of moving a large Egyptian obelisk" (26). Thinking of the wheel metaphors from the passage I quoted from "Of Fortune," she explains that "Bacon defines *disinvol-tura*, or disengagement, in terms of the well-oiled gears of a machine" (26). Although she doesn't put it this way, Wolfe's discussion of disinvoltura directs attention to the experience of unfeigned effortlessness. For the relationship between machinery and courtliness, see 29–87.

83. R. S. Crane, "The Relation of Bacon's *Essays* to His Program for the Advance-ment of Learning," in *Essential Articles for the Study of Francis Bacon*, ed. Brian Vickers (London: Sidgwick and Jackson, 1972), 272–92.

84. For the end of "interdisciplinarity" with the rise of Newtonian physics, see Robin Valenza, *Literature, Language, and the Rise of the Intellectual Disciplines in Britain, 1680–1820* (Cambridge: Cambridge University Press, 2009), 10.

85. *The Major Works*, 667n278.

86. See *OED*, s.v. "reluctation," n. Julian Martin observes that "Bacon used the sin-gle term 'experiment' to describe very different sorts of activity by the natural histo-rian: the passive reporting both of observed craft practices and techniques, and of par-ticular inquiries conducted by other men; the 'artificial' investigations he carried out himself; and any subsequent, 'more subtle,' investigations" (*Francis Bacon, the State, and the Reform of Natural Philosophy* [Cambridge: Cambridge University Press, 1992], 155).

87. Rebhorn makes the following tantalizing observation about Adamic sprezzatura: "Unmistakably, this 'gracious' courtier recalls the innocent Adam, who, before the Fall, likewise did not err, enjoyed harmony among his various faculties and with the natural world, and stood erect in his God-given dignity. To be sure, Castiglione's courtier lives in Urbino, not the Garden of Eden, embodies all the characteristics peculiar to his culture, and appears as a consummate artist and performer, not a 'natural' man. Nevertheless, during his inspired performance, the ghostly presence of Adam hovers over him, animating the image he projects with the affective energy that only recollection of man's first father could generate" (*Courtly Performances*, 44).

88. In her brilliant study of Bacon's "politics of inquiry," Julie Robin Solomon defines "objectivity" as "the holding in abeyance, or erasure, of the individual mind's desires, interests, assumptions, and intents while that mind is in the process of knowing the material world" (*Objectivity in the Making*, xix). "Disinterestedness," she goes on to explain, "is more made than found, more fabricated than natural" (xix). As I discussed in the introduction, my interest is not pristine contact with nature. Furthermore, I explore cases of absent desire rather than cases in which desire is deliberately "[held] in abeyance." See also her discussion of Bacon's appropriation of the concept of *metis*, "the crafty, temporarily self-distancing and world-attentive intelligence of travelers, traders, and counselors" (115, 119). When the effortlessness of Bacon's disinvoltura is a matter of skill, it comes close to craftiness or know-how.

89. In the *Redargutio Philosophiarum*, Bacon imagines evading mental athleticism by delegating responsibility to "machines": "Men apply their naked, or unaided, intellect to the task. From the mere number or quality of the minds engaged they hope great things. By dialectics, the athletic art of the mind, they strengthen their mental sinews. But they do not bring in machines to multiply and combine their individual efforts. And, as due aids are not supplied to the mind, so nature is studied without due attention" (*Redargutio Philosophiarum*, in *The Philosophy of Francis Bacon*, ed. Benjamin Farrington [Chicago: University of Chicago Press, 1964], 128–29).

90. It also calls to mind the *Advancement*'s metaphor of the "subtilty of the illaqueation" of reason (276). "Illaqueation" refers to "the action of catching or entangling in a noose or snare" (see *OED*, s.v. "illaqueation," n.).

91. Lisa Jardine's gloss of this procedure is clarifying: "Initiative methods are those which display the stages by which the author's conclusions were reached, so that the reader may both check that he would have reached the same conclusions on the same evidence, and pursue the investigation further if he so chooses. Since the inductive method is the sole means of arriving at sound conclusions from natural evidence, it is itself the ideal method for such presentation. Every stage in the derivation of scientific principles from sense data is recorded in the induction, so that the record itself will enable a future investigator to check the conclusions and pursue them further. Where such a rigorous method of inquiry has not been used, however, it is still possible for the author to set out his original assumptions and the way in which he arrived at his conclusions" (*Francis Bacon: Discovery and the Art of Discourse* [Cambridge: Cambridge University Press, 1974], 174).

92. Brian Vickers, *Francis Bacon and Renaissance Prose* (Cambridge: Cambridge University Press, 1968), 60–95.

93. See *OED*, s.v. "perambulation," n.

94. Vickers provides an organic metaphor that unifies book 2's wayward movement: "Here we constantly feel the onward movement peculiar to this kind of structure, the overlapping effect which at the same time that it takes you off on to a new branch remains connected to the stem and thus to everything which has gone before" (*Renaissance Prose*, 56).

95. Cicero, *On Duties*, 55.

96. Cicero, *On Duties*, 55.

97. Bacon, *New Organon*, in *Works*, 4:105–6.

98. Though C. W. Lemmi juxtaposes Bacon's account of Pan's discovery of Ceres (my interest here) with a passage from Natalis Comes, he is right to acknowledge in his note that the "passage is rather a suggestion than a source"; nothing like Bacon's version, with its emphasis on chance discovery, can be found in the lines from the *Mythologiae* (1567) to which Lemmi directs our attention (*The Classical Deities in Bacon: A Study in Mythological Symbolism* [Baltimore: Johns Hopkins University Press, 1933], 69n113, 69).

99. Francis Bacon, preface, *Wisdome of the ancients*, n.p.

100. For Bacon's borrowings in "Pan, or Nature," see Lemmi, *Classical Deities in Bacon*, 61–74. Though Bacon draws from Macrobius when he compares Pan's beard to sunbeams, Jardine points out that his emphasis on the beard's length, which he takes to illustrate the great distances traveled by the sun's rays, is original (*Discovery*, 182n2, 185). There is precedent for the interpretation of Pan's cloak as an image of mottled nature in Servius, Isidore of Seville, Albricus Philosophus, and Comes (Lemmi, *Classical Deities in Bacon*, 62, 58).

101. Francis Bacon, *Wisdome of the ancients*, 34. "And thought little of it" is the translator's interpolation, suggesting that he reads these lines the way I do. See Bacon, *De sapientia veterum*, in *Works*, 6:640.

102. Bacon, *Wisdome of the ancients*, 34; Bacon, *De sapientia veterum*, in *Works*, 6:640.

103. For our purposes, Bacon's "Orpheus, or Philosophy" is equally intriguing. There, he understands Orpheus's loss of Eurydice by looking back over his shoulder as a representation of the "curious diligence and untimely impatience" of the failed philosopher (Bacon, *Wisdome of the ancients*, 58). Bacon's vocabulary of eagerness as intellectual failure (in the original, "curiosam et intempestivam sedulitatem et impatientiam") challenges our ordinary expectations about the values of Baconian science (*De sapientia veterum*, in *Works*, 6:648).

104. Bacon, *Of the Dignity and Advancement of Learning*, in *Works*, 4:413, 420.

105. Bacon, *Of the Dignity and Advancement of Learning*, in *Works*, 4:413.

106. Bacon, *Of the Dignity and Advancement of Learning*, in *Works*, 4:413.

107. On "learned experience," see Jardine, *Discovery*, 144–49. "The rational method of inquiry by the Organon promises far greater things in the end," Bacon explains, but "this sagacity proceeding by learned experience will in the meantime present mankind with a number of inventions which lie near at hand" (*Of the Dignity and Advancement of Learning*, in *Works*, 9:78).

108. William Rawley, "To the Reader," in Francis Bacon, *Sylva Sylvarum* (London, 1670), sig. A4r; Bacon, *Sylva Sylvarum*, in *Works*, 2:529.

109. For commentary on the variety of meanings of *mundus*, see Pliny, *The Natural History of Pliny*, ed. and trans. John Bostock and H. T. Riley (London: Henry G. Bohn, 1855), 13n1.

110. Bacon, *Wisdome of the ancients*, 36; Bacon, *De sapientia veterum*, in *Works*, 6:640.

111. In identifying this concept as "ultimate[ly]" Platonic, I follow Lemmi, *Classical Deities in Bacon*, 70.

112. Although she does not offer an interpretation like mine, Jardine is the first to notice the novelty of this sequence: "As far as I am aware, this exploitation of the characteristics of echoes (rather than the fable of the nymph Echo) as a symbol for the true philosophy, which is the reflection of nature itself, is original to Bacon" (*Discovery*, 185).

113. One might also gloss the term as "carefully performed or prepared, studied, meticulous." See *Oxford Latin Dictionary*, s.v. "accuratus," adj.

2. The Angle of Thought

1. Robert Boyle, *New Experiments, to the number of 16, concerning the Relation between Light and Air*, in *The Works of Robert Boyle*, ed. Michael Hunter and Edward B. Davis (London: Pickering and Chatto, 1999–2000), 6:5, 13.

2. Boyle, *New Experiments*, in *Works*, 6:4.

3. Boyle, *New Experiments*, in *Works*, 6:5–6.

4. Though I do not share his evaluative judgment or his conjectures on motivation, I echo C. J. Horne's estimation of Boyle's style: "Though he aims, like Dryden, to write as a cultured man would talk, his style is hurried and careless, and his sentences rattle on without form or elegance" ("Literature and Science," in *A Guide to English Literature*, ed. Boris Ford [London: Cassell, 1957], 4:193).

5. Consider the following example: "Boyle was endeavouring to appear as a reliable purveyor of experimental testimony and to offer conventions by means of which others could do likewise. The provision of circumstantial details was a way of assuring readers that real experiments had yielded the findings stipulated" (Schaffer and Shapin, *Leviathan*, 64).

6. Boyle, *New Experiments*, in *Works*, 6:22.

7. Boyle foregrounds his confusion most explicitly when he mentions "*the trouble we found in managing the Engine in the dark*" (*New Experiments*, in *Works*, 6:13–14).

8. I thank Lawrence Principe for helping me understand this episode—and for directing me to Boyle's account of his examination of a glowing diamond in a bed "whose Curtains were carefully drawn," again suggesting the usefulness of this location for the darkness it affords (Boyle, *Experiments and Considerations Touching Colours*, in *Works*, 4:197). I am also grateful for Principe's clarifying remarks on the relationship between sociological and religious interpretations of Boyle's prose—which I discuss at the end of this introductory discussion (email message to author, July 3, 2017).

9. Boyle, *New Experiments*, in *Works*, 6:6.

10. Robert Boyle, *An Account of Philaretus in His Minority*, in *Robert Boyle: By Himself and His Friends*, ed. Michael Hunter (London: Pickering and Chatto, 1994), 15.

11. Boyle, *Account of Philaretus*, 15–16.

12. I intend no perfect analogy with Freud's account of the child's game—only the idea of a repetition that diminishes the terror of a remembered event: "At the outset he [the child] was in a *passive* situation—he was overpowered by the experience; but, by repeating it, unpleasurable though it was, as a game, he took on an *active* part.

These efforts might be put down to an instinct for mastery that was acting independently of whether the memory was in itself pleasurable or not" (*Beyond the Pleasure Principle*, in *The Standard Edition of the Complete Psychological Works of Sigmund Freud*, trans. and ed. James Strachey [London: Hogarth Press, 1953–74], 18:16).

13. Boyle, *New Experiments*, in *Works*, 6:7.

14. Boyle, *New Experiments*, in *Works*, 6:14.

15. Boyle, *New Experiments*, in *Works*, 6:14.

16. Boyle, *New Experiments*, in *Works*, 6:19.

17. For accounts of Restoration science in which Boyle plays a central role, see both Shapin, *A Social History of Truth*, and Schaffer and Shapin, *Leviathan and the Air-Pump*. For a sympathetic but acute critique of one version of "social construction" exemplified by Shapin and Schaffer, see Latour, *We Have Never Been Modern*. See my introduction for more on these lines of argument.

18. Richard Yeo shows that Boyle "came to be regarded as an exemplar of mismanagement," though he is speaking in particular of his note-taking practices (*Notebooks, English Virtuosi, and Early Modern Science* [Chicago: University of Chicago Press, 2014], 151). One of his anecdotal illustrations relates directly to our theme: John Evelyn writes a letter to William Wotton in which he remarks that the deceased Boyle's bedroom is "crowded with 'Boxes, Glasses, Potts, Chymicall & Mathematical Instruments; Bookes & Bundles of Papers" (151, citing Evelyn to Wotton, March 29, 1696). Wotton agrees in somewhat disorderly prose: "His Papers were truly, what he calls many Bundles of them himself a Chaos, rude & indigested many times God know's" (151, citing Wotton to Evelyn, August 8, 1699). Although I offer an alternative to his description of the *Reflections*, which affirms Michael Hunter's view that their purpose is the "pursuit of moral balance, self control, and piety," I have benefited from his illuminating account of Boyle's "promiscuous" notes (136, 169, citing Hunter, in *Robert Boyle: By Himself*, xvi).

19. "Way of thinking" is Boyle's refrain. Here is one example: "The way of Thinking, whose Productions begin to be known by the name of Occasional Meditations, is, if rightly practis'd, so advantageous, and so delightful, that 'tis Pity, the greatest part, ev'n of serious and devout Persons, should be so unacquainted with it" (*A Discourse Touching Occasional Meditations*, in *Works*, 5:21; hereafter abbreviated *Discourse* and cited parenthetically by page number through the end of this chapter). Yeo offers a useful account of the Aristotelian distinction between memory and recollection, which is still very much alive in the seventeenth century. "Recollection," he explains, "was understood as a process of searching, reviewing, and comparing ideas stored in memory; it was therefore considered as a rational and deliberative activity . . . Hobbes contrasted this deliberate searching with the undirected wandering of the mind, which he regarded as its default state" (28). In the *Reflections*, "recollection" is what Boyle doesn't do as often as we might expect.

20. For an illuminating discussion of affective disorientation, see Sianne Ngai, *Ugly Feelings* (Cambridge: Harvard University Press, 2005), 14. Interestingly, Ngai's cases suggest a "dysphoric" experience, whereas my Baconian-inflected cases are voluptuous.

21. "Diffidence" is the central motif of Rose-Mary Sargent's thoughtful account of Boyle's project. "His diffidence," she explains, "led him to present his results in a tentative manner, which has led one historian to speak of his 'curious inability' to present a complete and systematic treatment of any one subject. But it was not so

much an inability as a conscious decision about style" (*The Diffident Naturalist: Robert Boyle and the Philosophy of Experiment* [Chicago: University of Chicago Press, 1995], 185). Although Sargent is right to resist the verdict that Boyle simply fails to achieve systematicity, I suggest his extravagant disorderliness is different from sensible modesty. My view is likewise distinct from Peter R. Anstey's interesting discussion of Boyle's "nescience": an avowed state of unknowing (*The Philosophy of Robert Boyle* [New York: Routledge, 2000], 42 and 43).

22. As Anstey puts it, "What emerges is that the neutrality of the corpuscular hypothesis on the issues of the divisibility of matter, the *fuga vacui*, and so on is constitutive of the corpuscular philosophy itself" (*Philosophy*, 9).

23. René Descartes, *Meditations on First Philosophy*, trans. and ed. John Cottingham (Cambridge: Cambridge University Press, 1996), 43. Descartes offers an interesting counterpoint to Boyle's indifference. Making philosophical use of the meditation, he places value on exactly those features of the genre Boyle casts aside. Indeed, he dislikes indifference because he understands drifting uncertainty as the *opposite* of freedom. "The indifference I feel when there is no reason pushing me in one direction rather than another," he writes, "is the lowest grade of freedom; it is evidence not of any perfection of freedom, but rather of a defect in knowledge or a kind of negation. For if I always saw clearly what was true and good, I should never have to deliberate about the right judgment or choice; in that case, although I should be wholly free, it would be impossible for me ever to be in a state of indifference" (40).

24. For Boyle's "legendary" piety, see Michael Hunter, *Boyle: Between God and Science* (New Haven: Yale University Press, 2009), 6. For his "scrupulosity," see Hunter, *Robert Boyle, 1621–91: Scrupulosity and Science* (Woodbridge: Boydell Press, 2000), especially the discussion at 58–71. I do not share Geoffrey Cantor's view that Hunter pathologizes Boyle by speaking of his "religiosity" ("Boyling Over: A Commentary on the Preceding Papers," *British Journal for the History of Science* 32, no. 3 [1999]: 315–24).

25. Hunter, *Between God and Science*, 205.

26. If Hunter is correct that "only one as assiduous in his spiritual exercises as Boyle would have thought it appropriate to employ such [high] standards in the laboratory," we might also conclude that someone so committed to scrupulousness would also have an appetite for the respite offered by the observational mood (Hunter, *Scrupulosity*, 69).

27. For the importance of the meditation to the literary history of the seventeenth century, see Louis L. Martz, *The Poetry of Meditation: A Study in English Religious Literature of the Seventeenth Century* (New Haven: Yale University Press, 1954).

28. Edward B. Davis calls attention to Baxter's response to the *Reflections* in "Robert Boyle as the Source of an Isaac Watts Set for a William Billings Anthem," *The Hymn* 53, no. 1 (2002): 46–47. For a different interpretation of this exchange and of the *Reflections* as a whole, see Picciotto, *Labors of Innocence*, 273–83.

29. For Boyle's use of all three terms within the space of a few lines, see his *Occasional Reflections upon Several Subjects*, in *Works*, 5:10; hereafter abbreviated *OR* and cited parenthetically by page number through the end of this chapter.

30. *The Correspondence of Robert Boyle*, ed. Michael Hunter, Antonio Clericuzio and Lawrence Principe (London: Pickering and Chatto, 2001), 2:475–76.

31. *Correspondence of Robert Boyle*, 2:474.

32. Boyle actually uses the word in a somewhat different context, but for a similar purpose. It describes a habit of digression at once useful and inefficient. Note in addition Boyle's use of parentheses to distinguish passages that veer too far away from his present purpose, a strategy I discuss below with respect to *Of the High Veneration Man's Intellect Owes to God*: "Those that require more of Method than they will here find, may be Advertis'd, That much of this Scribble being design'd to serve Particular Acquaintances of Mine, 'twas fit it should Insist on those Points They were Concern'd in: and that (consequently) much of the Seeming Desultorinesse of my Method, and Frequency of my Rambling Excursions have been but Intentional and Charitable Digressions out of my Way, to bring some wandring Friends into theirs and may Closely enough pursue my Intentions, even when they seem most to deviate from my Theme. And as for the Longer Excursions which either You, or other Judicious Friends would needs have me leave here, and there, I have for the Ease of my Perusers Annex'd to them some Marks whereby they may be taken Notice of to be Digressions, that as I Submit to their Judgement, who think they may be Usefull to some Readers, so I may Comply with my own Unwillingnesse, to let them be Troublesome to others; who by this means have an Opportunity to Passe by if they please such as they shall not expect to find themselves (either upon their Own score, or that of their Acquaintances) Concern'd in" (*Some Considerations Touching the Style of the Scriptures* [1661], in *Works*, 2:390).

33. Joseph Hall, *Occasionall Meditations* (London, 1630), 122.

34. Hall, *Occasionall Meditations*, 237.

35. For the background to the composition of Rich's meditations, see "Introduction," in *The Occasional Meditations of Mary Rich, Countess of Warwick*, ed. Raymond A. Anselment (Tempe, AZ: American Center for Medieval and Renaissance Studies, 2009), 37.

36. Rich, *Occasional Meditations*, 147.

37. Rich, *Occasional Meditations*, 47.

38. St. Ignatius, *The Spiritual Exercises of St. Ignatius: A New Translation*, trans. Louis J. Puhl (Westminster, MD: Newman Press, 1962), 15; St. Ignatius, *Los Ejercicios Espirituales de San Ignacio de Loyola*, ed. M. R. P. Juan Roothaan (Zaragoza: Hechos y Dichos, 1959), 92.

39. Joseph Hall, *Arte of Divine Meditation* (London, 1605), 192.

40. For a century-spanning account of the fortunes of this (changing) word and concept, see Hamilton, *Security*.

41. Joseph Hall, *Meditations and Vowes, Divine and Morall* (London, 1605), A3v.

42. Hall, *Arte of Divine Meditation*, 17–18.

43. Bacon, *Advancement of Learning*, in *Works*, 3:439.

44. Boyle, *An Introductory Preface*, in *Works*, 5:20.

45. Montaigne, *Complete Essays*, 599.

46. Boyle, *Introductory Preface*, 8.

47. Montaigne, *Complete Essays*, 2.

48. Montaigne, *Complete Essays*, 2; Boyle, *Introductory Preface*, 8.

49. For the narrative of Boyle's turn to science in the late 1640s or early 1650s, see Michael Hunter, "How Boyle Became a Scientist," *History of Science* 33 (1995): 59–103; and Lawrence M. Principe, "Style and Thought of the Early Boyle: Discovery of the 1648 Manuscript of *Seraphic Love*," *Isis* 85 (1994): 247–60. I suggest we describe

his transformation as an *expansion* of the project as well as a *change in emphasis* but not a *change in direction*: his habits of reflection (as well as his moral and theological speculations) continue unabated, but he spends increasing time on scientific experiment proper. One of the premises of my interpretation is that the story of Baconianism doesn't consist only of recognizably "scientific" pursuits. Scott Black rightly observes: "Instead of a break, I see a bridge across which what Boyle calls a 'way of thinking' spans his moral and natural philosophies"—but he doesn't attempt to demonstrate that continuity in Boyle's later writings (*Of Essays and Reading in Early Modern Britain* [New York: Palgrave, 2006], 68). I value his suggestion that Boyle "adopts" the genre of the essay for the purposes of natural philosophy, but that way of putting it risks understating how formally innovative Boyle's writing is (68). Lawrence Principe anticipates my line of thinking, though without my emphasis on the errancy of Boyle's meditative practice, when he suggests that "occasional reflection" persists in his later career: "It is not unwarranted to suggest that this meditative habitude of thinking . . . drawing out unseen implications from observables . . . carried over unchanged as a crucial part of Boyle's experimental career" ("Virtuous Romance and Romantic Virtuoso: The Shaping of Robert Boyle's Literary Style," *Journal of the History of Ideas* 56, no. 3 [1995]: 393). Principe also makes the intriguing claim that Boyle's style is influenced by French romance; I find this argument highly suggestive—and intriguingly amenable to my own, since romance is very much the genre of "wandering"—but I'm not sure the examples he gives of stylistic borrowings are quite precise enough to persuade me that they are more than points of departure for further investigation: "Boyle's mature style, syntactically considered, is not unlike contemporary French fictional prose . . . in terms of sentence length and complexity, its preference for balanced parallel constructions, and its high descriptive content" ("Virtuous Romance," 395).

50. Boyle, *Account of Philaretus*, 8.

51. Boyle, *Account of Philaretus*, 12.

52. For a comparative analysis of Ignatius and Boyle on attention, see David Marno, "Easy Attention: Ignatius of Loyola and Robert Boyle," *Journal of Medieval and Early Modern Studies* 44, no. 1 (2014): 135–61. I would go farther than Marno: Boyle's difference from Loyola is dramatic enough that it can be understood as countermeditation.

53. Interestingly, scholars tend either to understand the *Reflections* as "prescientific" or assert, as if this were a noncontroversial point, its relevance to Boyle's better-known work in natural philosophy. I am grateful for the latter perspective (see, e.g., Black, *Of Essays and Reading*, and Marno, "Easy Attention"), but I do think much remains to be said about how Boyle conveys to us that his meditations are meant to contribute to an emergent culture of experiment (since their explicit content is mostly unrelated). Picciotto is the one scholar who offers a thorough account of the scientific premises of occasional reflection, but I suggest an alternative to the "brisk spirit of exercise" she discovers in Boyle's book (*Labors of Innocence*, 274).

54. Marno interestingly describes "Attention" as an "independent faculty"— almost a personification ("Easy Attention," 155).

55. Boyle refers to Browne several times: in, for example, *Experiments and Notes about the Mechanical Origine or Production of Electricity*, in *Works*, 8:511. I am tentative only about the specific connection I draw. For a discussion of Browne's scientific

seriousness, and the seriousness with which his contemporaries (including Boyle) take his experimental findings, see Gordon Keith Chalmers, "Sir Thomas Browne, True Scientist," *Osiris* 2 (1936): 28–79.

56. See *OED*, s.v. "expatiate," v.

57. Bacon, *Sylva Sylvarum*, in *Works*, 2:570. Bacon's "transcur" denotes "roving to and fro" in aimless but receptive circulation. See *OED*, s.v. "transcur," v.

58. I offer one of myriad examples: "*Tycho* will have two distinct matters of Heaven and Ayre; but to say truth, with some small qualification, they [he has been discussing the competing theories of natural philosophers] have one and the self same opinion, about the Essence and matter of Heavens, that it is not hard and impenetrable as *Peripateticks* hold, transparent, of a *quinta essentia, but that it is penetrable and soft as the ayre it selfe is, and that the Planets move in it, as Birds in the ayre, Fishes in the sea*. This they prove by motion of Comets, & otherwise . . . which are not generated, as *Aristotle* teacheth, in the aëriall Region of an hot and dry exhalation, and so consumed" (Robert Burton, *The Anatomy of Melancholy*, ed. Thomas C. Faulkner, Nicolas K. Kiessling, and Rhonda L. Blair [Oxford: Clarendon Press, 1990], 2:47).

59. Burton, *Anatomy*, 2:33.

60. Browne, *Religio Medici*, in *Works*, 1:18.

61. Bacon, *Novum Organum*, in *Works*, 4:7.

62. For Cicero's account of "eloquence as the ability to practice *decorum*, defined . . . as the ability to accommodate the occasion, taking account of times, places, and persons," as well as later developments in this line of thinking, see Kathy Eden, *Hermeneutics and the Rhetorical Tradition: Chapters in the Ancient Legacy and Its Humanist Reception* (New Haven: Yale University Press, 1997), 26.

63. Viktor Shklovsky, "Art as Device," in *Theory of Prose*, trans. Benjamin Sher (Champaign: Dalkey Archive, 1990), 1–14.

64. Sir Philip Sidney, "The Defence of Poesy," in *Sir Philip Sidney: The Major Works*, ed. Katherine Duncan-Jones (Oxford: Oxford University Press, 1989), 227.

65. "As with children, when physicians try to administer rank wormwood, they first touch the rims about the cups with the sweet yellow fluid of honey, that unthinking childhood be deluded as far as the lips, and meanwhile may drink up the bitter juice of wormwood, and though beguiled be not betrayed, but rather by such means be restored and regain health, so now do I" (Lucretius, *De rerum natura*, book 1, lines 936–43). Cf. Plato's comparison of poetry to the rendering "pleasant" of healing "nutriments" in *Laws*, trans. R. G. Bury (Cambridge: Harvard University Press, 1926), 1:113.

66. Exactly when Boyle digested Lucretius and the philosophy of Epicurus is unclear. As Monte Johnson and Catherine Wilson explain, "there are hundreds of references to Epicurus and Lucretius in his writings. If Boyle was sincere in maintaining that he had read little of Lucretius and was not conversant with Epicureanism in 1663, he made up for his neglect later" ("Lucretius and the History of Science," in *The Cambridge Companion to Lucretius*, ed. Stuart Gillespie and Philip Hardie [Cambridge: Cambridge University Press, 2007], 131–48).

67. Charles Lamb, *The Letters of Charles and Mary Lamb*, ed. E. V. Lucas (London: Methuen, 1905), 1:20.

68. In this chapter and the next (with respect to Walton, Boyle, and Marvell), I speak of the fusion of georgic and pastoral, which we can understand as the

translation of my underlying argument about the (disorienting) coordination of effortlessness and labor into the language of genre. I have not attempted to do justice to the history of these conventions, partly because others have made the attempt (we can look to William Empson, for instance, for extraordinary readings of the pastoral dimension in Marvell and Milton) and partly because I tend to see generic formulae as among *many* formal resources with which my authors evoke moods. I admire Empson's formula for pastoral, upon which Paul Alpers expands: "putting the complex into the simple"—though I would speak, with respect to my (semipastoral) cases, of "finding" or perhaps "observing" the complex in the simple rather than deliberately placing it there (Empson, *Some Versions of Pastoral* [London: Chatto and Windus, 1950], 23; Alpers, *What Is Pastoral?* [Chicago: University of Chicago Press, 1996], 37–43).

69. Charles Lamb, *Letters of Charles and Mary Lamb*, 2:563.

70. Izaak Walton, *The Compleat Angler or the Contemplative Man's Recreation* (London, 1653), 2. I quote by default from the edition of 1653, concentrating on Walton's first possible encounter with the *Reflections*, but have also quoted from the last of his five revisions, the *Universal Angler* of 1676, in which he added a great deal of new material. On Walton's interpolations to this final version, see John R. Cooper, *The Art of* The Compleat Angler (Durham: Duke University Press, 1968), 177–84. In putting Walton into conversation with Boyle, I have had to set aside my interest in Charles Cotton's "continuation" of the work, as well as Robert Venable's sequel, both of which are included in *The Universal Angler*.

71. I refer to Marvell's *The Garden*: "Annihilating all that's made / To a green thought in a green shade" (*The Poems of Andrew Marvell*, ed. Nigel Smith [London: Pearson Longman, 2003], lines 47–48).

72. "He could have added," Oliver continues, "the kissing of the milkmaid by the amorous member of the party" (H. J. Oliver, "The Composition and Revisions of the 'Compleat Angler,'" *Modern Language Review* 42, no. 3 [1947]: 295–313). What Lloyd actually says is that "these resemblances [between the *Occasional Reflections* and *The Compleat Angler*] are purely conventional and as accidental as the baiting of hooks and the casting of flies," arguing that "it is rather in an occasional glimpse of nature and the reflective mood that pervades the whole composition that the kinship is revealed"—though perhaps in this case he is simply describing pastoral (Claude Lloyd, "An Obscure Analogue of the Compleat Angler," *PMLA* 42, no. 2 [1927]: 402).

73. Cooper, *Art*, 164.

74. In a similar vein, J. Paul Hunter writes: "Boyle largely deserves the place Swift [who, in his 1701 *Meditation on a Broomstick*, parodied Boyle's practice of occasional reflection] and history have given him, the specialist who strayed too far from his expertise: he is like a good, well-meaning actor who strays into politics (as distinguished from bad actors who stray into politics), and it is hard to feel sorry for him in spite of good intentions. But he deserves a larger niche in literary, cultural, and intellectual history, though not because he was a great or even a clear thinker. Rather, his pedestrian commitments make him important in the history of taste, desire, and ideas, for his fuzzy categories and refusals to make distinctions are in fact responsible for popularizing ways of thinking crucial to the reception of novels" ("Robert Boyle and the Epistemology of the Novel," *Eighteenth-Century Fiction* 2, no. 4 [1990]: 275–91). My view is that the very "fuzz[iness]" of Boyle's "categories" permits an

experience of insight on which he places enormous value. Elsewhere, Hunter comes much closer to my line of argument: "The 'method' he [Boyle] advocates is in fact quite casual, involving none of the controlled procedures of Scientific Method" (*Before Novels: The Cultural Contexts of Eighteenth-Century English Fiction* [New York: Norton, 1990], 202). He also remarks on Boyle's "blissful vagueness about how to discriminate among possible interpretations" (206).

75. Walton, *Compleat Angler*, 51.

76. Izaak Walton, *A Discourse of Rivers, Fish-Ponds, Fish & Fishing*, in Charles Cotton, Robert Venable, and Izaak Walton, *The Universal Angler, Made So, By Three Books of Fishing* (London, 1676), 56.

77. Walton, *Compleat Angler*, "To the Reader," sig. A5v.

78. Walton, *Compleat Angler*, "To the Reader," sig. A5v.

79. Walton, *Universal Angler*, 5–6. See Montaigne, *Complete Essays*, 331.

80. Walton, *Universal Angler*, 5. I thank Megan Heffernan for her interpretation of this passage. For Walton's practice of altering quotations, see Austin Dobson, "On Certain Quotations in Walton's 'Angler,'" in *Side-Walk Studies* (London: Chatto and Windus, 1902), 250–62.

81. Walton, *Compleat Angler*, 5.

82. Walton, *Compleat Angler*, 5.

83. See my discussion of this claim in the introduction.

84. Bacon, *Advancement of Learning*, in *Works*, 3:317.

85. Walton, *Compleat Angler*, 214–15.

86. Walton, *Compleat Angler*, 215.

87. Walton, *Compleat Angler*, 215.

88. Walton, *Universal Angler*, 262.

89. Walton, *Universal Angler*, 263.

90. Montaigne, *Complete Essays*, 817.

91. Walton, *Universal Angler*, 262.

92. Walton, *Compleat Angler*, 72–37 (irregular pag.).

93. Walton, *Compleat Angler*, 37–38.

94. Cooper writes, "There is no sign that Walton was even aware of the great intellectual changes of his time. With unconscious irony, he cites zoological information from Bacon, as though the author of *The Advancement of Learning* were another authority to be accepted uncritically, like Aelian and Pliny" (*Art.* 52). His insensitivity to Walton's scientific interests has much to do with his negative view of early modern natural philosophy, of which he offers the following description from Alfred North Whitehead: "a dull affair, soundless, scentless, colourless; merely the hurrying of material, endlessly, meaninglessly" (52, citing *Science and the Modern World* [New York: Macmillan, 1925], 77).

95. Oliver, "Composition and Revisions," 400–401.

96. Walton, *Universal Angler*, 30.

97. Walton, *Universal Angler*, 32.

98. Walton, *Compleat Angler*, 21.

99. Walton, *Compleat Angler*, 23.

100. Walton, *Compleat Angler*, 140, 251.

101. Walton, *Compleat Angler*, 52.

102. Walton, *Compleat Angler*, 51–52.

103. Dobson, "On Certain Quotations."

104. Walton, *Compleat Angler*, 128–29.

105. Walton, *Compleat Angler*, 129–30.

106. Walton, *Universal Angler*, "To the Reader," sig. A5v.

107. Walton, *Compleat Angler*, 35.

108. Walton, *Compleat Angler*, 16.

109. Walton, *Compleat Angler*, 16. The revision has "ingenuous" (*Universal Angler*, 28). For the relationship between these two words in the culture of experiment, see Picciotto, *Labors of Innocence*, 66–70. "A word that had been a class marker to distinguish the trustworthy from the common," she writes, "was now interchangeable with a word that denoted the knowhow of 'mechanicks,' Through a constant shuffling of the *i* and the *u*, an experimental hybrid was created that utterly inverted the relationship between trustworthiness and gentlemanliness" (66).

110. Walton, *Universal Angler*, 24–25.

111. Walton, *Compleat Angler*, 224.

112. Walton, *Universal Angler*, 116–17.

113. Walton, *Universal Angler*, 185.

114. Hunter relocates it so that it immediately preced Section IV (*OR*, 93).

115. Boyle confesses his poor mathematical skill, but he also makes a philosophical argument for the limitations of mathematics. As Sargent puts it, "Boyle argued that, despite its fallibility, a demonstration produced by experimental science is superior to one produced by the mathematical way of reasoning. There is a complexity and subtlety to the physical world that pure mathematics is not able to capture" (*Diffident Naturalist*, 57).

116. "But when I distinctly see where things come from and where and when they come to me, and when I can connect my perceptions of them with the whole of the rest of my life without a break, then I am quite certain that when I encounter these things I am not asleep but awake" (Descartes, *Meditations*, 62).

117. For a detailed account of primary and secondary qualities in Boyle's philosophy, which is better understood as a distinction between the "mechanical affections of matter and all of matter's other properties," see Anstey, *Philosophy*, 15–112. For an attempt to revive the theory (with reference to Locke rather than Boyle), see Quentin Meillassoux, *After Finitude: An Essay on the Necessity of Contingency*, trans. Ray Brassier (New York: Continuum, 2008).

118. Such disagreement is common, and Boyle explicitly reflects on it as an engine of "entertain[ing]" discourse: "I know not whether *Eugenius* imagin'd that *Lindamor* did in this Discourse make some little Reflection, upon what we had lately said on the behalf of Princes: But I afterwards suspected, that it was partly to reply to this Observation, as well as entertain the Company with a new one that he subjoyn'd" (*OR*, 125).

119. "I cannot praise a fugitive and cloister'd vertue, unexercis'd & unbreath'd, that never sallies out and sees her adversary" (John Milton, *Complete Prose Works of John Milton*, ed. Don M. Wolfe [New Haven: Yale University Press, 1953–82], 2:515).

120. Francis Bacon, "Of Truth," in *Works*, 6:377. For a persuasive discussion of Bacon's association of "giddiness" with Montaigne's *Essais*, see Hovey, "'Montaigny Saith Prettily,'" 76.

121. Walton, *Compleat Angler*, 94–95.

122. Walton, *Compleat Angler*, 97–98.

123. See Arthur O. Lovejoy's account of plenitude in the neo-Platonic tradition in his classic *The Great Chain of Being: A Study of the History of an Idea* (New York: Harper and Row, 1960), 61–66. His explanation of the necessity of variety in nature is especially illuminating: "This generation of the Many from the One cannot come to an end so long as any possible variety of being in the descending series is left unrealized" (62). For a discussion of the seventeenth century's aesthetics of variety, in which the rhetorical tradition (going back to Aristotle) and a distinct discourse of "Christian optimism" (going back to the Church Fathers) converge, see H. V. S. Ogden, "The Principles of Variety and Contrast in Seventeenth Century Aesthetics, and Milton's Poetry," *Journal of the History of Ideas* 10 (1949): 159–82.

124. Robert Boyle, *Of the Study of the Book of Nature*, in *Works*, 13:147; hereafter abbreviated *Book of Nature* and cited parenthetically by page number through the end of this chapter.

125. Robert Boyle, *Of the High Veneration Man's Intellect Owes to God*, in *Works*, 10:184; hereafter abbreviated *High Veneration* and cited parenthetically by page number through the end of this chapter. As the passage continues, it evokes the language of "occasional reflection." The creatures are inexact shadows of God "as a Picture of a Watch or Man, or the name of either of them written with Pen and Ink, does not exhibit a true or perfect *Idea* of a thing (whose internal constitution a surface cannot fully represent) but onely gives occasion to the mind to think of it, and to frame one.'"

126. See my discussion of wonder and curiosity in the introduction.

127. For exceptions to natural laws in Boyle (God's freedom to impose different laws at different times and places), see Anstey, *Philosophy*, 174–75.

128. The logic seems to hold even for the most important questions, as Boyle suggests in the following observation: "Indignation prompts me to this reflexion, that if [Since] even our Hymns and Praises of God the Supreme Being deserve our blushes and need His pardon, what confusion will one day cover the faces of those, that do not onely speak slightly and carlesly, but oftentimes contemptuously, and perhaps drollingly, of that Supreme and Infinitely Perfect Being, to whom they owe those very Faculties and that with which they so ungratefully, as well as impiously misemploy?" (*High Veneration*, 197). Notice that "contempt" and "droll[ery]" in response to God's perfection earn Boyle's reproach, while "slight" and "carles" speech is taken for granted. However we praise our Maker, we owe Him "blushes" and need his "pardon." The inevitable "carles[ness]" of our prayers is a consequence of finitude. "Carless" is what it's like to be mundane: ordinary and resolutely earthbound—no matter how "Heavenly Minded." The best we can do falls significantly short—and therein lies the generosity of a universe that endlessly exceeds our ever-expanding understanding.

129. Baxter, in Boyle, *Correspondence of Robert Boyle*, 2:476.

130. Montaigne, *Complete Essays*, 21.

131. Boethius, *The Consolation of Philosophy*, trans. S. J. Tester, Loeb Classical Library (Cambridge: Harvard University Press, 1973), 427.

132. Hunter tells us that the source of the Latin saying is unknown (189n). The translation is his.

3. The Microscope Made Easy

1. Henry Baker, *The Microscope Made Easy* (London, 1742), v.

2. Baker, *Microscope*, 51.

3. Baker, *Microscope*, iii.

4. Robert Hooke, "Discourse concerning Telescopes and Microscopes," in *Philosophical Experiments and Observations of the Late Eminent Dr. Robert Hooke*, ed. W. Derham (London, 1729), 261. For an interesting discussion of this passage as evidence of the "microscope's fall from favour in scientific circles," see Tita Chico, "Minute Particulars: Microscopy and Eighteenth-Century Narrative," *Mosaic* 39, no. 2 (2006): 144.

5. Hooke, "Discourse," 261.

6. Hooke, "Discourse," 268.

7. Only the first section of this volume concerns the microscope, while the remainder describes experiments with mercury and magnets. This chapter's theme limits my discussion to Power's research in microscopy.

8. The scholarship describes "a deliberate effort to separate observation from conjecture" as constitutive of the "epistemic genre of the *observationes*," but I show how extravagantly experimental natural philosophers continue to explore the speculative and conjectural, even in the midst of firsthand accounts (Lorraine Daston, "The Empire of Observation, 1600–1800," in *Histories of Scientific Observation*, ed. Lorraine Daston and Elizabeth Lunbeck [Chicago: University of Chicago Press, 2011], 81).

9. Scholars have often made short work of Power by treating his book as a botched or unsuccessful *Micrographia*. "The *Micrographia*," writes Margaret 'Espinasse, "could almost be called a second and extremely improved edition of Power. Hooke took this amateur work and showed how a professional could handle it" (*Robert Hooke* [Berkeley: University of California Press, 1956], 54–55). She goes on to fault Power's book for lacking images and, on the basis of a single comparison (their descriptions of butterflies), concludes that Hooke "is more accurate" (55). Although I don't comment on the precision of Power's observations, I suggest that such cursory judgments authorize scholarly neglect of his philosophical and literary experiments.

10. *The Diary of Samuel Pepys*, ed. Henry B. Wheatley (London: George Bell and Sons, 1904), 4:316 (cited and discussed in 'Espinasse, *Robert Hooke*, 58).

11. James Grantham Turner, *The Politics of Landscape: Rural Scenery and Society in English Poetry, 1630–1660* (Cambridge: Harvard University Press, 1979), 61; Andrew Legouis, *Andrew Marvell: Poet, Puritan, Patriot* (Oxford: Oxford University Press, 1965), 82.

12. Andrew Marvell, *Upon Appleton House*, in *The Poems of Andrew Marvell*, ed. Nigel Smith (London: Pearson Longman, 2003), lines 761–68; hereafter cited parenthetically by line number through the end of this chapter.

13. "Your" and "you" might refer equally well to Fairfax or his estate—at least until the image of the "map" fixes our attention on the landscape.

14. "The ancient opinion that man was Microcosmus," writes Bacon, "an abstract or model of the world, hath been fantastically strained by Paracelsus and the alchemists, as if there were to be found in man's body certain correspondences and parallels, which should have respect to all varieties of things, as stars, planets, minerals, which are extant in the great world" (*Advancement of Learning*, in *Works*, 3:370).

15. Ben Jonson's "To Penshurst," for instance, describes a world of endlessly dependable comfort, as much a product of man's ministrations as a guarantor of his invulnerability (the masculine pronoun is the appropriate one, since the figures presiding over the estate are Robert Sidney and the poet): "The painted partridge lies in every field, / And for thy mess is waiting to be killed" ("To Penshurst" in *Ben Jonson: A Critical Edition of the Major Works*, ed. Ian Donaldson [Oxford: Oxford University Press, 1985], lines 29–30).

16. Harry Berger Jr., "Andrew Marvell: The Poem as Green World," in *Second World and Green World: Studies in Renaissance Fiction Making*, ed. John Patrick Lynch (Berkeley: University of California Press, 1988), 300.

17. Picciotto, *Labors of Innocence*, 14–15.

18. Nicolson, *Breaking of the Circle*, 181.

19. Picciotto, *Labors of Innocence*, 375.

20. Despite Andrew Shifflett's compelling discussion in *Stoicism, Politics and Literature in the Age of Milton: War and Peace Reconciled* (Cambridge: Cambridge University Press, 1998), 36–74, Marvell's wild abandon discourages me from taking Stoicism as the rubric for reading the poem.

21. By translating affect into artifice, we render the poet needlessly mysterious. Thus the most astute book-length study of Marvell's poems declares him a "poet-without-persona" who is also a "poet-with-too-many-personas" (Rosalie Colie, *"My Ecchoing Song": Andrew Marvell's Poetry of Criticism* [Princeton: Princeton University Press, 1970], 5). The scholar who succeeds at explaining the protean quality of his public face with great persuasive force nonetheless speaks of the "disconcerting dexterity" with which he "could tack between opposing sides" in his poems of the Commonwealth period (David Norbrook, *Writing the English Republic: Poetry, Rhetoric and Politics, 1627–1660* [Cambridge: Cambridge University Press, 1999], 244). (I quote from a description of Marchamont Nedham, but the comparison with Marvell is Norbrook's own.) I'm sympathetic with his disinclination to impose ideological uniformity on these poems, but I'm pressing the point that we need not be "disconcert[ed]" by disunity. When scholars descend to the details of Marvell's poems, they often transmute the duplicity of his public self into the indecision of the private one. Berger speculates, in a loosely psychoanalytic mode, that *Upon Appleton House* "is longer [than *The Garden*] in part because it dramatizes the mind's vacillation at closer range," as if the very engine of Marvell's poetic production were anxious ambivalence (Berger, "Andrew Marvell," 320). Derek Hirst and Steven Zwicker, who give us the most illuminating historicist account of the poem, single out the same psychic phenomenon, but redefine it in political terms, exploring Marvell's "ambivalence" with respect to military action, about which some decision was a "cruel necessity" ("High Summer at Nun Appleton, 1651: Andrew Marvell and Lord Fairfax's Occasions," *Historical Journal* 36 [1993]: 268, 263). Occasions for equivocation are easy to find, given Marvell's habit of standing astride categorical distinctions. Whether conceived as an ex-royalist, an ex-Cromwellian, or a semiconscious member of a sexual minority, Marvell seems a good candidate for the psychic pain of inconsistency. If we suspend our distrust of *disinvoltura*, we might attend instead to an experience of effortless complexity for which self-contradiction is no synonym. For a thoughtful argument for ideological continuity across Marvell's varied career, see John Wallace, *Destiny His Choice: The Loyalism of Andrew Marvell* (Cambridge: Cambridge University Press, 1968).

22. Michel Foucault, "Of Other Spaces," trans. Jan Miskowiec, *Diacritics* 16, no. 1 (Spring 1986): 24.

23. Bacon, *The Great Instauration*, in *Works*, 4:32.

24. Here, my argument resonates with Nigel Smith's description of the poem's "liberated subjectivity" in *Literature and Revolution in England, 1640–1660* (New Haven: Yale University Press, 1994), 325.

25. See Hirst and Zwicker, "High Summer," for a richly detailed account of the poem's historical context, especially the circumstances of Fairfax's retirement.

26. For this evidence, see Helen Darbishire, ed., *The Early Lives of Milton* (New York: Barnes and Noble, 1932), 175–76; and Hugh Brogan, "Marvell's *Epitaph on* —," *Renaissance Quarterly* 32 (Summer 1979): 197–99 (both cited in Picciotto, *Labors of Innocence*, 346–47).

27. Aubrey, *Brief Lives*, 54 (cited in Hodge, 94).

28. Picciotto, *Labors of Innocence*, 356. On Marvell's Baconianism, see 344–77. R. I. V. Hodge makes the important observation that "Marvell was contemporary with Newton and Hooke" in a conceptual rather than merely chronological sense (*Foreshortened Time: Andrew Marvell and 17th Century Revolutions* [Cambridge: D. S. Brewer, 1978], 68–95). I follow Smith, in Marvell, *Upon Appleton House*, in *Poems of Andrew Marvell*, 152, and Allan Pritchard, "Marvell's 'The Garden': A Restoration Poem?," *Studies in English Literature 1500–1900* 23 (1983): 371–88, in dating *The Garden* to 1668, but I share Smith's unwillingness to take Pritchard's argument as fact.

29. It's for this reason that I abstain from the vocabulary of disinterest; disinterest implies a will to stave off affective investment in order to preserve fairness, which is nearly the opposite of Marvell's unguardedness. The careful delimitations of the Kantian discourse on disinterest, which retains much of its force in late modernity, confirms my sense that the difference should be noted. I've benefited from the work of Kathryn Murphy and Anita Traninger, who offer an illuminating discussion of a distinct but related concept in the seventeenth and eighteenth centuries. Their account of the "enigmatic" dimension of "impartiality" is especially telling, since both poles of the term's semantic range imply a gesture Marvell does not perform in this poem: the deliberate suspension of a personal stake. "What seems characteristic is that [impartiality] oscillates semantically between a *refusal* to join or support one of two parties, or, figuratively, a suspension of judgement; and a certain *quality of judgement*, one that is informed by putting aside personal preferences and foregrounding the arguments at stake" ("Introduction: Instances of Impartiality" in *The Emergence of Impartiality* [Leiden: Brill, 2014], 4).

30. See James Grantham Turner, "Marvell's Warlike Studies," *Essays in Criticism* 28 (1978): 288–301. He extends his interpretation of the poem in his *The Politics of Landscape*, but my interest here is the lucid accuracy of that first interpretation.

31. Turner, "Warlike Studies," 300.

32. See, for instance, Hirst and Zwicker, "High Summer," 268, on Marvell's "ambivalence."

33. Turner, "Warlike Studies," 295.

34. Turner speaks eloquently of "a retirement in which the very apparatus of war is pacific" ("Warlike Studies," 291).

35. William Shakespeare, *Twelfth Night*, in *Complete Pelican Shakespeare*, act 3, scene 1, lines 11–13.

36. I'm indebted here and throughout this section to Gurton-Wachter, who explores nuanced poetic responses to the "militarization of attention" that attends the Napoleonic Wars (*Watchwords*). Her interest in the slackening of attention in Romantic poetry as it responds to the imperative to wartime vigilance is a source of inspiration.

37. I follow Smith in identifying Sir Thomas Fairfax, the son of William Fairfax and Isabel Thwaites, as the subject of this sentence, but it isn't crystal clear. See Marvell, *Upon Appleton House*, in *Poems of Andrew Marvell*, 223n281.

38. Boyle, *Occasional Reflections*, in *Works*, 5:95.

39. See Marvell, *Upon Appleton House*, in *Poems of Andrew Marvell*, 226n368.

40. As Smith points out, the verb is specifically used to speak of a "beam of light" (Marvell, *Upon Appleton House*, in *Poems of Andrew Marvell*, 226n368). See OED, s.v. "graze," v.

41. Edmund Spenser, *The Faerie Queene*, ed. A. C. Hamilton (San Francisco: Longman, 2001), book 3, canto 6, stanza 44, lines 8–9.

42. See Marvell, *Upon Appleton House*, in *Poems of Andrew Marvell*, 231n507–8.

43. I refer to the speaker's first use of the first person singular, though Marvell's silver-tongued Prioress beats him to it (137). The fact of the first person's belated emergence isn't news, but I call attention to the simultaneous appearance of a characteristic pattern of speech—one that we find is intimately linked to the speaker's first-person perspective.

44. See note 29 on the concept of disinterest.

45. Turner anticipates my line of interpretation, though I draw a distinction between affective minimalism and "wonder": "Slow eyes and pleasant footsteps produce a vivid mixture of clarity and lethargy. Marvell represents the state of suspended wonder in which we discover things" (*Politics of Landscape*, 67). See also Christopher Ricks, "Its Own Resemblance," in *Approaches to Marvell: The York Tercentenary Lectures*, ed. C. A. Patrides (London: Routledge and Kegan Paul, 1978), in particular his discussion of the poem's conclusion: "Marvell's lines here, characteristically, at once expand and contract, in a double perspective which calmly blinks" (117).

46. Marvell, *The Garden*, in *Poems of Andrew Marvell*, lines 47–48.

47. Abiezer Coppe, *A fiery flying rolle: a word from the Lord to all the great ones of the Earth, whom this may concerne: being the last warning piece at the dreadfull day of judgement* (London, 1650), 6.

48. J. C. Davis doubts the existence of Ranters, understanding them instead as a product of antisectarian caricature. See *Fear, Myth, and History: The Ranters and the Historians* (Cambridge: Cambridge University Press, 1986). I'm far from the first to suggest that Davis takes his argument too far, but Coppe's rhetoric is certainly excessive enough to lend fuel to antisectarian caricature—the fact of which says nothing about the unreality of Ranters. My understanding of Coppe's millenarianism is informed by Nigel Smith, who resists the temptation to cordon Ranter rhetoric off from "the general prophetic climate" of the historical moment (*Perfection Proclaimed: Language and Literature in English Radical Religion, 1640–1660* [Oxford: Oxford University Press, 1989], 56).

49. Margarita Stocker, *Apocalyptic Marvell: The Second Coming in Seventeenth Century Poetry* (Sussex: Harvester Press, 1986), 2. Especially instructive is her analysis of the way apocalyptic rhetoric confuses active and passive political stances, directing

God's human servants to pursue His ends with vigor but also affirming that He achieves his objectives without need of human assistance (12). For an insightful and technically precise account of Marvell's vision of apocalypse (but one distant from my emphasis on Marvell's use of millenarianism as a kind of poetics), see Catherine Gimelli Martin, "The Enclosed Garden and the Apocalypse: Immanent Versus Transcendent Time in Milton and Marvell," in *Milton and the Ends of Time*, ed. Juliet Cummins (Cambridge: Cambridge University Press, 2003).

50. John Rogers, *The Matter of Revolution: Science, Poetry, and Politics in the Age of Milton* (Ithaca: Cornell University Press, 1996) is of interest here because he explores the same set of historical contexts to which I direct our attention: natural philosophy, millenarianism, and war. I have learned a great deal from his account, but I do not follow him in arguing that Marvell accepts a quasi-Aristotelian teleology (see 54–55).

51. My turn to matters of religion calls to mind the debate around *adiaphora*, one of early modern England's most contentious discourses on "indifference." Because the controversy concerns the distinction between what scripture mandates and what remains open to human decision making, it doesn't immediately present itself as a resource for thinking about Marvell's disposition. Precise questions about permissibility seem less than apposite for an unregulated state of feeling in which nothing is excluded from possibility. For a particularly edifying introduction to this issue, which brings out its ideological and theological complexities and connects Elizabethan controversies to later developments, see John S. Coolidge, *The Pauline Renaissance: Puritanism and the Bible* (Oxford: Oxford University Press, 1970).

52. For the most detailed account of millenarianism's impact on the development of English science, see Webster, *Great Instauration*.

53. See *OED*, s.v. "apocalypse."

54. Bacon, *Great Instauration*, in *Works*, 4:33.

55. For a version of this argument about apocalypse, see Picciotto, *Labors of Innocence*, 356–77.

56. I allude here to Christopher Hill, *The World Turned Upside Down: Radical Ideas during the English Revolution* (London: Penguin, 1972).

57. Picciotto, *Labors of Innocence*, 358.

58. Picciotto, *Labors of Innocence*, 359.

59. Bacon, *Novum Organum*, in *Works*, 4:54; *Advancement of Learning*, in *Works*, 3:265. I echo Catherine Wilson's perceptive observations on this passage: "Yet this same Baconian investigator can at times be a pure and brilliant mirror, an angel and not a torturer" ("Visual Surface and Visual Symbol," 95). I've benefited from her description of the interlacing but distinct strands of thought that run through Bacon's work.

60. Leonardo da Vinci, *Leonardo da Vinci's Note-Books*, ed. and trans. Edward McCurdy (London: Duckworth & Co., 1908), 163.

61. I'm not persuaded by Edith and Norriss Hetherington's claim that these lines do not refer to the microscope. They argue instead for a closed tube in which a flea is examined through a lens at one end; the only advantage of this view, as far as I can tell, is that the insect *literally* "approach[es] the eye" as the viewer lifts it to her face ("Andrew Marvell, 'Upon Appleton House,' and Fleas in Multiplying Glasses," *English Language Notes* 13 [December 1, 1975]: 124). Since it's fair to describe this device as

a rudimentary microscope, and my interest is the figure of the lens rather than the specificity of the compound microscope, the stakes here are low.

62. Mary Carruthers, *The Book of Memory: A Study of Memory in Medieval Culture* (Cambridge: Cambridge University Press, 1990), 206–7. For further reflection on this figure, see Raymond B. Waddington, *Looking into Providences: Designs and Trials in Paradise Lost* (Toronto: University of Toronto Press, 2012), 45.

63. For a perceptive reading of Marvell's placement of Davenant among "th'universal herd" (456), see Picciotto, *Labors of Innocence*, 357–58, and her extended discussion of *Gondibert* (1651), 380–99.

64. I echo Stocker's wonderful use of this adjective (*Apocalyptic Marvell*, 46).

65. I do not imply that this meaning is specific to Montaigne, but simply that evocative examples can be found throughout the *Essais*. See, for instance, "Of the Resemblance of Children to Fathers," 3:578, and "De la ressemblance des enfans aux peres," 3:426.

66. Stephen Greenblatt has given this theory of cultural nostalgia its most memorable image in the ghost of Hamlet's father. Greenblatt famously argues that the stage ghost embodies, in its disembodied way, the problem of the demolition of Purgatory by the reformers: "With the doctrine of Purgatory and the elaborate practice that grew up around it," he writes, "the church had provided a powerful method of negotiating with the dead . . . The Protestant attack on the 'middle state of souls' and the middle place those souls inhabited destroyed this method for most people in England, but it did not destroy the longings and fears that Catholic doctrine had focused and exploited. Instead . . . the space of Purgatory becomes the space of the stage where old Hamlet's Ghost is doomed for a certain term to walk the night" (*Hamlet in Purgatory* [Princeton: Princeton University Press, 2002], 256–57). I recall this well-known argument as a point of contrast with Marvell's gambol through time, which does not represent a quest for lost comfort but rather an exploratory journey in which the speaker declines to fend off the unreformed past.

67. For Ovidian echoes in the sequence, see Marvell, *Upon Appleton House*, in *Poems of Andrew Marvell*, 230n477–80.

68. Walton, *Universal Angler*, 32.

69. Colie, *Ecchoing Song*, 205.

70. Colie, *Ecchoing Song*, 205.

71. As Anne Cotterill puts it in her thoughtful and darkly lyrical interpretation: "The scene is the meadows before the flood, yet already we are at sea" ("Marvell's Watery Maze: Digression and Discovery at Nun Appleton," *English Literary History* 69, no. 1 [2002]: 105). See also the larger argument to which her interpretation of the poem belongs in *Digressive Voices in Early Modern English Literature* (Oxford: Oxford University Press, 2004).

72. Jonson, "To Penshurst," line 33.

73. Picciotto writes: "The close association between the refining fire, the glass, and the eye that employs it in turn shapes Marvell's figuration of Maria. A purgatorial fire whose flames have themselves been tried by heaven as 'pure, and spotless as the Eye,' she is also the agent of nature's transformation into an equally spotless glass, the vitrified nature of Revelation" (*Labors of Innocence*, 363–64).

74. Cited in Marvell, *Upon Appleton House*, in *Poems of Andrew Marvell*, 239n688.

75. Bacon, *Novum Organum*, in *Works* 4:50.

76. Webster, *Great Instauration*, 6.

77. *The Diary of Robert Hooke, M.A., M.D., F.R.S., 1672–1680*, ed. Henry W. Robinson and Walter Adams (London: Taylor and Francis, 1935); discussed in 'Espinasse, *Robert Hooke*, 6.

78. See, for instance, *The Diaries of Robert Hooke, the Leonardo of London, 1635–1703*, ed. Richard Nichols (Sussex: Book Build, 1994).

79. Robert Hooke, *Micrographia: Or Some Physical Descriptions of Minute Bodies Made by Magnifying Glasses, with Observations and Inquiries Thereupon* (London, 1665), sig. b1r.

80. Hooke, *Micrographia*, 1.

81. Hooke, *Micrographia*, 1.

82. She does not precisely anticipate the movement of his thought, but she similarly juxtaposes a "Mathematical point" and a "natural" one—with a playful attention to the materiality of language Hooke likewise shows when he reflects on the unevenness of a printed period: "Concerning your Question, *Whether a Point be something, or nothing, or between both*; My opinion is, that a natural point is material; but that which the learned name a Mathematical point, is like their Logistical Egg, whereof there is nothing in Nature any otherwise, but a word, which word is material, as being natural; for concerning immaterial beings, it is impossible to believe there be any in Nature; and though witty Students, and subtil Arguers have both in past, and this present age, endeavoured to prove something, nothing; yet words and disputes have not power to annihilate any thing that is in Nature, no more then to create something out of nothing; and therefore they can neither make something, nothing; nor nothing to be something: for the most witty student, nor the subtilest disputant, cannot alter Nature, but each thing is and must be as Nature made it" (*Philosophical Letters* [London, 1664], 498).

83. Hooke, *Micrographia*, 2.

84. Hooke, *Micrographia*, 2.

85. John Milton, *Paradise Lost*, in *Milton's Poetry and Major Prose*, ed. Merritt Y. Hughes (Indianapolis: Prentice Hall, 1957), book 5, line 112.

86. Marvell, *The Last Instructions to a Painter*, in *Poems of Andrew Marvell*, lines 16–18.

87. Campbell, *Wonder and Science*, 183–202. In his lively account of the visual artistry of Restoration science, which gives special attention to Hooke, Matthew C. Hunter offers the following formulation, which nicely counters scholarly emphasis on observational aggression: "Microscopic things disclosed themselves partially, in time and in fragmentary aspect—and this only when parsed with observational vigilance, technical skill, and a patient, flexible posture of inquiry" (*Wicked Intelligence: Visual Art and the Science of Experiment in Restoration London* [Chicago: University of Chicago Press, 2013], 42). Still, the frame of his interpretation, "a ruthless cleverness" linked with *metis*, shifts the emphasis back in the other direction (7).

88. Hooke, *Micrographia*, 175.

89. Daston and Park, *Wonders and the Order of Nature*, 313.

90. I draw a contrast here with Barbara J. Shapiro, *A Culture of Fact: England, 1550–1720* (Ithaca: Cornell University Press, 2003), 105–38. Her account of Restoration science and its English prehistory is not unusual in emphasizing credibility of witnessing: "honesty, sharp eyes, and an ability to describe or illustrate what had been viewed" (118).

91. Daston and Park, *Wonders and the Order of Nature*, 313.

92. On the theme of sensation as inundation, I'm indebted to Amanda Goldstein for many years of illuminating conversation. See her visionary discussion of "susceptibility to influence" in "Growing Old Together: Lucretian Materialism in Shelley's Poetry of Life," *Representations* 128 (Fall 2014): 60–92. See also Passannante, *Lucretian Renaissance*, for an intriguing and intriguingly complementary account of the "pervasiveness" of Lucretian "influence" (154–97).

93. Daston and Park, *Wonders and the Order of Nature*, 313–14. For the passage they quote, see Hooke, *Micrographia*, sig. d2r.

94. Literary scholars rarely take up the case of Power, but there is a modern edition: Henry Power, *Experimental Philosophy*, ed. Marie Boas Hall (New York: Johnson Reprint Corporation, 1966). Charles Webster's harsh review, which focuses on Hall's introduction to this version, peremptorily dismisses the importance of Power's microscopic observations (Webster, review of *Experimental Philosophy*, by Henry Power, *British Journal for the History of Science* 4, no. 3 [1969]: 299–300). Interestingly, however, he elsewhere (unintentionally?) reverses the scholarly habit of minimizing Power's importance (by drawing a negative comparison with Hooke), describing the *Micrographia* as an *"aggrandized* version of *Experimental Philosophy"* ("Henry Power's Experimental Philosophy," *Ambix* 14 [1967]: 161; emphasis added). He also offers a thoughtful portrait of Power's intellectual life, which discusses the importance of both Cartesian and neo-Platonic philosophies to his thought.

95. Francis Godwin, *The Man in the Moon* (London: 1638). See also John Wilkins, *A Discourse Concerning a New World and Another Planet: The First Book, Discovery of a New World, or A Discourse tending to prove, that 'tis probable there may be another habitable World in the Moon* (London, 1638).

96. I'm thinking here both of his reflections on the telescope in the "Preface" and his concluding observation, "Of the Moon."

97. Henry Power, *Experimental Philosophy, In Three-Books: Containing New Experiments Microscopical, Mercurial, Magnetical* (London, 1664), 37; hereafter cited parenthetically through the end of this chapter. For Power's relationship to Browne, see Thomas Cowles, "Dr. Henry Power, Disciple of Sir Thomas Browne," *Isis* 20, no. 2 (1934): 344–66. See also their correspondence in Browne, *Works*, 4:253–70.

98. Catherine Wilson is perceptive about the confusions I tend to explore under the rubric of carelessness. Indeed, she offers a formulation that is by far the closest any scholar has come to my theme, though it is not aimed squarely at emotional life: "Without, then, undertaking the impossible and misguided task of defending objectivity, I will be content to present early modern science under a more benign aspect than has recently been customary, showing how engagement exists alongside detachment, and receptivity along with a desire for mastery" (*The Invisible World: Early Modern Philosophy and the Invention of the Microscope* [Princeton: Princeton University Press, 1995], 38). Such willingness to stay with complexity yields insight into the disorderliness of microscopy. "And why should science not have emerged from a haphazard and chaotic empiricism, from experience with making, mixing, and measuring, simply under pressure, as Aristotle says, from the truth?" (15).

99. Hooke, *Micrographia*, 3.

100. Hooke, *Micrographia*, 3. It's easy to assume that the Royal Society Fellows cast aspersions on the forms of artificial bodies in order then to defend the perfection

of natural ones, and it's true that Hooke likes to undermine the apparent elegance of art: at a different scale, he shows his readers, the beautiful products of human making are no longer beautiful. However, as Passannante points out, Hooke also speculates about the imperfection of the *natural* world (*Catastrophizing*). To be sure, we don't find that he denies the *underlying* (ontological) perfection of nature, but he seems *not* to assume that it appears that way to the human eye—that the manner of its perfection necessarily produces aesthetic pleasure. The following passage illustrates this point, beginning with a negative comparison of artifice to nature but finally speculating about the irregularity of natural forms: "Now though this point [the needle's] be commonly accounted the sharpest (whence when we would express the sharpness of a point the most *superlatively*, we say, As sharp as a Needle) yet the *Microscope* can afford us hundreds of Instances of Points many times sharper: such as those of the *hairs*, and *bristles*, and *claws* of multitudes of *Insects*; the *thorns*, or *crooks*, or *hairs* of *leaves*, and other small vegetables; nay, the ends of the *stiriae* or small *parallelipipeds* of *Amianthus*, and *alumen plumosum*; of many of which, though the Points are so sharp as not to be visible, though view'd with a *Microscope* (which magnifies the Object, in bulk, above a million of times) yet I doubt not, but were we able *practically* to make *Microscopes* according to the *theory* of them, we might find hills, and dales, and pores, and a sufficient bredth, or expansion, to give all those parts elbow-room, even in the blunt top of the very Point of any of these so very sharp bodies. For certainly the *quantity* or extension of any body may be *Divisible in infinitum*, though perhaps not the *matter*" (*Micrographia*, 2). The metaphor of "elbow room" projects the human observer into the microscopic landscape at the blunted end of the apparently sharp hair, bristle, or claw.

101. Hooke, *Micrographia*, 210.

102. Boyle, *Works*, 10:194.

103. Cavendish, *Philosophical Letters*, 133.

104. Wilson, *Invisible World*, 29.

105. Bacon, *Advancement of Learning*, in *Works*, 3:340.

106. I suggest he understands himself as a defender of atomism in a loose, corpuscularian sense. See chapter 2 for my discussion of this question in Boyle's philosophy.

107. "Anyone who considers himself in this way will be terrified at himself, and, seeing his mass, as given him by nature, supporting him between these two abysses of infinity and nothingness, will tremble at these marvels. I believe that with his curiosity changing into wonder he will be more disposed to contemplate them in silence than investigate them with presumption" (Blaise Pascal, *Pensées*, trans. A. J. Krailsheimer [New York: Penguin, 1995], fragment 72).

108. We might here compare Power's sentiment with that of Nehemiah Grew: "As Travellers sometimes amongst Mountains, by gaining the top of one, are so far from their Journies end, that they only come to see another lies before them: so the way of Nature is so impervious, and, as I may say, down hill and up hill, that how far soever we go, yet the surmounting of one difficulty, is wont still to give us the prospect of another" (*The Idea of a Phytological History Propounded* [London, 1673], 52). If this image strikes us as intimidating, the conclusion Grew draws from it is that we should celebrate the inconclusiveness of the enterprise: "For although a man shall never be able to hit Stars by shooting at them, yet he shall come much nearer to them, than another that throws at apples" (53).

109. Svetlana Alpers argues that Leeuwenhoek explores the "conditions of sight" by studying the eyes of various creatures, "call[ing] attention to the fact that the world is known not through being visible, but through the particular instruments that mediate what is seen" (*The Art of Describing: Dutch Art in the Seventeenth Century* [Chicago: University of Chicago Press, 1983], 83). Discussing this passage in Alpers, Picciotto offers the following formulation: "to focus on the eyes of organisms under the microscope intensified the experience of looking through a lens twice over; one looked through an optical instrument only to find another" (246).

110. The passage concerns the speed with which water evaporates in air. See Bacon, *Sylva Sylvarum*, in *Works*, 2:377.

111. Thomas Cowles, "Dr. Henry Power's Poem on the Microscope," *Isis* 21, no. 1 (1934): 73.

112. Wilson's formulations lend support to my argument: "To Glanvill's charge that fallen humanity is helpless to see the causes of things," she explains, "Power asserts that he does not agree. . . . It is not a question of restoring a human intellectual empire but of creating one for the first time" (*Invisible World*, 65). Furthermore, "the idol-defeating power of the aids to the mind is taken over, in a way Bacon would not have approved of, by aids to the eye"; what Power gives up, in other words, is the disciplinary method that should keep the mind in check (65).

113. Aït-Touati is right to notice that Hooke is not the mythographer Power is, but I do not second her judgment that the distinction between them is best understood as an opposition between mediation and directness. Hooke, she argues, referring to Power's anecdote (borrowed from Muffet) about the flea pulling the chariot, "produced such a life-like effect by drawings of living animals, rather than by lively written description" (150). Yet even the few examples I've given show that Hooke *shares* Power's predilection for elaborate metaphors and imagined scenes. Aït-Touati goes on to argue that Hooke "shows" in images what Power merely describes (152). If she meant only to observe that Hooke includes illustrations in his book, the point would be obvious, but she makes the larger point that Hooke, rhetorically speaking, strips layers of mediation away from Power's version. My discussion of Hooke's "armed mites" demonstrates that her assessment is incomplete.

114. Hooke ruminates on "the mark of a *full stop*, or *period*," exploring "both *printed* ones and *written*," before straying into a discussion of microscopic writing (*Micrographia*, 3). "But to come again to the point," he writes, when he returns to the matter at hand (3).

115. For the physiological advantages animals enjoy over human beings, see Laurie Shannon, *The Accommodated Animal: Cosmopolity in Shakespearean Locales* (Chicago: University of Chicago Press, 2013).

116. John Milton, *Samson Agonistes*, in *Miltons Poetry and Major Prose*, ed. Merritt Y. Hughes (Indianapolis: Prentice Hall, 1957), lines 93–97.

117. Cavendish, *Philosophical Letters*, 112.

118. Andrew Marvell, *The Mower to the Glow-worms*, in *Poems of Andrew Marvell*, lines 9–16.

119. For the anticlimactic joke in Leonardo da Vinci, see Passannante, *Catastrophizing*.

120. Here, I recall François's description of "the mildness of the disappointed lover who bears his disappointer no ill will" (*Open Secrets*, xix).

121. Marvell, *Mower to the Glow-worms*, in *Poems of Andrew Marvell*, lines 7–8.

4. The Paradise Without

1. I refer of course to Max Weber, *The Protestant Ethic and the Spirit of Capitalism* (New York: Scribner's, 1930).

2. Milton, *Paradise Lost*, book 5, line 16; hereafter cited parenthetically by book and line number.

3. For this allusion to the Song of Solomon, see Northrop Frye, "The Revelation to Eve," in *Paradise Lost: A Tercentenary Tribute*, ed. Balachandra Rajan (Toronto: University of Toronto Press, 1967), 23–24. Howard Schultz notes that the satanic voice Eve hears in her dream likewise echoes the Song, "changing the dove for the nightingale, keeping only the sensuous delights" ("Satan's Serenade," *Philological Quarterly* 27 [1948]: 24). Jane M. Petty develops this point by showing that the voice in Eve's dream echoes Adam's beckoning quite precisely, which explains their shared echo of the Song—while also, she argues, implying a natural as opposed to prophetic dream: a distortion of what sleeping Eve is *actually hearing* much like "the alarm clock bell becom[ing] church bells in the procrastinating dreamer's subconscious" ("The Voice at Eve's Ear in *Paradise Lost*," *Milton Quarterly* 19, no. 2 [1985]: 46). One difficulty here is that her theory diminishes the importance of the satanic toad found whispering at her ear. Though Petty knows very well that the distortion of dreams needs no explanation other than the strange filtration of the sleeper's mind, she observes: "Perhaps the distortion, or twisted rendering of the words, is wrought by the influence of the puff of smoke left after Ithuriel has speared the toad" (46). For an account of the genre of the invitation poem that takes the Song as its point of departure and culminates in a reading of *Paradise Lost*, including Adam's invitation in book 5, see Erik Gray, "Come Be My Love: The Song of Songs, *Paradise Lost*, and the Tradition of the Invitation Poem," *PMLA* 128, no. 2 (2013): 370–85. His observation that the invitation poem "devotes its attention not to the lovers themselves but to their destination—the *locus amoenus* ('pleasant place'), described in sensuous detail, to which the beloved is invited" is a precedent for my reading of this scene (371).

4. For myrrh as a second source of balm, see John Milton, *The Riverside Milton*, ed. Roy Flannagan (Boston: Houghton Mifflin, 1998), 476n12.

5. Marvell, *Upon Appleton House*, in *Poems of Andrew Marvell*, line 296.

6. Stanley Fish, *Surprised by Sin: The Reader in* Paradise Lost (Cambridge: Harvard University Press, 1967).

7. For a discussion of this tradition, culminating in an interpretation of Marvell's *The Garden*, see Stanley Stewart, *The Enclosed Garden: The Tradition and the Image in Seventeenth-Century Poetry* (Madison: University of Wisconsin Press, 1966).

8. For these resonances, see Stewart, *Enclosed Garden*, especially 1–59.

9. Bacon, *Advancement of Learning*, in *Works*, 3:264, 265.

10. Bacon, *Advancement of Learning*, in *Works*, 3:265.

11. Bacon, *Advancement of Learning*, in *Works*, 3:317.

12. Browne, *Religio Medici*, in *Works* 1:24. Browne elaborates the observation of Proverbs 6 on the natural diligence of the ant, reflecting on the magnificent works of ants, bees, and spiders.

13. Picciotto, *Labors of Innocence*, 402.

14. Picciotto, *Labors of Innocence*, 402.

15. Picciotto, *Labors of Innocence*, 435–42.

16. In Picciotto's apt observation that "long blocks of steeply enjambed lines provide a sense of headlong forward momentum while staving off a complete representation of what is happening," the word "headlong" signals her distinct line of reasoning, which leads to her conclusion that Milton "contrive[es] unfamiliar observational conditions and demand[s] repeated trials to resolve Boylean 'scruples'" (*Labors of Innocence*, 439).

17. Karen L. Edwards, *Milton and the Natural World: Science and Poetry in* Paradise Lost (Cambridge: Cambridge University Press, 1999), 10; see especially chapter 2. Edwards ably refutes an older generation of scholarship that insisted on Milton's scientific "backwardness" (3). One target of her critique is Kester Svendsen, *Milton and Science* (Cambridge: Harvard University Press, 1956). Svendsen treats the ongoing influence of ancient and medieval sources in Milton's poem as evidence that he essentially belongs to the past; he allegorizes the break by pointing to the publication of Thomas Sprat's *History of the Royal Society* and *Paradise Lost* in the same year (42). For a capacious evaluation of Milton's engagement with seventeenth-century science, see Harinder Singh Marjara, *Contemplation of Created Things: Science in* Paradise Lost (Toronto: University of Toronto Press, 1992).

18. Picciotto writes: "While Milton provides a very solid sense of the physical reality of the observational conditions he sets up, he leaves what is seen under these conditions largely up to us, literally making *Paradise Lost* a trial to read, its images only as vivid as the reader works to render them" (405). I would describe this "work" as the pleasure of investigation—another example of Boyle's "hare's law." I am persuaded by Picciotto's account of the technical requirements for reading the poem, but my interpretation emphasizes the importance of the reader's affinity—indeed, alacrity—for intellectual labor. When we speak colloquially of the "wandering mind," we take for granted that we pleasantly pass the time with imaginative inventions rather than having to bend them into shape through strenuous force of will.

19. For a persuasive account of Baconian influence in Milton's body of work, see Catherine Gimelli Martin, *Milton among the Puritans: The Case for Historical Revisionism* (Ashgate: Farnham, 2010).

20. Martin, *Milton among the Puritans*, 77, 231.

21. Martin, *Milton among the Puritans*, 236.

22. Picciotto, *Labors of Innocence*, 400–507. On the question of felix culpa, Picciotto acknowledges a debt to Diane Kelsey McColley, *A Gust for Paradise: Milton's Eden and the Visual Arts* (Urbana: University of Illinois Press, 1993), 401, 722n1. Needless to say, I do not share McColley's view that representations of Paradise like Milton's "suggest not an escape from difficulty, as so many unimaginative minds have supposed paradise to be, but a re-creation of the possibilities of Edenic consciousness and conduct" (xii). Indeed, "Edenic consciousness" enables an experience of effortless labor that seems to me more "imaginative" than the simplicity of an opposition between ease and difficulty. What I find most helpful about McColley's discussion is her rejection of a conception of felix culpa as a "binary dialectic that makes opposites interdependent and also plots them along an axis of good and evil," which she rightly argues "can . . . make sin seem inevitable, in violation of the doctrine of free will" (155). For a persuasive argument that the fortunate fall has become a misleading

"cliché of Milton criticism," see Dennis Richard Danielson, *Milton's Good God: A Study in Literary Theodicy* (Cambridge: Cambridge University Press, 1982), 202–27.

23. I take up this issue below when I discuss Milton's line of reasoning about exposure to vice.

24. We might think here of Gordon Teskey's language of "delirium," which invites us to understand Milton's art as a "flickering" between "incompatible perspectives" (*Delirious Milton: The Fate of the Poet in Modernity* [Cambridge: Harvard University Press, 2006], 4, 5). Likewise resonant with my interpretation is his description of an experience of poetic creation in which the one "continually depart[s] from a standard" or "guide" (4).

25. For a thoughtful discussion of the concept of "holy rest" in *Paradise Lost*, but one that takes Milton's interest in repose as an invitation to adopt a "divine perspective" on the poem rather than (as I do) the ordinary perspective of a human being, see Michael Lieb, "'Holy Rest': A Reading of *Paradise Lost*," *English Literary History* 39, no. 2 (1972): 253.

26. Theological contexts have been crucial to much of the most influential scholarship on *Paradise Lost*, but I have benefited in particular from the discussion of the role of Arminianism in Milton's epic in Waddington, *Looking into Providences*. For a lucid discussion of the relationship between Arminianism, Pelagianism, Calvinism, and other theological traditions, see Maurice Kelley, "Introduction," in *Complete Prose Works of John Milton*, 6:74–86.

27. For a useful précis of several critical reflections on the "fall before the fall," see John Leonard, *Faithful Labourers: A Reception History of Paradise Lost, 1667–1970* (Oxford: Oxford University Press, 2013), 2:558–70.

28. In his *De Doctrina Christiana*, Milton describes the Tree as a bare minimum for the demonstration of faith: "it was necessary [in Eden] that one thing at least should be either forbidden or commanded, and above all something which was in itself neither good nor evil, so that man's obedience might in this way be made evident" (*Christian Doctrine*, in *Complete Prose Works of John Milton*, 6:351–52). In *Paradise Lost*, Adam likewise emphasizes that the Tree is a single exception to otherwise unchecked freedom: "God hath pronounc't it death to taste that Tree, / The only sign of our obedience left / Among so many signs of power and rule / Conferr'd upon us" (4.427–30).

29. Adam develops his point at some length; here is another of his formulations: "[God] requires / From us no other service than to keep / This one, this easy charge" (4.419–21). Stanley Fish has pointed out the frequency with which scholars backproject Adam and Eve's "disobedience," arraigning no less accomplished a cohort of scholars than William Empson, W. B. C. Watkins, and Christopher Ricks, who make the following assertions: (1) "Adam and Eve were fated to fall" (Empson); (2) "Their disobedience, as we see it, is determined, partly by circumstances, partly by their own natures" (Watkins); and (3) "They were created with a propensity to fall" (Ricks) (Fish, *Surprised by Sin*, 209–10; I quote his glosses of Empson, Watkins, and Ricks). Whatever their differences, Fish is right to notice in these interpretations a shared belief that the Tree's magnetic field encompasses the whole of Paradise, distorting innocence by forecasting its demise.

30. Because I contend that Milton represents the default psychology of innocence as a state of blithe indifference to transgression, I inevitably raise another

Fishian question: how to narrate (how to explain) a needless journey from innocence to guilt. Fish's answer is frustrating because it isn't one, which he bluntly announces with the following pronouncement: "There is no cause of the Fall" (*Surprised by Sin*, 258). Although I reject his enthusiasm for mystification (if the poem is a machine for reminding of our distance from Paradise, as he contends, it need only confirm, rather than elucidate, the fact of the Fall and of our persistent fallenness), I agree that absolutely nothing about Paradise entails or even encourages the violation of God's law. Yet the Fall's contingency matters for reasons that Fish doesn't anticipate. Because Milton wishes to show us how eminently needless the Fall was (and thus how easily reversible it remains), he draws our attention to the faultiness of the logic on the basis of which Eve makes the several bad decisions that ultimately expel her from Paradise. If Milton's account fails to describe the causal mechanism of transgression, all the better! A causeless thing, motivated by nothing but circumstance (and justified with the most tendentious of rationalizations), should be as easy to annul as it is to avoid repeating.

31. There are additional reasons to entertain the possibility of dedramatized versions of the Fall. Stephen Fallon has argued that "Milton's monism makes it difficult, and perhaps impossible, to separate the strands of sin and punishment, since philosophical error and moral error go hand in hand" (*Milton among the Philosophers: Poetry and Materialism in Seventeenth-Century England* [Ithaca: Cornell University Press, 1991], 242). If Milton envisions a cosmos of material gradations extending from the ethereal to the fleshly, as Fallon argues, perhaps Adam and Eve's step "downward" is less the crossing of a stark divide between good and evil than an incremental change in an ever-changing cosmos. Needless to say, the poem often suggests otherwise.

32. John Rogers has argued perceptively that the expulsion from Paradise can be read as both sovereign decree and natural occurrence (*Matter of Revolution*, 144–76).

33. Fish, *How Milton Works*, 534.

34. Fish, *How Milton Works*, 529. He cites Barbara Lewalski, "Innocence and Experience in Milton's Eden," in *New Essays on Paradise Lost*, ed. Thomas Kranidas (Berkeley: University of California Press, 1969), 93; and J. M. Evans, *Paradise Lost and the Genesis Tradition* (Oxford: Oxford University Press, 1968), 269.

35. Grossman, *"Authors to Themselves,"* 139. For his interpretation of the "separation scene," see 138–43.

36. Grossman's reading is in fact powerfully dialectical; he doesn't indulge in a fantasy of unmediated self-awareness. Ultimately, he argues that Adam and Eve's fault is to "place their own doings in an atemporal context," "testing their love" through "doctrinal" dispute rather than attending to the truth of their "historical situation"; what they lose in this scene is a sense of contingent relations—not a sense of pristine isolation as freestanding selves (Grossman, *"Authors to Themselves,"* 131). Fish also ignores the significant extent to which Grossman's point is about the *generically* dramatic dimensions of book 9. On this question of Milton's climactic episode as a "closet drama," see 126–50, especially 127–28.

37. Grossman, *"Authors to Themselves,"* 139.

38. Fish, *How Milton Works*, 543.

39. John Martin Evans, *Milton's Imperial Epic:* Paradise Lost *and the Discourse of Colonialism* (Ithaca: Cornell University Press, 1996), 51.

40. For an insightful discussion of the unconventionally positive connotations of "wild" nature in *Paradise Lost*, see John R. Knott, "Milton's Wild Garden," *Studies in Philology* 102, no. 1 (2005): 66–82. My argument later in this chapter explains why I don't accept his claim that Eve's "absorption in gardening" at the moment of the Serpent's approach is "a form of self-absorption" connected with the "limitations of gardening in Eden" (79, 78). I nonetheless affirm Knott's central argument: "If rule and art were to prevail over nature in the Garden, it would lose its marvelous qualities and the capacity to offer extraordinary pleasures. Tidiness and control are not really the point" (78).

41. Picciotto shows that Walter Charleton and other Baconians describe Eve's interest in the Fruit as carnal desire, and she locates the motif of appetitive Eve among the misogynist tropes of experimentalist rhetoric (*Labors of Innocence*, 226–42).

42. Christopher Ricks, *Milton's Grand Style* (Oxford: Oxford University Press, 1963), 139.

43. Kevis Goodman, "'Wasted Labor'? Milton's Eve, the Poet's Work, and the Challenge of Sympathy," *English Literary History* 64, no. 2 (1997): 427.

44. Nicholas Billingsley, *Kosmobrephia or the infancy of the world* (London, 1658), 55.

45. For an extraordinary reading of Milton with Marx, see Marshall Grossman, "The Fruit of One's Labor in Miltonic Practice and Marxian Theory," *English Literary History* 59, no. 1 (1992): 77–105.

46. Of course, Eve's misprision is far from idiosyncratic; both common sense and literary tradition lend credence to her assumption—but it's also one to which the observational mood presents a compelling alternative. Goodman borrows a simile from Virgil's *Georgics* to explain the simultaneously active and passive meanings conveyed by the term *labor*: "both the weight of the current felt in the arms of the rower and his expenditure of effort while rowing against the stream"—"striving," then, as well as "suffering" ("Wasted Labor," 421). For Goodman, Milton's sense of "work" is likewise double. Eve's mistake, I suggest, is to demand that "labor" encompass both poles of its semantic range. Eve doesn't want to work so much as "earn" her "Supper"—and it's this specific desire that drives her from her home.

47. Smith, *Key of Green*, 136–37.

48. Joshua Scodel, "Edenic Freedoms," *Milton Studies* 56 (2015): 168–69. I am especially indebted to Scodel's argument for its discussion of the tension between, on the one hand, the freedom implicit in a variety of choices, and, on the other, the freedom to make the *best* choice through rational deliberation (158–71). My interpretation of innocence as the default psychology of effortlessness elaborates the meaning of the former.

49. Milton's sequel ends with a similar image, pivoting from the scene of trial to the homeward journey: "hee unobserv'd / Home to his Mother's house private return'd" (*Paradise Regained*, in *Milton's Poetry and Major Prose*, book 4, lines 638–39). Though this is not the place for a reading of *Paradise Regained*, it's worth pointing out that Milton here suggests both unselfconsciousness and antitheatrical self-presentation, offering us one last glimpse of our "unobserv'd" protagonist en route to domestic "priva[cy]."

50. See Daston and Park, *Wonders and the Order of Nature*, 307.

51. Bacon, *Advancement of Learning*, in *Works*, 3:317.

52. Milton also compares Adam's desire for knowledge to someone "whose drouth / Yet scarce allay'd still eyes the current stream, / Whose liquid murmur heard new thirst excites" (7.66–69). Hughes directs our attention to *Purgatorio*, where Dante professes an "urgent" "thirst" for "more revelation of theological truth," but I suggest that Milton's metaphor invites a different interpretation (347n66). An "allay'd" desire is precisely *not* a strong one, and the "new" "excit[ation]" of a satisfied wish should be relatively subdued; this is *not* the thirst of the parched and desperate. Later, Raphael explains to Adam that "Knowledge is as food, and needs no less / Her Temperance over Appetite, to know / In measure what the mind may well contain, / Oppresses else with Surfeit, and soon turns / Wisdom to Folly, as Nourishment to Wind" (7.127–31). Raphael only recommends self-discipline in a highly qualified sense. "Appetite" is self-regulating (its "Temperance" is automatic) insofar as the natural development of the feeling of fullness keeps it in check; we do not have to tell ourselves not to eat if our bodies have already found gratification for that desire.

53. See *OED*, s.v. "repast," n. For meanings of "sweet" other than the "characteristic flavor" of "sugar," see *OED*, s.v. "sweet," adj., which presents us with a number of revealing possibilities: "free from offensive or disagreeable taste," "fresh" (as in [unsalty] water or [unspoiled] milk), and "pleasing (in general); yielding pleasure or enjoyment"—"to the mind or feelings" as well as "to the senses."

54. Editors gloss "Fruits of Palm-Tree" as a reference to either coconuts or dates, but the one scholar who has given the question sustained attention believes she has settled the question in favor of date honey (Roselyn A. Farren, "Milton's Sweet Palm: A Note on *Paradise Lost* 8.210–16," *Milton Quarterly* 46, no. 1 [2012]: 61–63). Her argument assumes that Milton repeats what he finds in his sources—and yet, as Lucy Hutchinson disapprovingly points out, Milton is more than happy to deviate from Scripture (Lucy Hutchinson, *Order and Disorder*, ed. David Norbrook [Oxford: Blackwell, 2001], 5). However runny, date honey doesn't satisfy "thirst / And hunger both" the way coconuts, of which Milton would probably have had some knowledge, quite obviously do. For a discussion of European knowledge of the coconut in early modernity, see "II.e.2. Coconut," in *The Cambridge World History of Food*, ed. Kenneth F. Kiple and Kriemhild Coneè Ornelas (Cambridge: Cambridge University Press, 2000). Though here I am taking the liberty of interjecting subjective experience, what's wonderful about the comparison is that coconuts are almost savory; they are refreshing and flavorful without being sugary sweet. The liquid they contain is a perfect metaphor for cool (intellectual) satisfaction.

55. W. B. C. Watkins, *An Anatomy of Milton's Verse* (Baton Rouge: Louisiana State University Press, 1955), 12.

56. Picciotto is insightful on this theme: "Milton," she writes, "suggests that the escape from the prison house of the senses begins by using those senses—letting one's eyes wander over creation and beyond their fixed abode on earth" (408). I suggest that Milton's "begin[ning]" is more than that—that his epic dilates on the pleasures and epistemological affordances of waywardness. Needless to say, many readers assume a straightforwardly negative connotation: "The word *wander*," writes Isabel Gamble MacCaffrey, "has almost always a pejorative, or melancholy, connotation in *Paradise Lost*" (*Paradise Lost as "Myth"* [Cambridge: Harvard University Press, 1959], 188).

57. "And as she lay vpon the durtie ground, / Her huge long taile her den all ouerspred, / Yet was in knots and many boughtes vpwound, / Pointed with mortall sting" (Spenser, *Faerie Queene*, book 1, canto 1, stanza 15, lines 1–4).

58. For a thoughtful (but divergent) account of this tradition as it bears on Milton's poem, see Waddington, *Looking into Providences*, 92–100.

59. Petty, "Voice at Eve's Ear in *Paradise Lost*," 46.

60. Because Satan is "fraught / With envy against the Son of God, that day / Honor'd by his great Father," he doesn't sleep (5.657–58). We might suppose that he chooses to stay awake and scheme against God, for this is how he uses these extra waking hours—but, to my mind, his anxiety suggests insomnia. For an insightful discussion of sleep as a form of care (and insomnia as a deleterious absence of care), see Benjamin Parris, "'Watching to Banish Care': Sleep and Insomnia in Book 1 of *The Faerie Queene*," *Modern Philology* 113, no. 2 (2015): 151–77.

61. By the end of the tale, of course, the Serpent has more than his share of worries; as God tells him at Genesis 3:15 (Milton repeats these lines, almost verbatim, at 10.179–81), "I will put enmity between thee and the woman, and between thy seed and her seed; it shall bruise thy head, and thou shalt bruise his heel" (KJV).

62. Richard Baxter sees all worldly pleasure as a will o' wisp, leading humankind off the righteous path that leads to salvation in Christ—a perfect counterpoint to Milton's scenario, in which Eve follows the night fire *away* from immersive attention to the natural world and so into error: "Is it not rather the worlds delights, that are all meer dreams and shadows? Is not all its glory, as the light of a Glow-worm, a wandering fire, yielding but small directing light, and as little comforting heat in all our doubtful, and sorrowful darkness?" (*The Saints Everlasting Rest* [London, 1650], 580).

63. See *OED*, s.v. "straight," adv.

64. Browne, *Pseudodoxia Epidemica*, in *Works*, 2:69.

65. Empson, *Some Versions of Pastoral*, 180–81.

66. Empson, *Some Versions of Pastoral*, 181.

67. Waddington, *Looking into Providences*, 141–42. David Quint observes: "In . . . contingent freedom lies their [Adam and Eve's] real human happiness. . . . It is opposed to an excessive sense of security or assurance that may lead to their fall—especially once they make the mistake, like Eve, in the separation scene of Book 9, of identifying security with happiness" (*Epic and Empire: Politics and Generic Form from Virgil to Milton* [Princeton: Princeton University Press, 1993], 301).

68. David Quint, *Inside* Paradise Lost: *Reading the Designs of Milton's Epic* (Princeton: Princeton University Press, 2014), 154. For Eve as teacher's pet, see 164.

69. Bacon, *Advancement of Learning*, in *Works*, 3:285.

70. In his thought-provoking essay on composure, the psychoanalyst Adam Phillips discusses this precise moment in *Paradise Lost*: "The composure is itself ominous, a sign of elaborate calculation, an implicit acknowledgment that there are now parts of Eve that need to be composed" ("On Composure," in *On Kissing*, 42). Yet Eve errs, I suggest, in setting her sights on "composure": it's her very concern with self-presentation and thus with the agon of an encounter with an adversary that undermines her observational mood. Phillips calls composure "a paradoxical self-cure for the experience of traumatic excitement," but composure's *other* is at the opposite end of the spectrum: disregard, relaxation, or underanimation.

71. Yet the perspective of *Areopagitica* is not identical to Eve's; more can be said about the possible compatibility of Milton's critique of Eve's heroism and the view he defends in the tract. Milton writes: "I cannot praise a fugitive and cloister'd vertue,

unexercis'd & unbreath'd, that never sallies out and sees her adversary" (*Complete Prose Works*, 2:515). The polemicist distinguishes cowardice from heroism, but perhaps *neither* is a good description of the default psychology of innocence. Before Eve's departure, she and Adam are neither "cloister'd" nor braced for combat. In *Areopagitica*, of course, Milton's purpose is to argue against the worry that mere exposure to evil (in printed books) is harmful. This too chimes strangely with Adam's argument that Eve, however strong she is, should nonetheless steer clear of temptation. Yet rushing to meet danger is different from simply acknowledging and accepting its proximity. In both *Areopagitica* and *Paradise Lost*, we might conclude, facing evil just means opening your eyes to it—but only sometimes, Milton warns us, do we wisely avail ourselves of our native capacity to do so without fearful retreat or equally fearful aggression.

72. See *OED*, s.v. "Satan," n.

73. Fish draws the connection with Milton's *Mask* (*How Milton Works*, 544).

74. I am again indebted to Gurton-Wachter's reflections on styles of attention as responses to conflict (*Watchwords*).

75. Hall, *Meditations and Vowes*, A3v.

76. Elaborating the analogy between self-cultivation and the cultivation of the garden, Lewalski observes that Eve "goes forth to work in the external garden but is 'mindless' of her prior responsibility toward the paradise within" ("Innocence and Experience in Milton's Eden," 94).

77. Fish takes up the language of "mindlessness" from Watkins, disputing the idea that Adam and Eve "do not really know the good" because they do not know evil (*Surprised by Sin*, 144). When he refers to "the reader's inability to simulate the mindlessness of innocence by reading passively (without implications)," he misses the possibility of reading *with* implications but taking them in stride—as occasions for reflection rather than challenges to virtue (148).

78. Nathanael Culverwel, *An elegant and learned discourse of the light of nature, with several other treatises* (London, 1652), 189.

79. For the importance throughout *Paradise Lost* of "the temporal seams of noon and twilight" as moments in which "events and choices whose outcome is undecided or contingent on forces that are impartial, evenly balanced or subject to the turn and counter-turn of 'revolving' minds," see Jessica Wolfe, *Homer and the Question of Strife from Erasmus to Hobbes* (Toronto: University of Toronto Press, 2015), 349–53.

80. Milton compares Eve to Pomona (with her pruning hook) and Ceres (with her plough): in this moment, she is quite explicitly an emblem of labor (9.393–96; see Hughes's discussion of these emblematic representations at 387n394 and 387n395).

81. For a complementary argument that "spontaneity" is a distinctive feature of innocence, and that Adam and Eve originally enjoy freedom from the burden of moral deliberation, see Richard Strier, "Milton's Fetters, or, Why Eden Is Better Than Heaven," *Milton Studies* 39 (2000): 169–97.

82. For a discussion of these alternatives, see Scodel, "Edenic Freedoms," 186.

83. John Leonard, *Naming in Paradise: Milton and the Language of Adam and Eve* (Oxford: Clarendon Press, 1990), 217. For a version of this argument that takes the typological identity of Christ as "second Adam" as encouragement to anticipate the first Adam's self-sacrifice on Eve's behalf, see Dennis Danielson, "Through the Telescope of Typology: What Adam Should Have Done," *Milton Quarterly* 23 (1989): 121–27.

84. Martin, *Milton among the Puritans*, 242.

85. For a sensitive appraisal of Milton's use of "maze words" (*maze, amazement, mazy, labyrinth*), see Kathleen M. Swaim, "The Art of the Maze in Book IX of *Paradise Lost*," *Studies in English Literature 1500–1900* 12 (1972): 129–40. I have presented an alternative to her interpretation of the sleeping Serpent as an image of "circularity, selfishness, and the subtle intellect as generating center" (135).

86. Adam describes the experience of conversation with Raphael as follows: "What thanks sufficient, or what recompense / Equal have I to render thee, Divine / Historian, who thus largely hast allay'd / The thirst I had of knowledge, and voutsaf't / Things else by me unsearchable, now heard / With wonder, but delight, and, as is due, / With glory attributed to the high / Creator" (8.5–13). For a discussion of the Miltonic resonances of the concept of *vaghezza*, an experience of wonder that at once suggests "grace, charm, or beauty" (features of the wondrous object) and "loving," "wandering," or "desiring" (features of the experience itself), see N. K. Sugimura, "Eve's Reflection and the Passion of Wonder in *Paradise Lost*," *Essays in Criticism* 64, no. 1 (2014): 9–10. Although I affirm her argument that Eve falls prey to the perils of wonder ("the rippling *vaghezza* of Satan's serpentine form"), I qualify her claim that Milton associates Adam's "wonder" with the "Platonic-cum-Aristotelian tradition according to which 'philosophy beings in wonder'"; he does so with the significant Baconian qualifications this book has described in detail (20, 19).

87. Edwards offers the following witty formulation of the (experimentalist) course Eve should have taken when confronted with the Serpent's rhetorical ploys: "Natural philosophy functions as sex education does: greater understanding is more effective than ignorance as a defense against the consequences of seduction" (*Milton and the Natural World*, 39).

88. My argument here echoes Wolfe's point that Satan "ponders" the wrong questions, corrupting the process of deliberation through which he might have chosen a better course. She writes: "Here [at the beginning of Book 4], as at the beginning of Book 9, where 'after long debate, irresolute / Of thoughts resolved' (9.87–88), he opts to assume the form of a serpent in order to tempt Eve, Satan's deliberative faculty is shown to be defective: were his will not corrupted by Sin, Satan might instead be deliberating *whether* to tempt Eve, rather than how to do so most effectively. Satan's irresolute 'revolving' of this question—the wrong question—shows just how feebly his fallen conscience is able to 'judge between the alternatives which persistently confron[t] it" (*Homer and the Question of Strife*, 328).

89. Edwards, *Milton and the Natural World*, 21.

90. The Serpent was likewise "blithe" at 9.625. Note that "blithe" doesn't yet convey "heedlessness" (not until the twentieth century); Eve feigns happiness. See *OED*, s.v. "blithe," adj.

91. David Loewenstein points out that Milton's "one just man" is not just alienated from the social world around him but also "helpless to alter history's degenerative and repetitive course" (*Milton and the Drama of History: Historical Vision, Iconoclasm, and the Literary Imagination* [Cambridge: Cambridge University Press, 2006], 104). Milton's vision combines utopianism with an awareness of the widespread propensity to accept the world as it is.

92. I quote at some length from a passage (from *The Doctrine and Discipline of Divorce*, 1643) in which he describes this bad habit: "Many men, whether it be their

fate, or fond opinion, easily perswade themselves, if God would but be pleas'd a while to withdraw his just punishments from us, and to restraine what power either the devil, or any earthly enemy hath to worke us woe, that then mans nature would find immediate rest and releasement from all evils. But verily they who think so, if they be such as have a minde large anough to take into their thoughts a general survey of humane things, would soone prove themselves in that opinion farre deceiv'd. For though it were granted us by divine indulgence to be exempt from all that can be harmfull to us from without, yet the perversnesse of our folly is so bent, that we should never lin [cease] hammering out of our owne hearts, as it were out of a flint, the seeds and sparkles of new miseries to our selves, till all were in a blaze againe. And no marvell if out of our own hearts, for they are evill; but ev'n out of those things which God meant us, either for a principall good, or a pure contentment, we are still hatching and contriving up our selves matter of continuall sorrow and perplexitie" (*Complete Prose Works*, 2:234–35). I'm aware that Milton's talk of "evill" "hearts" does not at first seem to sit well with my view, but I'm also struck by his suggestion that we "hammer out . . . miseries" from it—as if whatever evil we find inside ourselves remains inactive until we eagerly breathe life into it. Indeed, we "hatch" and "contriv[e]" "sorrow and perplexitie" "ev'n out of" sources of "pure contentment."

93. Norbrook, *Writing the English Republic*, 491.

94. Ricks, *Milton's Grand Style*, 147. Although the word "delicacy" does different work here, John Peter's interpretation resonates with Ricks's when he describes Milton's "absence of false delicacy, of poetic self-consciousness" as "the positive side of . . . tactlessness"; he rightly praises *Paradise Lost* for its willingness to describe what an idealizing impulse would erase (*A Critique of* Paradise Lost [New York: Columbia University Press, 1960], 92).

95. Samuel Johnson, "Milton," in *The Works of Samuel Johnson*, ed. Francis Pearson Walesby and Arthur Murphy (Oxford: Talboys and Wheeler, 1825), 7:130–31. I do not share Johnson's view of the limits of realism: "Milton's delight was to sport in the wide regions of possibility; reality was a scene too narrow for his mind" (131).

Postscript

1. For alternatives to this trajectory (aimed at the French Enlightenment), see Pierre Saint-Amand, *The Pursuit of Laziness: An Idle Interpretation of the Enlightenment*, trans. Jennifer Curtis Gage (Princeton: Princeton University Press, 2011); and David W. Bates, *Enlightenment Aberrations: Error and Revolution in France* (Ithaca: Cornell University Press, 2002). Interestingly, Saint-Amand's first case study is Pierre Carlet de Marivaux's French adaptation of Joseph Addison and Richard Steele's *The Spectator*; his itinerary begins in England (17). Though he does not observe the same strange compatibility of intense productivity and careless disregard I discover in the Baconian context, his interest in "light moments of distraction or detachment, poised within a bubble of leisure, of vaporous atmosphere" resonates intriguingly with my account (12).

2. Valenza, *Intellectual Disciplines*, 10.

3. 'Espinasse neatly sums up the transition, emphasizing the attenuation of the practical (or "applied") component of experimental natural philosophy: "The Royal

Society of the early eighteenth century was dominated by Newton himself, by rar-efied astronomy and physics, and the universities began to labour to interpret the Newtonian laws; while the development of the steam-engine was left to a blacksmith, Newcomen, and the chronometer for ascertaining the longitude was constructed by Harrison, a carpenter. But in the earlier period the same man, Robert Hooke, both contributed to theoretical physics and astronomy and also constructed the air-pump and important new clock- and watch-mechanisms. Pure scientist, inventor, and tech-nician were one and the same person" (*Robert Hooke*, 40–41). For the difference New-ton makes to the Royal Society, see also Picciotto, *Labors of Innocence*, 583–91.

4. Picciotto writes: "Released from the corporate logics and the identification with artisanal laborer, experimentalist-identified intellectuals could now define themselves in relation to an icon of otherworldly genius" (*Labors of Innocence*, 584).

5. William Wordsworth, "*The Prelude* of 1850, in Fourteen Books," in *The Prelude: 1799, 1805, 1850*, ed. Jonathan Wordsworth, M. H. Abrams, and Stephen Gill (New York: Norton, 1979), book 3, lines 58–63.

6. For the alliance between poetry and science in the Romantic era, see Amanda Jo Goldstein, *Sweet Science: Romantic Materialism and the New Logics of Life* (Chicago: University of Chicago Press, 2017).

7. For the concept of postcritical reading, see Felski, *Limits*, 151–85. For the ongo-ing debate around suspicion and its alternatives, see also Stephen Best and Sharon Marcus, "Surface Reading: An Introduction," *Representations* 108 (2009): 1–21; Elaine Freedgood and Cannon Schmitt, "Denotatively, Technically, Literally," *Represen-tations* 125 (2014): 1–14; Heather Love, "Close but Not Deep: Literary Ethics and the Descriptive Turn," *New Literary History* 41 (2010): 371–91; and Michael Warner, "Uncritical Reading," in *Polemic: Critical or Uncritical*, ed. Jane Gallop (New York: Routledge, 2004), 13–36.

8. Paul Ricoeur, *Freud and Philosophy: An Essay on Interpretation* (New Haven: Yale University Press, 1977), 32–36. For an account that takes Ricoeur as its point of depar-ture, see Felski, *Limits*.

9. Bacon, *Advancement of Learning*, in *Works*, 3:287.

10. Felski acknowledges the flexibility of interpretive affect by speaking of a "crit-ical mood" that encompasses a range of different methods, and she rightly observes that "mood is not synonymous with method" (*Limits*, 6, 152). Nonetheless, she rou-tinely attributes "suspicion" to a flat, unreflective character type (a caricature) with predictable methodological commitments: "The suspicious person is sharp-eyed and hyperalert; mistrustful of appearances, fearful of being duped, she is always on the lookout for concealed threats and discredited motives" (*Limits*, 33).

11. I'm thinking broadly of Marxian and psychoanalytic traditions, but also of Sedgwick's critique of Judith Butler, which seems to me confused. After proposing that Nietzsche, Marx, and Freud profess a "seeming faith, inexplicable in their own terms, in the effects of [exposure]," she cites "the influential final pages of *Gender Trouble*" as an "example" of such warrantless trust in the virtues of demystification, and then goes on to quote passages from these pages in which Butler uses metaphors of unveiling (Sedgwick italicizes them for us): "reveals," "denaturalized," "reveals," "will enact and reveal," "exposes," "exposes," "reveal," "exposing" (Eve Kosofsky Sedgwick, "Paranoid Reading and Reparative Reading, or, You're so Paranoid, You Probably Think This Essay Is about You," in *Touching Feeling: Affect, Pedagogy,*

Performativity [Durham: Duke University Press, 2003], 139). I accept Sedgwick's critique of the taken-for-granted-ness of these procedures, especially under the influence of arguments like Butler's—but who could read Butler and actually believe that her "faith" in "exposure" is "inexplicable in [her] own terms"? Her investment in a theory of ideology, which works by enforcing exactly the taken-for-granted-ness Sedgwick here intends to question, is clear enough. Heather Love has rightly pointed out that Sedgwick, in the very midst of her critique, performs a kind of paranoid reading, and I do not intend to fault her for it. However, I do want to observe that Sedgwick thinks Butler fails to justify something for which the justification is clear ("Truth and Consequences: On Paranoid Reading and Reparative Reading," *Criticism* 52, no. 2 [2010]: 235–41).

12. Cleanth Brooks, "The Heresy of Paraphrase," in *The Well-Wrought Urn: Studies in the Structure of Poetry* (Orlando: Harcourt, 1942), 203.

13. Roland Barthes, *The Neutral: Lecture Course at the College de France [1977–1978]* (New York: Columbia University Press, 2007), 61.

WORKS CITED

Adorno, Theodor, and Max Horkheimer. *Dialectic of Enlightenment: Philosophical Fragments*. Translated by Edmund Jephcott. Edited by Gunzelin Schmid Noerr. Stanford: Stanford University Press, 2002.

——. "The Essay as Form." In *Notes to Literature*, vol. 1, edited by Rolf Tiedemann, translated by Shierry Weber Nicholson, 3–23. New York: Columbia University Press, 1991.

——. *Minima moralia: Reflections on a Damaged Life*. New York: Verso, 2005.

Aït-Touati, Frédérique. *Fictions of the Cosmos: Science and Literature in the Seventeenth Century*. Translated by Susan Emanuel. Chicago: University of Chicago Press, 2011.

Alpers, Paul. *What Is Pastoral?* Chicago: University of Chicago Press, 1996.

Alpers, Svetlana. *The Art of Describing: Dutch Art in the Seventeenth Century*. Chicago: University of Chicago Press, 1983.

Anstey, Peter R. *The Philosophy of Robert Boyle*. New York: Routledge, 2000.

Aristotle, *The Complete Works of Aristotle: Revised Oxford Translation*. Edited by Jonathan Barnes. 2 vols. Princeton: Princeton University Press, 1984.

Aubrey, John. *Brief Lives*. Edited by Andrew Clark. Oxford: Oxford University Press, 1898.

Bacon, Francis. *Francis Bacon: The Major Works*. Edited by Brian Vickers. Oxford: Oxford University Press, 1996.

——. *Refutation of Philosophies*. In *The Philosophy of Francis Bacon*, edited by Benjamin Farrington, 103–33. Chicago: University of Chicago Press, 1964.

——. *Sylva Sylvarum*. London, 1670.

——. *The wisdom of the ancients*. Translated by Arthur Gorges. London, 1619.

——. *The Works of Francis Bacon*. Edited by J. Spedding, R. L. Ellis, and D. D. Heath. 14 vols. London: Longman and Co., 1857–74.

Baker, Henry. *The Microscope Made Easy*. London, 1742.

Barbour, Reid. "Bacon, Atomism, and Imposture: The True and the Useful in History, Myth and Theory." In *Francis Bacon and the Refiguring of Modern Thought: Essays to Commemorate* The Advancement of Learning *(1605–2005)*, edited by Julie Robin Solomon and Catherine Gimelli Martin, 17–43. Aldershot: Ashgate, 2005.

——. *English Epicures and Stoics: Ancient Legacies in Early Stuart Culture*. Amherst: University of Massachusetts Press, 1998.

Barnes, Barry. *About Science*. Oxford: Blackwell, 1985.

Barthes, Roland. *The Neutral: Lecture Course at the College de France [1977–1978]*. New York: Columbia University Press, 2007.

Bates, David W. *Enlightenment Aberrations: Error and Revolution in France*. Ithaca: Cornell University Press, 2002.

Baxter, Henry. *The Saints Everlasting Rest*. London, 1650.

Becker, Daniel Levin. *Many Subtle Channels: In Praise of Potential Literature*. Cambridge: Harvard University Press, 2012.

Belsey, Catherine. "Iago the Essayist." In *Shakespeare in Theory and Practice*, 157–71. Edinburgh: Edinburgh University Press, 2008.

Berger, Harry, Jr. *The Absence of Grace: Sprezzatura and Suspicion in Two Renaissance Courtesy Books*. Stanford: Stanford University Press, 2000.

——. "Andrew Marvell: The Poem as Green World." In *Second World and Green World: Studies in Renaissance Fiction Making*, edited by John Patrick Lynch, 252–323. Berkeley: University of California Press, 1988.

Berlant, Lauren. *Cruel Optimism*. Durham: Duke University Press, 2011.

Best, Stephen, and Sharon Marcus. "Surface Reading: An Introduction." *Representations* 108 (2009): 1–21.

Billingsley, Nicholas. *Kosmobrephia or the infancy of the world*. London, 1658.

Black, Scott. *Of Essays and Reading in Early Modern Britain*. New York: Palgrave, 2006.

Blumenberg, Hans. *Care Crosses the River*. Translated by Paul Fleming. Stanford: Stanford University Press, 2010.

——. *The Legitimacy of the Modern Age*. Translated by Robert M. Wallace. Cambridge: MIT Press, 1993.

——. *Paradigms for a Metaphorology*. Translated by Robert Savage. Ithaca: Cornell University Press.

——. *Shipwreck with Spectator: Paradigm of a Metaphor for Existence*. Cambridge: MIT Press, 1997.

Boethius. *The Consolation of Philosophy*. Translated by S. J. Tester. Loeb Classical Library. Cambridge: Harvard University Press, 1973.

Bouwsma, William J. "The Two Faces of Humanism: Stoicism and Augustinianism in Renaissance Thought." In *Itinerarium Italicum: The Profile of the Italian Renaissance in the Mirror of Its European Transformations. Dedicated to Paul Oskar Kristeller on the Occasion of His 70th Birthday*, edited by Heiko A. Oberman and Thomas A. Brady Jr., 3–60. Leiden: E. J. Brill, 1975.

Boyle, Robert. *An Account of Philaretus in His Minority*. In *Robert Boyle: By Himself and His Friends*. Edited by Michael Hunter. London: Pickering and Chatto, 1994.

——. *The Correspondence of Robert Boyle*. Edited by Michael Hunter, Antonio Clericuzio, and Lawrence Principe. 7 vols. London: Pickering and Chatto, 2001.

——. *The Works of Robert Boyle*. Edited by Michael Hunter and Edward B. Davis. 14 vols. London: Pickering and Chatto, 1999–2000.

Briggs, John C. *Francis Bacon and the Rhetoric of Nature*. Cambridge: Harvard University Press, 1989.

Brogan, Hugh. "Marvell's *Epitaph on* —." *Renaissance Quarterly* 32 (Summer 1979): 197–99.

Brooks, Cleanth. "The Heresy of Paraphrase." In *The Well-Wrought Urn: Studies in the Structure of Poetry*, 192–214. Orlando: Harcourt, 1942.

Browne, Thomas. *The Works of Thomas Browne*. Edited by Geoffrey Keynes. 2nd ed. 4 vols. Chicago: University of Chicago Press, 1964.

Burdach, Konrad. "Faust und die Sorge." *Deutsche Vierteljahrsschrift für Literaturwissenschaft und Geiesesgeschichte* 1 (1923): 1–60.

Burke, Peter. *The Fortunes of the Courtier: The European Reception of Castiglione's Cortegiano*. University Park: Pennsylvania State University Press, 1996.

Burton, Robert. *The Anatomy of Melancholy*. Edited by Thomas C. Faulkner, Nicolas K. Kiessling, and Rhonda L. Blair. 2 vols. Oxford: Clarendon Press, 1990.

Campbell, Mary Baine. *Wonder and Science: Imagining Worlds in Early Modern Europe*. Ithaca: Cornell University Press, 1999.

Cantor, Geoffrey. "Boyling Over: A Commentary on the Preceding Papers." *British Journal for the History of Science* 32, no. 3 (1999): 315–24.

Carruthers, Mary. *The Book of Memory: A Study of Memory in Medieval Culture*. Cambridge: Cambridge University Press, 1990.

Castiglione, Baldassare. *The Courtyer of Count Baldessar Castilio*. Translated by Thomas Hoby. London, 1561.

———. *Il Libro del Cortegiano*. Edited by Bruno Maier. Torino: Unione Tipografico-Editrice Torinese, 1964.

Cavell, Stanley. *The Claim of Reason: Wittgenstein, Skepticism, Morality, and Tragedy*. Oxford: Oxford University Press, 1999.

Cavendish, Margaret. *Philosophical Letters*. London, 1664.

Chalmers, Gordon Keith. "Sir Thomas Browne, True Scientist." *Osiris* 2 (1936): 28–79.

Chico, Tita. "Minute Particulars: Microscopy and Eighteenth-Century Narrative." *Mosaic* 39, no. 2 (2006): 143–61.

Cicero. *On Duties*. Translated by Walter Miller. Loeb Classical Library. Cambridge: Harvard University Press, 1913.

Colie, Rosalie. *"My Ecchoing Song": Andrew Marvell's Poetry of Criticism*. Princeton: Princeton University Press, 1970.

Coolidge, John S. *The Pauline Renaissance: Puritanism and the Bible*. Oxford: Oxford University Press, 1970.

Cooper, John R. *The Art of* The Compleat Angler. Durham: Duke University Press, 1968.

Coppe, Abiezer. *A fiery flying rolle: a word from the Lord to all the great ones of the Earth, whom this may concerne: being the last warning piece at the dreadfull day of judgement*. London, 1650.

Cotterill, Anne. *Digressive Voices in Early Modern English Literature*. Oxford: Oxford University Press, 2004.

———. "Marvell's Watery Maze: Digression and Discovery at Nun Appleton." *English Literary History* 69, no. 1 (2002): 103–32.

Cowles, Thomas. "Dr. Henry Power, Disciple of Sir Thomas Browne." *Isis* 20, no. 2 (1934): 344–66.

———. "Dr. Henry Power's Poem on the Microscope." *Isis* 21, no. 1 (1934): 71–80.

Crane, R. S. "The Relation of Bacon's *Essays* to His Program for the Advancement of Learning." In *Essential Articles for the Study of Francis Bacon*, edited by Brian Vickers, 272–92. London: Sidgwick and Jackson, 1972.

Culverwel, Nathanael. *An elegant and learned discourse of the light of nature, with several other treatises*. London, 1652.

Danielson, Dennis Richard. *Milton's Good God: A Study in Literary Theodicy*. Cambridge: Cambridge University Press, 1982.

———. "Through the Telescope of Typology: What Adam Should Have Done." *Milton Quarterly* 23 (1989): 121–27.

Darbishire, Helen, ed. *The Early Lives of Milton*. New York: Barnes and Noble, 1932.

Daston, Lorraine. "Baconian Facts, Academic Civility, and the Prehistory of Objectivity." In *Rethinking Objectivity*, edited by Allan Megill, 37–63. Durham: Duke University Press, 1994.

——. "The Empire of Observation, 1600–1800." In *Histories of Scientific Observation*, edited by Lorraine Daston and Elizabeth Lunbeck, 81–113. Chicago: University of Chicago Press, 2011.

Daston, Lorraine, and Peter Galison. *Objectivity*. New York: Zone Books, 2007.

Daston, Lorraine, and Elizabeth Lunbeck, "Introduction: Observation Observed." In *Histories of Scientific Observation*, edited by Lorraine Daston and Elizabeth Lunbeck, 1–10. Chicago: University of Chicago Press, 2011.

Daston, Lorraine, and Katharine Park. *Wonders and the Order of Nature, 1150–1750*. New York: Zone Books, 1998.

Davis, Edward B. "Robert Boyle as the Source of an Isaac Watts Set for a William Billings Anthem." *The Hymn* 53, no. 1 (2002): 46–47.

Davis, J. C. *Fear, Myth, and History: The Ranters and the Historians*. Cambridge: Cambridge University Press, 1986.

Dear, Peter. *Revolutionizing the Sciences: European Knowledge and Its Ambitions, 1500–1700*. 2nd ed. Princeton: Princeton University Press, 2009.

Descartes, René. *Meditations on First Philosophy*. Translated and edited by John Cottingham. Cambridge: Cambridge University Press, 1996.

Dobson, Austin. "On Certain Quotations in Walton's 'Angler.'" In *Side-Walk Studies*, 250–62. London: Chatto and Windus, 1902.

Eden, Kathy. *Hermeneutics and the Rhetorical Tradition: Chapters in the Ancient Legacy and Its Humanist Reception*. New Haven: Yale University Press, 1997.

Edwards, Karen L. *Milton and the Natural World: Science and Poetry in* Paradise Lost. Cambridge: Cambridge University Press, 1999.

Empson, William. *Some Versions of Pastoral*. London: Chatto and Windus, 1950.

'Espinasse, Margaret. *Robert Hooke*. Berkeley: University of California Press, 1956.

Evans, John Martin. *Milton's Imperial Epic:* Paradise Lost *and the Discourse of Colonialism*. Ithaca: Cornell University Press, 1996.

——. Paradise Lost *and the Genesis Tradition*. Oxford: Oxford University Press, 1968.

Fallon, Stephen. *Milton among the Philosophers: Poetry and Materialism in Seventeenth-Century England*. Ithaca: Cornell University Press, 1991.

Farren, Roselyn A. "Milton's Sweet Palm: A Note on *Paradise Lost* 8.210–16." *Milton Quarterly* 46, no. 1 (2012): 61–36.

Felski, Rita. *The Limits of Critique*. Chicago: University of Chicago Press, 2015.

Feyerabend, Paul. *Against Method*. 4th ed. London: Verso, 2010.

Fish, Stanley. *How Milton Works*. Cambridge: Harvard University Press, 2001.

——. *Self-Consuming Artifacts: The Experience of Seventeenth-Century Literature*. Berkeley: University of California Press, 1972.

——. *Surprised by Sin: The Reader in* Paradise Lost. Cambridge: Harvard University Press, 1967.

Fisher, Philip. *The Vehement Passions*. Princeton: Princeton University Press, 2001.

Fleck, Ludwik. "The Problem of Epistemology [1936]." In *Cognition and Fact—Materials on Ludwik Fleck*, edited by R. S. Cohen and T. Schnelle, 79–112. Dordrecht: D. Reidel, 1986.

Foucault, Michel. *The History of Sexuality*. Vol. 2: *The Use of Pleasure*. Translated by Robert Hurley. New York: Vintage Books, 1990.

——. *The History of Sexuality*. Vol. 3: *The Care of the Self*. Translated by Robert Hurley. New York: Vintage Books, 1988.

———. "Of Other Spaces." Translated by Jan Miskowiec. *Diacritics* 16, no. 1 (Spring 1986): 22–27.

François, Anne-Lise. *Open Secrets: The Literature of Uncounted Experience.* Stanford: Stanford University Press, 2008.

Freedgood, Elaine, and Cannon Schmitt. "Denotatively, Technically, Literally." *Representations* 125 (2014): 1–14.

Freud, Sigmund. *The Standard Edition of the Complete Psychological Works of Sigmund Freud.* Translated and edited by James Strachey. 24 vols. London: Hogarth Press, 1953–74.

Fried, Gregory. "The King is Dead: Heidegger's *Black Notebooks.*" *Los Angeles Review of Books*, September 13, 2014. http://lareviewofbooks.org/review/king-dead-heideggers-black-notebooks.

Fried. Michael. *Absorption and Theatricality: Painting and Beholder in the Age of Diderot.* Chicago: University of Chicago Press, 1980.

Frye, Northrop. "The Revelation to Eve." In *Paradise Lost: A Tercentenary Tribute*, edited by Balachandra Rajan, 8–47. Toronto: University of Toronto Press, 1967.

Gal, Ofer, and Raz Chen-Morris. *Baroque Science.* Chicago: University of Chicago Press, 2013.

Gale, Monica R. *Virgil on the Nature of Things: The* Georgics, Lucretius *and the Didactic Tradition.* Cambridge: Cambridge University Press, 2000.

Gaukroger, Stephen. *Francis Bacon and the Transformation of Early-Modern Philosophy.* Cambridge: Cambridge University Press, 2001.

Giglioni, Guido. "From the Woods of Experience to the Open Fields of Metaphysics: Bacon's Notion of *Silva.*" *Renaissance Studies* 28, no. 2 (2014): 242–61.

Glanvill, Joseph. *Scepsis scientifica.* London, 1665.

Godwin, Francis. *The Man in the Moon.* London, 1638.

Goldberg, Jonathan. *The Seeds of Things: Theorizing Sexuality and Materiality in Renaissance Representations.* New York: Fordham University Press, 2009.

Goldstein, Amanda. "Growing Old Together: Lucretian Materialism in Shelley's Poetry of Life." *Representations* 128 (Fall 2014): 60–92.

———. *Sweet Science: Romantic Materialism and the New Logics of Life.* Chicago: University of Chicago Press, 2017.

Goodman, Kevis. "'Wasted Labor'? Milton's Eve, the Poet's Work, and the Challenge of Sympathy." *English Literary History* 64, no. 2 (1997): 415–46.

Goodstein, Elizabeth S. *Experience without Qualities: Boredom and Modernity.* Stanford: Stanford University Press, 2005.

Gray, Erik. "Come Be My Love: The Song of Songs, *Paradise Lost*, and the Tradition of the Invitation Poem." *PMLA* 128, no. 2 (2013): 370–85.

Green, Felicity. *Montaigne and the Life of Freedom.* Cambridge: Cambridge University Press, 2012.

Greenblatt, Stephen. *Hamlet in Purgatory.* Princeton: Princeton University Press, 2002.

———. *Renaissance Self-Fashioning from More to Shakespeare.* Chicago: University of Chicago Press, 1980.

———. *The Swerve: How the World Became Modern.* New York: W. W. Norton, 2011.

Grew, Nehemiah. *The Idea of a Phytological History Propounded.* London, 1673.

Grossman, Marshall. *"Authors to Themselves": Milton and the Revelation of History.* Cambridge: Cambridge University Press, 1987.

——. "The Fruit of One's Labor in Miltonic Practice and Marxian Theory." *English Literary History* 59, no. 1 (1992): 77–105.

Gumbrecht, Hans Ulrich. *Atmosphere, Mood, Stimmung: On a Hidden Potential of Literature.* Stanford: Stanford University Press, 2012.

Gurton-Wachter, Lily. *Watchwords: Romanticism and the Poetics of Attention.* Stanford: Stanford University Press, 2016.

Guyer, Paul. "Editor's Introduction." In *Critique of the Power of Judgment,* by Immanuel Kant, xii-lii. Translated and edited by Paul Guyer. Cambridge: Cambridge University Press, 2000.

Hadot, Pierre. *Philosophy as a Way of Life: Spiritual Exercises from Socrates to Foucault.* Edited by Arnold Davidson. Translated by Michael Chase. Oxford: Blackwell, 1995.

Hall, Joseph. *Arte of Divine Meditation.* London, 1605.

——. *Meditations and Vowes, Divine and Morall.* London, 1605.

——. *Occasionall Meditations.* London, 1630.

Hamilton, John T. *Security: Politics, Humanity, and the Philology of Care.* Princeton: Princeton University Press, 2013.

Hampton, Timothy. "Difficult Engagements: Private Passion and Public Service in Montaigne's *Essais.*" In *Politics and the Passions, 1500–1800,* edited by Victoria Kahn, Neil Saccamano, and Daniela Coli, 30–48. Princeton: Princeton University Press, 2006.

——. *Fictions of Embassy: Literature and Diplomacy in Early Modern Europe.* Ithaca: Cornell University Press, 2009.

——. *Writing from History: The Rhetoric of Exemplarity in Renaissance Literature.* Ithaca: Cornell University Press, 1990.

Harrison, Peter. *The Bible, Protestantism and the Rise of Natural Science.* Cambridge: Cambridge University Press, 1998.

——. *The Fall of Man and the Foundations of Science.* Cambridge: Cambridge University Press, 2008.

Hartle, Ann. *Michel de Montaigne: Accidental Philosopher.* Cambridge: Cambridge University Press, 2007.

Heidegger, Martin. *Being and Time.* Translated by Joan Stambaugh. Revised by Dennis J. Schmidt. Albany: State University of New York Press, 2010.

Hetherington, Edith W., and Norriss A. Hetherington. "Andrew Marvell, 'Upon Appleton House,' and Fleas in Multiplying Glasses." *English Language Notes* 13 (December 1975): 122–24.

Hill, Christopher. *The World Turned Upside Down: Radical Ideas during the English Revolution.* London: Penguin, 1972.

Hirst, Derek, and Steven Zwicker. "High Summer at Nun Appleton, 1651: Andrew Marvell and Lord Fairfax's Occasions." *Historical Journal* 36 (1993): 247–69.

Hodge, R. I. V. *Foreshortened Time: Andrew Marvell and 17th Century Revolutions.* Cambridge: D. S. Brewer, 1978.

The Holy Bible. King James Version. Oxford: Oxford University Press, 2010.

Hooke, Robert. *The Diaries of Robert Hooke, the Leonardo of London, 1635–1703.* Edited by Richard Nichols. Sussex: Book Build, 1994.

——. *The Diary of Robert Hooke, M.A., M.D., F.R.S., 1672–1680.* Edited by Henry W. Robinson and Walter Adams. London: Taylor and Francis, 1935.

———. "Discourse Concerning Telescopes and Microscopes." In *Philosophical Experiments and Observations of the Late Eminent Dr. Robert Hooke*, edited by W. Derham, 257–68. London, 1726.

———. *Micrographia: Or Some Physical Descriptions of Minute Bodies Made by Magnifying Glasses, with Observations and Inquiries Thereupon*. London, 1665.

Horkheimer, Max. *Critique of Instrumental Reason*. Translated by Matthew J. O'Connell et al. London: Verso, 2012.

———. *Eclipse of Reason*. New York: Continuum, 2004.

Horne, C. J. "Literature and Science." In *A Guide to English Literature*, edited by Boris Ford, 4:188–202. London: Cassell, 1957.

Houghton, Walter E., Jr. "The History of Trades: Its Relation to Seventeenth-Century Thought: As Seen in Bacon, Petty, Evelyn, and Boyle." *Journal of the History of Ideas* 2, no. 1 (1941): 33–60.

Hovey, Kenneth Alan. "'Montaigny Saith Prettily': Bacon's French and the Essay." *PMLA* 106, no. 1 (1991): 71–82.

Hunter, J. Paul. *Before Novels: The Cultural Contexts of Eighteenth-Century English Fiction*. New York: Norton, 1990.

———. "Robert Boyle and the Epistemology of the Novel." *Eighteenth-Century Fiction* 2, no. 4 (1990): 275-91.

Hunter, Matthew C. *Wicked Intelligence: Visual Art and the Science of Experiment in Restoration London*. Chicago: University of Chicago Press, 2013.

Hunter, Michael. *Boyle: Between God and Science*. New Haven: Yale University Press, 2009.

———. "How Boyle Became a Scientist." *History of Science* 33 (1995): 59–103.

———. *Robert Boyle, 1621–91: Scrupulosity and Science*. Woodbridge: Boydell Press, 2000.

———. *Science and Society in Restoration England*. Cambridge: Cambridge University Press, 1981.

Ignatius, Saint. *Los Ejercicios Espirituales de San Ignacio de Loyola*. Edited by M. R. P. Juan Roothaan. Zaragoza: Hechos y Dichos, 1959.

———. *The Spiritual Exercises of St. Ignatius: A New Translation*. Translated by Louis J. Puhl. Westminster, MD: Newman Press, 1962.

James, Susan. *Passion and Action: The Emotions in Seventeenth-Century Philosophy*. Oxford: Oxford University Press, 1997.

Jameson, Fredric. "Postmodernism, or The Cultural Logic of Late Capitalism." *New Left Review* 146 (1984): 53–92.

Jardine, Lisa. *Francis Bacon: Discovery and the Art of Discourse*. Cambridge: Cambridge University Press, 1974.

Johnson, Monte, and Catherine Wilson. "Lucretius and the History of Science." In *The Cambridge Companion to Lucretius*, edited by Stuart Gillespie and Philip Hardie, 131–48. Cambridge: Cambridge University Press, 2007.

Johnson, Samuel. "Milton." In *The Works of Samuel Johnson*, edited by Francis Pearson Walesby and Arthur Murphy, 7:66–142. Oxford: Talboys and Wheeler, 1825.

Jones, Matthew L. *The Good Life in the Scientific Revolution: Descartes, Pascal, Leibniz, and the Cultivation of Virtue*. Chicago: University of Chicago Press, 2006.

Jonson, Ben. *Ben Jonson: A Critical Edition of the Major Works*. Edited by Ian Donaldson. Oxford: Oxford University Press, 1985.

Joubert, Laurent. *Treatise on Laughter*. Translated and edited by Gregory David de Rocher. Tuscaloosa: Alabama University Press, 1990.

Kiple, Kenneth F., and Kriemhild Coneè Ornelas, eds. *The Cambridge World History of Food*. Cambridge: Cambridge University Press, 2000.

Knott, John R. "Milton's Wild Garden." *Studies in Philology* 102, no. 1 (2005): 66–82.

Lamb, Charles. *The Letters of Charles and Mary Lamb*. Edited by E. V. Lucas. 2 vols. London: Methuen, 1905.

Latour, Bruno. *The Pasteurization of France*. Translated by Alan Sheridan and John Law. Cambridge: Harvard University Press, 1988.

———. *We Have Never Been Modern*. Translated by Catherine Porter. Cambridge: Harvard University Press, 1993.

Lear, Jonathan. *Aristotle: The Desire to Understand*. Cambridge: Cambridge University Press, 1988.

Legouis, Andrew. *Andrew Marvell: Poet, Puritan, Patriot*. Oxford: Oxford University Press, 1965.

Leonard, John. *Faithful Labourers: A Reception History of Paradise Lost, 1667–1970*. 2 vols. Oxford: Oxford University Press, 2013.

———. *Naming in Paradise: Milton and the Language of Adam and Eve*. Oxford: Clarendon Press, 1990.

Leonardo da Vinci. *Leonardo da Vinci's Note-Books*. Translated and edited by Edward McCurdy. London: Duckworth & Co., 1908.

Lemmi, C. W. *The Classical Deities in Bacon: A Study in Mythological Symbolism*. Baltimore: Johns Hopkins University Press, 1933.

Levao, Robert. "Francis Bacon and the Mobility of Science." *Representations* 40 (1992): 1–32.

Lewalski, Barbara. "Innocence and Experience in Milton's Eden." In *New Essays on Paradise Lost*, edited by Thomas Kranidas, 80–117. Berkeley: University of California Press, 1969.

Lieb, Michael. "'Holy Rest': A Reading of *Paradise Lost*." *English Literary History* 39, no. 2 (1972): 238–54.

Lipsius, Justus. *Concerning Constancy*. Translated and edited by R. V. Young. Tempe, AZ: Center for Medieval and Renaissance Studies, 2011.

Lloyd, Claude. "An Obscure Analogue of the Compleat Angler." *PMLA* 42, no. 2 (1927): 400–403.

Loewenstein, David. *Milton and the Drama of History: Historical Vision, Iconoclasm, and the Literary Imagination*. Cambridge: Cambridge University Press, 2006.

Love, Heather. "Close but Not Deep: Literary Ethics and the Descriptive Turn." *New Literary History* 41 (2010): 371–91.

———. "Truth and Consequences: On Paranoid Reading and Reparative Reading." *Criticism* 52, no. 2 (2010): 235–41.

Lovejoy, Arthur O. *The Great Chain of Being: A Study of the History of an Idea*. New York: Harper and Row, 1960.

Lucretius. *De rerum natura*. Translated by W. H. D. Rouse. Revised by Martin Ferguson Smith. Loeb Classical Library. Cambridge: Harvard University Press, 1982.

Lunbeck, Elizabeth. "Empathy as Psychoanalytic Mode of Observation: Between Sentiment and Science." In *Histories of Scientific Observation*, edited by Lorraine Daston and Elizabeth Lunbeck, 255–75. Chicago: University of Chicago Press, 2011.

MacCaffrey, Isabel Gamble. *Paradise Lost as "Myth."* Cambridge: Harvard University Press, 1959.

MacPhail, Eric. "Montaigne and the Trial of Socrates." *Bibliothèque d'Humanisme et Renaissance* 63, no. 3 (2001): 457–75.

Marjara, Harinder Singh. *Contemplation of Created Things: Science in* Paradise Lost. Toronto: University of Toronto Press, 1992.

Marno, David. "Easy Attention: Ignatius of Loyola and Robert Boyle." *Journal of Medieval and Early Modern Studies* 44, no. 1 (2014): 135–61.

Martin, Catherine Gimelli. "The Enclosed Garden and the Apocalypse: Immanent Versus Transcendent Time in Milton and Marvell." In *Milton and the Ends of Time*, edited by Juliet Cummins, 144–68. Cambridge: Cambridge University Press, 2003.

———. *Milton among the Puritans: The Case for Historical Revisionism.* Farnham: Ashgate, 2010.

Martin, Julian. *Francis Bacon, the State, and the Reform of Natural Philosophy.* Cambridge: Cambridge University Press, 1992.

Martz, Louis L. *The Poetry of Meditation: A Study in English Religious Literature of the Seventeenth Century.* New Haven: Yale University Press, 1954.

Marvell, Andrew. *The Poems of Andrew Marvell.* Edited by Nigel Smith. London: Pearson Longman, 2003.

Massumi, Brian. *Parables for the Virtual: Movement, Affect, Sensation.* Durham: Duke University Press, 2002.

McColley, Diane Kelsey. *A Gust for Paradise: Milton's Eden and the Visual Arts.* Urbana: University of Illinois Press, 1993.

Merchant, Carolyn. *The Death of Nature: Women, Ecology and the Scientific Revolution.* New York: HarperCollins, 1980.

Meillassoux, Quentin. *After Finitude: An Essay on the Necessity of Contingency.* Translated by Ray Brassier. New York: Continuum, 2008.

Michaels, Walter Benn. "Neoliberal Aesthetics: Fried, Rancière and the Form of the Photograph." *nonsite* 1 (January 25, 2011). http://nonsite.org/article/neoliberal-aesthetics-fried-ranciere-and-the-form-of-the-photograph.

Milton, John. *Complete Prose Works of John Milton.* Edited by Don M. Wolfe. 8 vols. New Haven: Yale University Press, 1953–82.

———. *Milton's Poetry and Major Prose.* Edited by Merritt Y. Hughes. Indianapolis: Prentice Hall, 1957.

———. *The Riverside Milton.* Edited by Roy Flannagan. Boston: Houghton Mifflin, 1998.

Montaigne, Michel de. *The Complete Essays of Montaigne.* Translated by Donald Frame. Stanford: Stanford University Press, 1948.

———. *Essais.* Edited by Alexandre Micha. 3 vols. Paris: Garnier-Flammarion, 1969.

———. *Essays.* Translated by Charles Cotton. 3 vols. London, 1685.

———. *Essays.* Translated by Charles Cotton. Edited by William Carew Hazlitt. 3 vols. London, 1877.

———. *Montaigne's Annotated Copy of Lucretius: A Transcription and Study of the Manuscript, Notes, and Pen-Marks.* Edited by M. A. Screech. Geneva: Droz, 1998.

Morreale, Margherita. "*Desenvoltura, Suelto* y *Soltura* en Boscán." *Revista de Filología Española* 38 (1954): 257–64.

Murphy, Kathryn, and Anita Traninger. "Introduction: Instances of Impartiality." In *The Emergence of Impartiality*, edited by Kathryn Murphy and Anita Traninger, 1–32. Leiden: Brill, 2014.

Ngai, Sianne. *Our Aesthetic Categories: Zany, Cute, Interesting.* Cambridge: Harvard University Press, 2012.

——. *Ugly Feelings.* Cambridge: Harvard University Press, 2005.

Nicolson, Marjorie Hope. *The Breaking of the Circle: Studies in the Effect of the "New Science" upon Seventeenth-Century Poetry.* Rev. ed. New York: Columbia University Press, 1950.

Norbrook, David. *Writing the English Republic: Poetry, Rhetoric and Politics, 1627–1660.* Cambridge: Cambridge University Press, 1999.

North, Paul. *The Problem of Distraction.* Stanford: Stanford University Press, 2012.

Ochs, Kathleen H. "The Royal Society of London's History of Trades Programme: An Early Episode in Applied Science." *Notes and Records of the Royal Society of London* 39, no. 2 (1985): 129–58.

Ogden, H. V. S. "The Principles of Variety and Contrast in Seventeenth Century Aesthetics, and Milton's Poetry." *Journal of the History of Ideas* 10 (1949): 159–82.

Oliver, H. J. "The Composition and Revisions of the 'Compleat Angler.'" *Modern Language Review* 42, no. 3 (1947): 295–313.

Oxford English Dictionary Online. 2nd ed. Oxford: Oxford University Press, 1989. www.oed.com.

Oxford Latin Dictionary. Edited by P. G. W. Glare. 2nd ed. Oxford: Oxford University Press, 2012.

Palmer, Ada. *Reading Lucretius in the Renaissance.* Cambridge: Harvard University Press, 2014.

Parris, Benjamin. "'Watching to Banish Care': Sleep and Insomnia in Book 1 of *The Faerie Queene.*" *Modern Philology* 113, no. 2 (2015): 151–77.

Pascal, Blaise. *Pensées.* Translated by A. J. Krailsheimer. New York: Penguin, 1995.

Passannante, Gerard. *Catastrophizing: Materialism and the Making of Disaster.* Chicago: University of Chicago Press, forthcoming.

——. *The Lucretian Renaissance: Philology and the Afterlife of Tradition.* Chicago: University of Chicago Press, 2011.

Paster, Gail Kern. *The Body Embarrassed: Drama and the Disciplines of Shame in Early Modern England.* Ithaca: Cornell University Press, 1993.

——. *Humoring the Body: Emotions and the Shakespearean Stage.* Chicago: University of Chicago Press, 2004.

Pepys, Samuel. *The Diary of Samuel Pepys.* Edited by Henry B. Wheatley and Mynors Bright. 8 vols. London: George Bell and Sons, 1904.

Pérez-Ramos, Antonio. *Francis Bacon's Idea of Science and the Maker's Knowledge Tradition.* Oxford: Oxford University Press, 1989.

Pesic, Peter. "Francis Bacon, Violence, and the Motion of Liberty: The Aristotelian Background." *Journal of the History of Ideas* 75, no. 1 (2014): 69–90.

Peter, John. *A Critique of Paradise Lost.* New York: Columbia University Press, 1960.

Petty, Jane M. "The Voice at Eve's Ear in *Paradise Lost.*" *Milton Quarterly* 19, no. 2 (1985): 42–47.

Pfau, Thomas. *Romantic Moods: Paranoia, Trauma, and Melancholy, 1790–1840.* Baltimore: Johns Hopkins University Press, 2005.

Phillips, Adam. *On Kissing, Tickling, and Being Bored: Psychoanalytic Essays on the Unexamined Life*. Cambridge: Harvard University Press, 1994.

Picciotto, Joanna. *Labors of Innocence in Early Modern England*. Cambridge: Harvard University Press, 2010.

Plato. *Laws*. Translated by R. G. Bury. 2 vols. Loeb Classical Library. Cambridge: Harvard University Press, 1926.

Pliny the Elder. *The Natural History of Pliny*. Translated and edited by John Bostock and H. T. Riley. 6 vols. London: Henry G. Bohn, 1855–57.

Pomata, Gianna. "Observation Rising: Birth of an Epistemic Genre, 1500–1650." In *Histories of Scientific Observation*, edited by Lorraine Daston and Elizabeth Lunbeck, 45–80. Chicago: University of Chicago Press, 2011.

Popper, Karl. *The Logic of Scientific Discovery*. London: Routledge, 2002.

Porter, James I. "Lucretius and the Poetics of Void." In *Le jardin romain: Épicurisme et poésie à Rome. Mélanges offerts à Mayotte Bollack*, edited by Annick Monet, 197–226. Villeneuve d'Ascq: Presses de l'Université Charles-de-Gaulle, 2003.

Power, Henry. *Experimental Philosophy, In Three-Books: Containing New Experiments Microscopical, Mercurial, Magnetical*. London, 1664.

——. *Experimental Philosophy*. Edited by Marie Boas Hall. New York: Johnson Reprint Corporation, 1966.

Principe, Lawrence M. "Style and Thought of the Early Boyle: Discovery of the 1648 Manuscript of *Seraphic Love*." *Isis* 85 (1994): 247–60.

——. "Virtuous Romance and Romantic Virtuoso: The Shaping of Robert Boyle's Literary Style." *Journal of the History of Ideas* 56, no. 3 (1995): 377–97.

Pritchard, Allan. "Marvell's 'The Garden': A Restoration Poem?" *Studies in English Literature 1500–1900* 23 (1983): 371–88.

Quint, David. "Courtier, Prince, Lady: The Design of the *Book of the Courtier*." *Italian Quarterly* 37, nos. 143–46 (2000): 186–95.

——. *Epic and Empire: Politics and Generic Form from Virgil to Milton*. Princeton: Princeton University Press, 1993.

——. *Inside* Paradise Lost: *Reading the Designs of Milton's Epic*. Princeton: Princeton University Press, 2014.

——. *Montaigne and the Quality of Mercy: Ethical and Political Themes in the* Essais. Princeton: Princeton University Press, 1998.

Rebhorn. Wayne A. *Courtly Performances: Masking and Festivity in Castiglione's* Book of the Courtier. Detroit: Wayne State University Press, 1978.

Rich, Mary. *The Occasional Meditations of Mary Rich, Countess of Warwick*. Edited by Raymond A. Anselment. Tempe, AZ: Center for Medieval and Renaissance Studies, 2009.

Ricks, Christopher. "Its Own Resemblance." In *Approaches to Marvell: The York Tercentenary Lectures*, edited by C. A. Patrides, 108–35. London: Routledge and Kegan Paul, 1978.

——. *Milton's Grand Style*. Oxford: Oxford University Press, 1963.

Ricoeur, Paul. *Freud and Philosophy: An Essay on Interpretation*. New Haven: Yale University Press, 1977.

Rogers, John. *The Matter of Revolution: Science, Poetry, and Politics in the Age of Milton*. Ithaca: Cornell University Press, 1996.

Saint-Amand, Pierre. *The Pursuit of Laziness: An Idle Interpretation of the Enlighten-ment.* Translated by Jennifer Curtis Gage. Princeton: Princeton University Press, 2011.

Sargent, Rose-Mary. *The Diffident Naturalist: Robert Boyle and the Philosophy of Experi-ment.* Chicago: University of Chicago Press, 1995.

Schaffer, Simon, and Steven Shapin. *Leviathan and the Air-Pump: Hobbes, Boyle, and the Experimental Life.* Princeton: Princeton University Press, 1985.

Schoenfeldt, Michael. *Bodies and Selves in Early Modern England.* Cambridge: Cam-bridge University Press, 1999.

Schultz, Howard. "Satan's Serenade." *Philological Quarterly* 27 (1948): 17–26.

Scodel, Joshua. "The Affirmation of Paradox: A Reading of Montaigne's 'De la Phis-ionomie' (III: 12)." *Yale French Studies* 64 (1983): 209–37.

——. "Edenic Freedoms." *Milton Studies* 56 (2015): 153–200.

——. *Excess and the Mean in Early Modern English Literature.* Princeton: Princeton Uni-versity Press, 2002.

Sedgwick, Eve Kosofsky. "Paranoid Reading and Reparative Reading, or, You're so Paranoid, You Probably Think This Essay Is about You." In *Touching Feeling: Affect, Pedagogy, Performativity,* 123–52. Durham: Duke University Press, 2003.

Seneca the Younger. *Epistles 1–65.* Translated by Richard M. Gummere. Loeb Classi-cal Library. Cambridge: Harvard University Press, 1917.

Sextus Empiricus. *Outlines of Skepticism.* Edited by Julia Annas and Jonathan Barnes. 2nd ed. Cambridge: Cambridge University Press, 2000.

Shakespeare, William. *The Complete Pelican Shakespeare.* Edited by Stephen Orgel and A. R. Braunmuller. New York: Penguin, 2002.

Shannon, Laurie. *The Accommodated Animal: Cosmopolity in Shakespearean Locales.* Chi-cago: University of Chicago Press, 2013.

Shapin, Steven. *A Social History of Truth: Civility and Science in Seventeenth-Century England.* Chicago: University of Chicago Press, 1994.

Shapiro, Barbara J. *A Culture of Fact: England, 1550–1720.* Ithaca: Cornell University Press, 2003.

Shifflett, Andrew. *Stoicism, Politics and Literature in the Age of Milton: War and Peace Reconciled.* Cambridge: Cambridge University Press, 1998.

Shklovsky, Viktor. "Art as Device." In *Theory of Prose,* 1–14. Translated by Benjamin Sher. Champaign: Dalkey Archive, 1990.

Sidney, Sir Philip. "The Defence of Poesy." In *Sir Philip Sidney: The Major Works,* edited by Katherine Duncan-Jones, 212–50. Oxford: Oxford University Press, 1989.

Simon, David Carroll. "The Anatomy of *Schadenfreude*; or, Montaigne's Laughter." *Critical Inquiry* 43 (Winter 2017): 250–80.

Smith, Barbara Herrnstein. *Scandalous Knowledge: Science, Truth and the Human.* Dur-ham: Duke University Press, 2005.

Smith, Bruce R. *The Key of Green.* Chicago: University of Chicago Press, 2009.

Smith, Nigel. *Literature and Revolution in England, 1640–1660.* New Haven: Yale Uni-versity Press, 1994.

——. *Perfection Proclaimed: Language and Literature in English Radical Religion, 1640–1660.* Oxford: Oxford University Press, 1989.

Solomon, Julie Robin. *Objectivity in the Making: Francis Bacon and the Politics of Inquiry.* Baltimore: Johns Hopkins University Press, 1998.

Sorabji, Richard. *Emotion and Peace of Mind: From Stoic Agitation to Christian Temptation*. Oxford: Oxford University Press, 2000.

Spacks, Patricia Meyer. *Boredom: The Literary History of a State of Mind*. Chicago: University of Chicago Press, 1995.

Spenser, Edmund. *The Faerie Queene*. Edited by A. C. Hamilton. San Francisco: Longman, 2001.

Spiller, Elizabeth. *Science, Reading, and Renaissance Literature: The Art of Making Knowledge, 1580–1670*. Cambridge: Cambridge University Press, 2004.

Spitzer, Leo. "Milieu and Ambiance." *Philosophy and Phenomenological Research* 3, no. 1 (1942): 169–218.

Stewart, Kathleen. "The Point of Precision." *Representations* 135 (Summer 2016): 31–44.

Stewart, Stanley. *The Enclosed Garden: The Tradition and the Image in Seventeenth-Century Poetry*. Madison: University of Wisconsin Press, 1966.

Stocker, Margarita. *Apocalyptic Marvell: The Second Coming in Seventeenth Century Poetry*. Sussex: Harvester Press, 1986.

Strier, Richard. "Milton's Fetters, or, Why Eden Is Better Than Heaven." *Milton Studies* 39 (2000): 169–97.

———. *The Unrepentant Renaissance: From Petrarch to Shakespeare to Milton*. Chicago: University of Chicago Press, 2011.

Sugimura, N. K. "Eve's Reflection and the Passion of Wonder in *Paradise Lost*." *Essays in Criticism* 64, no. 1 (2014): 1–28.

Svendsen, Kester. *Milton and Science*. Cambridge: Harvard University Press, 1956.

Swaim, Kathleen M. "The Art of the Maze in Book IX of *Paradise Lost*." *Studies in English Literature 1500–1900* 12 (1972): 129–40.

Terada, Rei. *Feeling in Theory: Emotion after the "Death of the Subject."* Cambridge: Harvard University Press, 2001.

———. *Looking Away: Phenomenality and Dissatisfaction, Kant to Adorno*. Cambridge: Harvard University Press, 2009.

Teskey, Gordon. *Delirious Milton: The Fate of the Poet in Modernity*. Cambridge: Harvard University Press, 2006.

Tetel, Marcel. "The Humanistic Situation: Montaigne and Castiglione." *Sixteenth Century Journal* 10, no. 3 (1979): 69–84.

Tribbi, Jay. "Cooking (with) Clio and Cleo: Eloquence and Experiment in Seventeenth-Century Florence." *Journal of the History of Ideas* 52, no. 3 (1991): 417–39.

Turner, Henry S. "Lessons from Literature for the Historian of Science (and Vice Versa): Reflections on 'Form.'" *Isis* 101 (2010): 578–89.

Turner, James Grantham. "Marvell's Warlike Studies." *Essays in Criticism* 28 (1978): 288–301.

———. *The Politics of Landscape: Rural Scenery and Society in English Poetry, 1630–1660*. Cambridge: Harvard University Press, 1979.

Valenza, Robin. *Literature, Language, and the Rise of the Intellectual Disciplines in Britain, 1680–1820*. Cambridge: Cambridge University Press, 2009.

Vickers, Brian. *Francis Bacon and Renaissance Prose*. Cambridge: Cambridge University Press, 1968.

Villey, Pierre. *Montaigne et François Bacon*. Paris: Revue de la Renaissance, 1913.

Virgil. *Georgics*. Edited by R. A. B. Mynors. Oxford: Clarendon Press, 1990.

Waddington, Raymond B. *Looking into Providences: Designs and Trials in* Paradise Lost. Toronto: University of Toronto Press, 2012.

Wallace, John. *Destiny His Choice: The Loyalism of Andrew Marvell*. Cambridge: Cambridge University Press, 1968.

Walton, Izaak. *The Compleat Angler or the Contemplative Man's Recreation*. London, 1653.

———. *A Discourse of Rivers, Fish-Ponds, Fish & Fishing*. In Charles Cotton, Robert Venable, and Izaak Walton, *The Universal Angler, Made So, By Three Books of Fishing*. London, 1676.

Warner, Michael. "Uncritical Reading." In *Polemic: Critical or Uncritical*, edited by Jane Gallop, 13–38. New York: Routledge, 2004.

Watkins, W. B. C. *An Anatomy of Milton's Verse*. Baton Rouge: Louisiana State University Press, 1955.

Weber, Max. *The Protestant Ethic and the Spirit of Capitalism*. New York: Scribner's, 1930.

Webster, Charles. Review of *Experimental Philosophy*, by Henry Power. *British Journal for the History of Science* 4, no. 3 (1969): 299–300.

———. *The Great Instauration: Science, Medicine, and Reform, 1626–1660*. Oxford: Peter Lang, 2002.

———. "Henry Power's Experimental Philosophy." *Ambix* 14 (1967): 150–78.

Whigham, Frank. *Ambition and Privilege: The Social Tropes of Elizabethan Courtesy Theory*. Berkeley: University of California Press, 1984.

Whitehead, Alfred North. *Science and the Modern World*. New York: Macmillan, 1925.

Whitney, Charles. *Francis Bacon and Modernity*. New Haven: Yale University Press, 1986.

Wilkins, John. *A Discourse Concerning a New World and Another Planet: The First Book, Discovery of a New World, or A Discourse tending to prove, that 'tis probably there may be another habitable World in the Moon*. London, 1638.

Williams, Raymond. *Marxism and Literature*. Oxford: Oxford University Press, 1977.

Wilson, Catherine. *Epicureanism at the Origins of Modernity*. Oxford: Clarendon Press, 2008.

———. *The Invisible World: Early Modern Philosophy and the Invention of the Microscope*. Princeton: Princeton University Press, 1995.

———. "Visual Surface and Visual Symbol: The Microscope and the Occult in Early Modern Science." *Journal of the History of Ideas* 49 (January–March 1988): 85–108.

Wolfe, Jessica. *Homer and the Question of Strife from Erasmus to Hobbes*. Toronto: University of Toronto Press, 2015.

———. *Machinery, Humanism, and Renaissance Literature*. Cambridge: Cambridge University Press, 2004.

Wordsworth, William. *The Prelude: 1799, 1805, 1850*. Edited by Jonathan Wordsworth, M. H. Abrams, and Stephen Gill. New York: Norton, 1979.

Yeo, Richard. *Notebooks, English Virtuosi, and Early Modern Science*. Chicago: University of Chicago Press, 2014.

Zerilli, Linda M. G. "The Turn to Affect and the Problem of Judgment." *New Literary History* 46 (2015): 261–86.

INDEX

Page numbers followed by n or nn indicate notes.

CPSIA information can be obtained
at www.ICGtesting.com
Printed in the USA
BVHW04*0333010518
514898BV00002B/4/P